Uncertainty and Operations Research

Editor-in-chief

Xiang Li, Beijing University of Chemical Technology, Beijing, China

Decision analysis based on uncertain data is natural in many real-world applications, and sometimes such an analysis is inevitable. In the past years, researchers have proposed many efficient operations research models and methods, which have been widely applied to real-life problems, such as finance, management, manufacturing, supply chain, transportation, among others. This book series aims to provide a global forum for advancing the analysis, understanding, development, and practice of uncertainty theory and operations research for solving economic, engineering, management, and social problems.

More information about this series at http://www.springer.com/series/11709

Huchang Liao · Zeshui Xu

Hesitant Fuzzy Decision Making Methodologies and Applications

 Springer

Huchang Liao
Business School
Sichuan University
Chengdu, Sichuan
China

Zeshui Xu
Business School
Sichuan University
Chengdu, Sichuan
China

ISSN 2195-996X   ISSN 2195-9978   (electronic)
Uncertainty and Operations Research
ISBN 978-981-10-9829-1   ISBN 978-981-10-3265-3   (eBook)
DOI 10.1007/978-981-10-3265-3

This Springer imprint is published by Springer Nature
The registered company is Springer Nature Singapore Pte Ltd.
The registered company address is: 152 Beach Road, #22-06/08 Gateway East, Singapore 189721, Singapore

Preface

Decision making always happens in our daily life, for example, choosing a car to buy, or selecting an electronic product from Amazon or Ebay. In the traditional multiple criteria decision making models, all evaluation values are precise, which is too ideal to match our daily life. To make it more applicable and efficient, the decision making models should depict the situation as close as possible to the real-world, but sometimes it is very difficult or impossible due to the incomplete information or knowledge and the complexity and uncertainty involved in the practical decision making problems. Many different theories and tools were proposed in the realm of decision making, such as the probability theory. However, in many cases, uncertainty is not probabilistic in nature but rather imprecise or vague, for example, "fast" speed, "cheap" price, "good" student, and so forth. The fuzzy set theory, which was originally introduced by Zadeh (1965), is one of the most efficient decision aid techniques providing the ability to deal with imprecise and vague information. Nonetheless, to cope with imperfect or imprecise information that two or more sources of vagueness appear simultaneously, the traditional fuzzy set shows some limitations. Hence, it has been extended into several different forms, such as the type 2 fuzzy set, the type n fuzzy set, the interval-valued fuzzy set, the fuzzy multisets, and so on. All these extensions are based on the same rationale that it is not clear to assign the membership degree of an element to a fixed set. Recently, Torra (2010) proposed a new generalized type of fuzzy set called hesitant fuzzy set. The motivation of introducing such a set is that sometimes the uncertain membership degree is not due to possibility distribution (as in type 2 fuzzy set), or a margin of error (as in interval fuzzy set), but because of a set of possible values. The hesitant fuzzy set shows many advantages over the traditional fuzzy set and its extensions, especially in group decision making with anonymity. It opens new perspectives for research on decision making under hesitant fuzzy environments.

In this book, we give a thorough and systematic introduction to the latest research results on hesitant fuzzy decision making theory, which include the operational laws of hesitant fuzzy sets, the correlation and entropy measures of hesitant fuzzy sets, the hesitant fuzzy hybrid weighted aggregation operators, the hesitant fuzzy multiple criteria decision making methods with complete or incomplete weights, the hesitant fuzzy preference relation theory, etc. We apply these methodologies to various fields such as decision making, medical diagnosis, cluster analysis, service quality management, e-learning management, environmental management, etc. The book is constructed into six chapters that deal with different but related issues, which are listed as follows:

Chapter 1 mainly introduces the state of the art of hesitant fuzzy sets. The chapter first defines the concept of hesitant fuzzy set. The mean and hesitant degrees of a hesitant fuzzy element are also defined. Then the chapter defines the operational laws of hesitant fuzzy elements, especially the subtraction and division operations. A theorem is given to show that the dimension of the derived hesitant fuzzy element may increase as the addition or multiplication operations are done, and thus, some adjusted operations are given. The comparison laws of hesitant fuzzy elements are given based on the score function and variance function of the hesitant fuzzy element. We also introduce the extensions of hesitant fuzzy sets, including the interval-valued hesitant fuzzy set, the dual hesitant fuzzy set and the hesitant fuzzy linguistic term set.

Chapter 2 introduces some novel correlation and entropy measures of hesitant fuzzy sets and applies them to hesitant fuzzy decision making. The chapter first points out the weakness of the existing correlation measures of hesitant fuzzy sets, and then introduces a novel correlation coefficient formula to measure the relationship between two hesitant fuzzy sets. The definitions of mean and variance of a hesitant fuzzy set are introduced. The weighted correlation coefficients are also defined. This chapter then applies the correlation coefficients to medical diagnosis and cluster analysis. After analyzing the existing entropy measures of hesitant fuzzy sets, this chapter introduces some novel two-tuple entropy measures of hesitant fuzzy sets.

Chapter 3 mainly introduces the hesitant fuzzy hybrid weighted aggregation operators for hesitant fuzzy information. The chapter first introduces the hesitant fuzzy weighted aggregation operators, such as the hesitant fuzzy weighted averaging operator, the hesitant fuzzy weighted geometric operator, the adjusted hesitant fuzzy weighted averaging operator, the adjusted hesitant fuzzy weighted geometric operator, the hesitant fuzzy ordered weighted averaging operator, the hesitant fuzzy ordered weighted geometric operator, the hesitant fuzzy hybrid averaging operator and the hesitant fuzzy hybrid geometric operator. Then the chapter points out the drawbacks of the existing hesitant fuzzy hybrid operators that they do not satisfy the desirable property, i.e., idempotency. To circumvent this flaw, a sort of new hesitant fuzzy hybrid weighted aggregation operators are introduced, such as the hesitant fuzzy hybrid weighted averaging operator, the hesitant fuzzy hybrid weighted geometric operator, the quasi hesitant fuzzy hybrid weighted averaging operator, the quasi hesitant fuzzy hybrid weighted geometric operator, and their generalized and induced forms. The properties of these operators are investigated in-depth. Finally, we apply these hesitant fuzzy weighted aggregation operators to multiple criteria decision making with hesitant fuzzy information.

Chapter 4 introduces the hesitant fuzzy multiple criteria decision making methods with complete weight information. After describing the hesitant fuzzy multiple criteria decision making problem and the basic idea of the VIKOR method, the chapter introduces the procedure of hesitant fuzzy VIKOR method to handle the problems where the assessments of alternatives on different criteria are given as hesitant fuzzy element and the weights of criteria are completely given as crisp values. This chapter applies the hesitant fuzzy VIKOR method to a service quality management problem. In addition, we also introduce the hesitant fuzzy ELECTRE methods, including the hesitant fuzzy ELECTRE I and the hesitant fuzzy ELECTRE II, for hesitant fuzzy multiple criteria decision making and apply these two methods to solve practical decision making problems.

Chapter 5 introduces the hesitant fuzzy multiple criteria decision making methods with incomplete weight information. Based on the definitions of hesitant fuzzy positive ideal solution and the hesitant fuzzy negative ideal solution, the satisfaction degree of an alternative is introduced. Then, we construct several optimization models to derive the weights of criteria, and discuss the interactive method for multiple criteria decision making problems with hesitant fuzzy information. In addition, we introduce the minimum deviation methods for hesitant fuzzy multiple criteria decision making with incomplete weight information and address the corresponding interval-valued cases. This chapter also presents how to solve the hesitant fuzzy multiple stages multiple criteria decision making problems where the weights of different stages are unknown.

Chapter 6 introduces the hesitant fuzzy preference relation and its multiplicative consistency as well as its consistency index. The chapter defines the concept of hesitant fuzzy preference relation and investigates its desirable properties. The concepts of multiplicative consistency, perfect multiplicative consistency and acceptable multiplicative consistency of a hesitant fuzzy preference relation are defined. Then the chapter introduces two algorithms to improve the consistency level of a hesitant fuzzy preference relation. The chapter provides a method to determine the values of the consistency index of hesitant fuzzy preference relations with different orders. Afterwards, we investigate the consensus reaching process of group decision making based on the hesitant fuzzy preference relations. Finally, the chapter presents how to use interval-valued hesitant fuzzy preference relation in group decision making.

This book is suitable for the engineers, technicians, and researchers in the fields of fuzzy mathematics, operations research, information science, management science and engineering, etc. It can also be used as a textbook for postgraduate and senior-year undergraduate students of the relevant professional institutions of higher learning.

This work was supported in part by the National Natural Science Foundation of China under Grants 71501135, 61273209, and 71571123, the China Postdoctoral Science Foundation under Grants 2016T90863 and 2016M602698, the Scientific Research Foundation for Excellent Young Scholars at Sichuan University under Grant 2016SCU04A23, and the Scientific Research Foundation for Scholars at Sichuan University under Grant 1082204112042.

Special thanks to Dr. Xiao-Jun Zeng at the University of Manchester and Dr. Meimei Xia at the Beijing Jiaotong University for lots of insightful ideas and great suggestions.

Chengdu, China   Huchang Liao
May 2016   Zeshui Xu

Contents

Abstract

The hesitant fuzzy set, which permits the membership degree of an element to a set presented by several possible values between 0 and 1, is an extension of the fuzzy set. Hesitant fuzzy set shows many advantages over traditional fuzzy set and its other extensions, especially in group decision making with anonymity. It opens new perspectives for research on decision making under hesitant environments. Since it was first introduced by Torra in 2010, hesitant fuzzy set theory has been widely investigated and applied to a variety of fields. In this book, we give a thorough and systematic introduction to the latest research results on hesitant fuzzy decision making theory, which include the operational laws of hesitant fuzzy sets, the correlation and entropy measures of hesitant fuzzy sets, the hesitant fuzzy hybrid weighted aggregation operators, the hesitant fuzzy multiple criteria decision making methods with complete or incomplete weight information, the hesitant fuzzy preference relation theory, etc. We apply these methodologies to various fields such as decision making, medical diagnosis, cluster analysis, service quality management, e-learning management, environmental management, etc. This book is suitable for the engineers, technicians, and researchers in the fields of fuzzy mathematics, operations research, information science, management science and engineering, etc. It can also be used as a textbook for postgraduate and senior-year undergraduate students of the relevant professional institutions of higher learning.

Chapter 1
Hesitant Fuzzy Set and Its Extensions

As uncertainty takes place almost everywhere in our daily life, many different tools have been developed to recognize, represent, manipulate, and tackle such uncertainty. Among the most popular theories to handle uncertainty include the probability theory and the fuzzy set theory, which are proposed to interpret statistical uncertainty and fuzzy uncertainty, respectively. These two types of models possess philosophically different kinds of information: the probability theory conveys information about relative frequencies, while the fuzzy set theory represents similarities of objects to the imprecisely defined properties (Bezdek 1993). Since it was originally introduced by Zadeh (1965), the fuzzy set has turned out to be one of the most efficient decision aid techniques providing the ability to deal with uncertainty and vagueness. After the pioneering work of Zadeh (1965), the fuzzy set theory has been extended in a number of directions, the most impressive one of which relates to the representation of the membership grades of the underlying fuzzy set (Yager 2014). Recently, on the basis of the extensional forms of fuzzy set, Torra (2010) proposed a new generalized type of fuzzy set called hesitant fuzzy set (HFS), which opens new perspectives for further research on decision making under hesitant environments.

HFS shows many advantages over traditional fuzzy set and its other extensions, especially in group decision making with anonymity. The HFS has attracted many scholars' attentions. Torra (2010) firstly gave the concept of HFS, and defined the complement, union and intersection of HFSs. Furthermore, Torra and Narukawa (2009) presented an extension principle permitting to generalize the existing operations on fuzzy sets to HFSs, and described the application of this new type of set in the framework of decision making. Xu and Xia (2011a, b) originally gave the mathematical expressions of HFS, and investigated the distance, similarity and correlation measures for HFSs. Torra (2010) also established the relationship between HFS and intuitionistic fuzzy set (IFS), based on which, Xia and Xu (2011a) gave some operational laws for HFSs, such as the addition and multiplication operations. Afterwards, Liao and Xu (2014a) introduced the subtraction and division operations over HFSs.

© Springer Nature Singapore Pte Ltd. 2017
H. Liao and Z. Xu, *Hesitant Fuzzy Decision Making Methodologies
and Applications*, Uncertainty and Operations Research,
DOI 10.1007/978-981-10-3265-3_1

In this chapter, we first introduce the HFS and its operations, and then give the subtraction and division operations over HFSs. The motivation of introducing these operations for HFSs is based on the relationship between HFS and IFS: HFS encompasses IFS as a particular case and the envelope of a HFS is an IFS (Torra 2010). Several operational laws of these two operations over HFSs are given. The relationship between IFS and HFS is further verified in terms of these two operations. In addition, the relationships between these two operations are established. We also discuss the comparison laws for HFSs. HFS has been extended into different forms, such as the interval-valued hesitant fuzzy set (IVHFS) (Chen et al. 2013b), the dual hesitant fuzzy set (DHFS) (Zhu et al. 2012) and the hesitant fuzzy linguistic term set (Rodríguez et al. 2012). In this chapter, we also introduce the definitions, the operational laws and the comparison laws of these extended HFSs.

1.1 Hesitant Fuzzy Set

1.1.1 Introduction to Hesitant Fuzzy Set

Zadeh (1965) introduced the concept of fuzzy set, which leads to a completely new and very active research area today named as fuzzy logic.

Definition 1.1 (*Zadeh* 1965). An ordinary fuzzy set F in a set X is characterized by a membership function μ_F which takes the values in the interval $[0, 1]$, i.e., $\mu_F : X \rightarrow [0, 1]$. The value of μ_F at x, $\mu_F(x)$, named fuzzy number, represents the grade of membership (grade, for short) of x in F and is a point in $[0, 1]$.

For example, we can use the fuzzy set

$$
\begin{aligned}
F &= \mu_F(x_1)/x_1 + \mu_F(x_2)/x_2 + + \mu_F(x_3)/x_3 + \mu_F(x_4)/x_4 \\
&= 1/0 + 0.9/0.1 + 0.7/0.2 + 0.4/0.3
\end{aligned}
\tag{1.1}
$$

to represent the linguistic term "low", where the operation "$+$" stands for logical sum (or).

As the membership grades in a fuzzy set are expressed as precise values drawn from the unit interval $[0, 1]$, the fuzzy set cannot capture the human ability in expressing imprecise and vague membership grades of a fuzzy set. On the one hand, in realistic decision making, imprecision may arise due to the unquantifiable information, incomplete information, unobtainable information, partial ignorance, and so forth. To cope with imperfect and imprecise information that two or more sources of vagueness appear simultaneously, the traditional fuzzy set shows some limitations. It uses a crisp number in unit interval [0,1] as a membership degree of an element to a set; however, very often, such a crisp number is difficult to be determined by a decision maker (or an expert). On the other hand, if a group of decision makers (or experts) are asked to evaluate the candidate alternatives, they often find some disagreements among themselves. Since the decision makers (or experts) may have different opinions over the alternatives and they cannot persuade each other easily, a consensus result is hard to be obtained but a set of possible values. In such a case, the traditional fuzzy set cannot be used to depict the group's opinions. Hence, the classical fuzzy set has been extended into several different forms, such as the IFS (Atanassov 1986), the interval-valued IFS (Atanassov and Gargov 1989), the type 2 fuzzy set (Mizumoto and Tanaka 1976), the type n fuzzy set (Dubois and Prade 1980), and the fuzzy multisets (also named the fuzzy bags) (Yager 1986). All these extensions are based on the same rationale that it is not clear to assign the membership degree of an element to a fixed set.

The IFS, which assigns to each element a membership degree, a non-membership degree and a hesitancy degree, is more powerful than fuzzy set in dealing with vagueness and uncertainty.

Definition 1.2 (*Atanassov* 1983, 2012). Let a crisp set X be fixed and let $A \subset X$ be a fixed set. An IFS A^* on X is an object of the following form:

$$A^* = \{ <x, \mu_A(x), v_A(x) > |x \in X\} \tag{1.2}$$

where the functions $\mu_A : A \rightarrow [0, 1]$ and $v_A : A \rightarrow [0, 1]$ define the degree of membership and the degree of non-membership of the element $x \in X$ to the set A, respectively, and for every $x \in X$

$$0 \leq \mu_A + v_A \leq 1 \tag{1.3}$$

Obviously, every ordinary fuzzy set has the form:

$$A^* = \{ <x, \mu_A(x), 1 - \mu_A(x) > |x \in X\} \tag{1.4}$$

That is to say, the ordinary fuzzy set is a special case of IFS.
For each IFS A^* on X

$$\pi_A(x) = 1 - \mu_A(x) - v_A(x) \tag{1.5}$$

is called the degree of non-determinacy (uncertainty) of the membership of the element $x \in X$ to the set A. In the case of ordinary fuzzy sets, $\pi_A(x) = 0$ for every $x \in X$.

However, when giving the membership degree of an element to a set, the difficulty of establishing the membership degree is not because we have some possibility distribution (as in type 2 fuzzy set), or a margin of error (as in interval fuzzy set and IFS), but because we have a set of possible values. In such cases, HFS, as a generalization of fuzzy set, permits the membership degree of an element to a set presented by several possible values between 0 and 1. It can better describe the situations where people have hesitancy in providing their preferences over objects in the process of decision making. The HFS was originally proposed by Torra (2010).

Definition 1.3 (*Torra* 2010). Let X be a fixed set, a HFS on X is in terms of a function h that when applied to X returns a subset of $[0, 1]$.

To be easily understood, Xia and Xu (2011a) represented the HFS in terms of the following mathematical symbol:

$$H = \{ <x, \ h_A(x) > |x \in X\} \tag{1.6}$$

where $h_A(x)$ is a set of values in $[0, 1]$, denoting the possible membership degrees of the element $x \in X$ to the set $A \subset X$. For convenience, Xia and Xu (2011b) called $h_A(x)$ a hesitant fuzzy element (HFE), which denotes a basic component of the HFS.

Since the possible values of the membership degree in a HFS are random, the HFS is, to some extent, more natural in representing the fuzziness and vagueness than all the other extensional forms of fuzzy set. On the one hand, it is very close to human's cognitive process by using HFS. It is noted that modeling fuzzy information by other extended forms of fuzzy set is based on the elicitation of single or interval values that should encompass and express the information provided by the decision makers (or experts) when determining the membership of an element to a given set. Nevertheless, in some cases, the decision makers (or experts) involved in the problem may have a set of possible values, and thus cannot provide a single or an interval value to express their preferences or assessments because they are thinking of several possible values at the same time. In such a case, the HFS, whose membership degree is represented by a set of possible values, can solve this problem perfectly, while the other extensions of fuzzy set are invalid.

On the other hand, due to the increasing complexity of socio-economic environments, it is less and less possible for single decision maker (or expert) to consider all relevant aspects of a problem when evaluating the considered objects. Hence, in order to get a more reasonable decision result, a decision organization, such as the board of directors of a company, which contains a collection of decision makers (or experts), is set up explicitly or implicitly to assess the alternatives. As pointed by Yu (1973), "*when a group of individuals intend to form a corporation with themselves as the shareholders or form a union to increase their total bargaining power, they usually find some disagreements among themselves. The disagreements come from the difference in their subjective evaluations of the decision making problems which arise.*" Since the decision makers (or experts) may have different opinions over the alternatives due to their different knowledge backgrounds or benefits and they cannot persuade each other easily, a consensus evaluation result is sometimes hard to obtain but several possible evaluation values. Then the HFS is suitable to handle this issue, and it is more powerful than all the other extended fuzzy sets. For example, suppose that a decision organization is asked to provide the degree to which an alternative is superior to another, and the decision makers prefer to use the values between 0 and 1 to express their preferences. Some decision makers in the organization provide 0.2, some provide 0.6, and the others provide 0.8. These three parts cannot persuade each other, and thus, the degree to which the alternative is superior to the other can be represented by the hesitant fuzzy element (HFE) $\{0.2, 0.6, 0.8\}$. Note that the HFE $\{0.2, 0.6, 0.8\}$ can describe the above situation more objectively than the crisp number 0.2 (or 0.6 or 0.8), or the interval-valued fuzzy number $[0.2, 0.8]$, or the intuitionistic fuzzy number $(0.2, 0.8)$, because the degrees to which an alternative is superior to another are not the convex combination of 0.2 and 0.8, or the interval between 0.2 and 0.8, but just three possible values 0.2, 0.6 and 0.8. If we use any of the extended fuzzy sets to represent the assessments given by these three parts of the decision organization, much useful information may be lost and this may lead to an unreasonable decision. Therefore, it is more suitable and powerful to describe the uncertain evaluation information by HFS.

The HFS encompasses IFS as a particular case, and it is a particular case of type 2 fuzzy set. The typical HFS is the one where $h(x)$ is finite. Torra (2010) gave some special HFEs for x in X:

(1) Empty set: $h(x) = \{0\}$, denoted as O^* for simplicity.
(2) Full set: $h(x) = \{1\}$, denoted as E^*.
(3) Complete ignorance (all is possible): $h(x) = [0, 1]$, denoted as U^*.
(4) Nonsense set: $h(x) = \oslash^*$.

Liao and Xu (2014a) made some deep clarifications on these special HFEs from the view points of the definition of HFS and also from the practical decision making process. As presented in the definition, the HFS on a reference set X is in terms of a function h that when applied to X returns a subset of $[0, 1]$. Hence, if the HFS h returns no value, it is adequate for us to assert that h is a nonsense set. Analogously, if it returns the set $[0, 1]$, which means all values between 0 and 1 are possible, we call it complete ignorance. Particularly, if it returns only one value $\gamma \in [0, 1]$, this certainly makes sense because single value $\gamma \in [0, 1]$ can also be seen as a subset of $[0, 1]$, i.e., we can take γ as $[\gamma, \gamma]$. When $\gamma = 0$, which means the membership degree is zero, then we call it the empty set; if $\gamma = 1$, then we call it the full set. Note that we shall not take the empty set as the set that there is no any value in it, and we also should not take the full set as the set of all possible values. This is the difference between the HFS and the traditional set. The interpretation of these four special HFEs in decision making process is obvious. Consider that an organization with several experts from different areas evaluates an alternative using HFS. The empty set depicts that all experts oppose the alternative. The full set means that all experts agree with it. The complete ignorance represents that all experts have no idea on the alternative, and the nonsense set implies nonsense.

Given an intuitionistic fuzzy number (IFN) (Xu 2007b)$(x, \mu_A(x), \nu_A(x))$, its corresponding HFE is straightforward: $h(x) = [\mu_A(x), 1 - \nu_A(x)]$ if $\mu_A(x) \neq 1 - \nu_A(x)$. But, the construction of IFN from HFE is not so easy when the HFE contains more than one value for each $x \in X$. As for this issue, Torra (2010) pointed out that the envelope of a HFE is an IFN, expressed in the following definition:

Definition 1.4 (*Torra* 2010). Given a HFE h, the IFN $A_{env}(h)$ is defined as the envelope of h, where $A_{env}(h)$ can be represented as $(h^-, 1 - h^+)$, with $h^- = \min\{\gamma | \gamma \in h\}$ and $h^+ = \max\{\gamma | \gamma \in h\}$.

Definition 1.5 (*Liao et al.* 2015b). For a reference set X, let $h(x) = \{\gamma_1, \gamma_2, \ldots, \gamma_l\}$ be a HFE with γ_k ($k = 1, 2, \ldots, l$) being the possible membership grades of $x \in X$ to a given set and l being the number of values in $h(x)$. The mean of the HFE $h(x)$ is defined as:

$$\bar{h}(x) = \frac{1}{l} \sum\nolimits_{k=1}^{l} \gamma_k \tag{1.7}$$

Definition 1.6 (*Liao et al.* 2015b). For a reference set X, let $h(x) = \{\gamma_1, \gamma_2, \ldots, \gamma_l\}$ be a HFE with γ_k ($k = 1, 2, \ldots, l$) being the possible membership grades of $x \in X$ to a given set and l being the number of values in $h(x)$. The hesitant degree of the HFE $h(x)$ is defined as:

$$\varphi_{h(x)} = \sqrt{\frac{1}{l} \sum\nolimits_{k=1}^{l} [\gamma_k - (\bar{h}(x))]^2} = \sqrt{\frac{1}{l} \sum\nolimits_{k=1}^{l} \left[\gamma_k - \left(\frac{1}{l} \sum\nolimits_{k=1}^{l} \gamma_k\right)\right]^2} \tag{1.8}$$

Example 1.1 (Liao et al. 2015b). For two HFEs $h_1 = \{0.1, 0.3, 0.5\}$ and $h_2 = \{0.1, 0.3, 0.8\}$, based on Eqs. (1.7) and (1.8), we have $\bar{h}_1 = 0.45$, $\bar{h}_2 = 0.6$, $\varphi_{h_1} = 0.2217$, and $\varphi_{h_2} = 0.3786$. Therefore, the HFE h_2 is more hesitant than the HFE h_1.

1.1.2 Operational Laws of Hesitant Fuzzy Elements

Torra (2010) defined some operations such as complement, union and intersection for HFEs:

Definition 1.7 (*Torra* 2010). For three HFEs h, h_1 and h_2, the following operations are defined:

(1) Lower bound: $h^-(x) = \min h(x)$.
(2) Upper bound: $h^+(x) = \max h(x)$.
(3) $h^c = \cup_{\gamma \in h} \{1 - \gamma\}$.
(4) $h_1 \cup h_2 = \{h \in h_1 \cup h_2 | h \geq \max(h_1^-, h_2^-)\}$.
(5) $h_1 \cap h_2 = \{h \in h_1 \cup h_2 | h \geq \min(h_1^+, h_2^+)\}$.

Afterwards, Xia and Xu (2011a) gave other forms of (4) and (5) as follows:

(6) $h_1 \cup h_2 = \cup_{\gamma_1 \in h_1, \gamma_2 \in h_2} \max\{\gamma_1, \gamma_2\}$.
(7) $h_1 \cap h_2 = \cup_{\gamma_1 \in h_1, \gamma_2 \in h_2} \min\{\gamma_1, \gamma_2\}$.

Torra (2010) further studied the relationships between HFEs and IFNs:

Proposition 1.1 (Torra 2010). *Let h, h_1 and h_2 be three HFEs. Then,*

(1) $A_{env}(h^c) = (A_{env}(h))^c$.
(2) $A_{env}(h_1 \cup h_2) = A_{env}(h_1) \cup A_{env}(h_2)$.
(3) $A_{env}(h_1 \cap h_2) = A_{env}(h_1) \cap A_{env}(h_2)$.

Proposition 1.2 (Torra 2010). *Let h_1 and h_2 be two HFEs with $h(x)$ being a nonempty convex set for all x in X, i.e., h_1 and h_2 are IFNs. Then,*

(1) *h_1^c is equivalent to IFS complement.*
(2) *$h_1 \cap h_2$ is equivalent to IFS intersection.*
(3) *$h_1 \cup h_2$ is equivalent to IFS union.*

Proposition 1.2 reveals that the operations defined for HFEs are consistent with the ones for IFNs. Based on the relationships between HFEs and IFNs, Xia and Xu (2011a) gave some operational laws for HFEs.

Definition 1.8 (*Xia and Xu* 2011a). Let h, h_1 and h_2 be three HFEs, and λ be a positive real number, then

(1) $h^\lambda = \cup_{\gamma \in h} \{\gamma^\lambda\}$.
(2) $\lambda h = \cup_{\gamma \in h} \{1 - (1 - \gamma)^\lambda\}$.
(3) $h_1 \oplus h_2 = \cup_{\gamma_1 \in h_1, \gamma_2 \in h_2} \{\gamma_1 + \gamma_2 - \gamma_1 \gamma_2\}$.
(4) $h_1 \otimes h_2 = \cup_{\gamma_1 \in h_1, \gamma_2 \in h_2} \{\gamma_1 \gamma_2\}$.

Let $h_j (j = 1, 2, \ldots, n)$ be a collection of HFEs, Liao et al. (2014a) generalized (3) and (4) in Definition 1.8 to the following forms:

(5) $\overset{n}{\underset{j=1}{\oplus}} h_j = \cup_{\gamma_j \in h_j} \left\{1 - \prod_{j=1}^{n} (1 - \gamma_j)\right\}$.
(6) $\overset{n}{\underset{j=1}{\otimes}} h_j = \cup_{\gamma_j \in h_j} \left\{\prod_{j=1}^{n} \gamma_j\right\}$.

It is noted that the number of values in different HFEs may be different. Let l_{h_j} be the number of the HFE h_j. Based on the above operational laws, the following theorem holds:

Theorem 1.1 (Liao et al. 2014a). *Suppose h_1 and h_2 are two HFEs, then*

$$l_{h_1 \oplus h_2} = l_{h_1} \times l_{h_2}, \; l_{h_1 \otimes h_2} = l_{h_1} \times l_{h_2} \tag{1.9}$$

Similarly, it also holds when there are n different HFEs, i.e.,

$$l_{\overset{n}{\underset{j=1}{\oplus}} h_j} = \prod_{j=1}^{n} l_{h_j}, \; l_{\overset{n}{\underset{j=1}{\otimes}} h_j} = \prod_{j=1}^{n} l_{h_j} \tag{1.10}$$

Example 1.2 (Liao and Xu 2013). Let $h_1 = (0.1, 0.2, 0.7)$ and $h_2 = (0.2, 0.4)$ be two HFEs, then by the operational laws of HFSs given in Definition 1.8, we have

$$\begin{aligned} h_1 \oplus h_2 &= \cup_{\gamma_1 \in h_1, \gamma_2 \in h_2} \{\gamma_1 + \gamma_2 - \gamma_1 \gamma_2\} \\ &= \{0.1 + 0.2 - 0.1 * 0.2, 0.1 + 0.4 - 0.1 * 0.4, 0.2 + 0.2 - 0.2 * 0.2, 0.2 + 0.4 - 0.2 * 0.4, \\ &\quad 0.7 + 0.2 - 0.7 * 0.2, 0.7 + 0.4 - 0.7 * 0.4\} = \{0.28, 0.36, 0.46, 0.52, 0.76, 0.82\} \end{aligned}$$

$$h_1 \otimes h_2 = \cup_{\gamma_1 \in h_1, \gamma_2 \in h_2} \{\gamma_1 \gamma_2\} = \{0.1 * 0.2, 0.1 * 0.4, 0.2 * 0.2, 0.2 * 0.4, 0.7 * 0.2, 0.7 * 0.4\}$$
$$= \{0.02, 0.04, 0.04, 0.08, 0.14, 0.28\}$$

Thus, $l_{h_1 \oplus h_2} = 6 = 3 \times 2 = l_{h_1} \times l_{h_2}$, $l_{h_1 \otimes h_2} = 6 = 3 \times 2 = l_{h_1} \times l_{h_2}$.

Theorem 1.1 and Example 1.2 reveal that the dimension of the derived HFE may increase as the addition or multiplication operations are done, which may increase the complexity of calculation. In order not to increase the dimension of the derived HFE in the process of calculation, Liao et al. (2014a) adjusted the operational laws of HFEs into the following forms:

Definition 1.9 (*Liao et al.* 2014a). Let $h_j (j = 1, 2, \ldots, n)$ be a collection of HFEs, and λ be a positive real number, then

(1) $h^\lambda = \{(h^{\sigma(t)})^\lambda, t = 1, 2, \ldots, l\}$.

(2) $\lambda h = \{1 - (1 - h^{\sigma(t)})^\lambda, t = 1, 2, \ldots, l\}$.

(3) $h_1 \oplus h_2 = \{h_1^{\sigma(t)} + h_2^{\sigma(t)} - h_1^{\sigma(t)} h_2^{\sigma(t)}, t = 1, 2, \ldots, l\}$.

(4) $h_1 \otimes h_2 = \{h_1^{\sigma(t)} h_2^{\sigma(t)}, t = 1, 2, \ldots, l\}$.

(5) $\overset{n}{\underset{j=1}{\oplus}} h_j = \{1 - \prod_{j=1}^{n} (1 - h_j^{\sigma(t)}), t = 1, 2, \ldots, l\}$.

(6) $\overset{n}{\underset{j=1}{\otimes}} h_j = \{\prod_{j=1}^{n} h_j^{\sigma(t)}, t = 1, 2, \ldots, l\}$.

where $h_j^{\sigma(t)}$ is the tth smallest value in h_j.

Example 1.3 (Liao and Xu 2013). Let $h_1 = \{0.2, 0.3, 0.5, 0.8\}$ and $h_2 = \{0.4, 0.6, 0.8\}$ be two HFEs respectively. Taking addition and multiplication operations as an example, by using Definition 1.9, we have

$$h_1 \oplus h_2 = \left\{ h_1^{\sigma(t)} + h_2^{\sigma(t)} - h_1^{\sigma(t)} h_2^{\sigma(t)} \middle| t = 1, 2, 3, 4 \right\}$$
$$= \{0.2 + 0.4 - 0.2 \times 0.4, 0.3 + 0.5 - 0.3 \times 0.5, 0.5 + 0.6 - 0.5 \times 0.6, 0.8 + 0.8 - 0.8 \times 0.8\}$$
$$= \{0.52, 0.65, 0.8, 0.96\}$$

$$h_1 \otimes h_2 = \left\{ h_1^{\sigma(t)} h_2^{\sigma(t)} \middle| t = 1, 2, \ldots, l \right\} = \{0.2 \times 0.4, 0.3 \times 0.5, 0.5 \times 0.6, 0.8 \times 0.8\}$$
$$= \{0.08, 0.15, 0.3, 0.64\}$$

It is noted that neither Torra (2010) nor Xia and Xu (2011a) paid any attention to the subtraction and division operations over HFEs. The subtraction and division operations are significantly important in forming the integral theoretical framework of HFS. Meanwhile, it is also an indispensable foundation in developing some well-known decision making method such as PROMETHEE with hesitant fuzzy information. Hence, in the following, we introduce these basic operations over HFEs.

Considering the relationships between IFS and HFS, to start our investigation, let us first review the subtraction and division operations over IFSs. The subtraction and division operations over IFSs were firstly proposed by Atanassov and Riečan (2006).

Later, Chen (2007) also introduced these operations for IFSs, which were derived from the deconvolution for equations using addition and multiplication operations of IFSs, and the forms of these two operations they proposed were similar to those of Atanassov and Riečan (2006). Based on the different versions of the operation "negation", Atanassov (2009) further developed a family of different kinds of subtraction operations for IFSs. Among all these different subtraction operations, Atanassov (2012) finally chose the following forms as the standard definitions for subtraction and division operations over IFSs in his recent published book:

Definition 1.10 (*Atanassov* 2012). For two given IFSs A and B, the subtraction and division operations have the forms:

$$A \ominus B = \left\{ \left(x, \mu_{A \ominus B}(x), v_{A \ominus B}(x) \right) | x \in X \right\} \tag{1.11}$$

where

$$\mu_{A \ominus B}(x) = \begin{cases} \frac{\mu_A(x) - \mu_B(x)}{1 - \mu_B(x)} & \begin{aligned} &\text{if } \mu_A(x) \geq \mu_B(x) \text{ and } v_A(x) \leq v_B(x) \\ &\text{and } v_B(x) > 0 \\ &\text{and } v_A(x)\pi_B \leq \pi_A(x)v_B(x) \end{aligned} \\ 0, & \text{otherwise} \end{cases} \tag{1.12}$$

and

$$v_{A \ominus B}(x) = \begin{cases} \frac{v_A(x)}{v_B(x)}, & \begin{aligned} &\text{if } \mu_A(x) \geq \mu_B(x) \text{ and } v_A(x) \leq v_B(x) \\ &\text{and } v_B(x) > 0 \\ &\text{and } v_A(x)\pi_B(x) \leq \pi_A(x)v_B(x) \end{aligned} \\ 1, & \text{otherwise} \end{cases} \tag{1.13}$$

and

$$A \oslash B = \left\{ \left(x, \mu_{A \oslash B}(x), v_{A \oslash B}(x) \right) | x \in X \right\} \tag{1.14}$$

where

$$\mu_{A \oslash B}(x) = \begin{cases} \frac{\mu_A(x)}{\mu_B(x)}, & \begin{aligned} &\text{if } \mu_A(x) \leq \mu_B(x) \text{ and } v_A(x) \geq v_B(x) \\ &\text{and } \mu_B(x) > 0 \\ &\text{and } \mu_A(x)\pi_B(x) \leq \pi_A(x)\mu_B(x) \end{aligned} \\ 0, & \text{otherwise} \end{cases} \tag{1.15}$$

and

$$v_{A \oslash B}(x) = \begin{cases} \frac{v_A(x) - v_B(x)}{1 - v_B(x)} & \begin{aligned} &\text{if } \mu_A(x) \leq \mu_B(x) \text{ and } v_A(x) \geq v_B(x) \\ &\text{and } \mu_B(x) > 0 \\ &\text{and } \mu_A(x)\pi_B(x) \leq \pi_A(x)\mu_B(x) \end{aligned} \\ 1 & \text{otherwise} \end{cases} \tag{1.16}$$

Inspired by Definition 1.10 and based on the relationships between IFSs and HFSs, the definitions of subtraction and division operations over HFEs can be introduced:

Definition 1.11 (*Liao and Xu* 2014a). Let h, h_1 and h_2 be three HFEs, then

(1) $h_1 \ominus h_2 = \cup_{\gamma_1 \in h_1, \gamma_2 \in h_2} \{t\}$, where

$$t = \begin{cases} \frac{\gamma_1 - \gamma_2}{1 - \gamma_2}, & \text{if } \gamma_1 \geq \gamma_2 \text{ and } \gamma_2 \neq 1 \\ 0, & \text{otherwise} \end{cases}$$

(2) $h_1 \oslash h_2 = \cup_{\gamma_1 \in h_1, \gamma_2 \in h_2} \{t\}$, where

$$t = \begin{cases} \frac{\gamma_1}{\gamma_2}, & \text{if } \gamma_1 \leq \gamma_2 \text{ and } \gamma_2 \neq 0 \\ 1, & \text{otherwise} \end{cases}$$

To make it more adequate, let $h \oslash U^* = O^*, h \oslash U^* = O^*$. According to Definition 1.11, it is obvious that for any HFE h, the following equations hold:

- $h \ominus h = O^*$; $h \ominus O^* = h$; $h \ominus E^* = O^*$.
- $h \oslash h = E^*$; $h \oslash E^* = h$; $h \oslash O^* = E^*$.

In addition, it follows from the above equations that some special cases hold:

- $E^* \ominus E^* = O^*$; $U^* \ominus E^* = O^*$; $O^* \ominus E^* = O^*$.
- $E^* \ominus U^* = O^*$; $U^* \ominus U^* = O^*$; $O^* \ominus U^* = O^*$.
- $E^* \ominus O^* = E^*$; $U^* \ominus O^* = U^*$; $O^* \ominus O^* = O^*$.
- $E^* \oslash E^* = E^*$; $U^* \oslash E^* = U^*$; $O^* \oslash E^* = O^*$.
- $E^* \oslash U^* = O^*$; $U^* \oslash U^* = O^*$; $O^* \oslash U^* = O^*$.
- $E^* \oslash O^* = E^*$; $U^* \oslash O^* = E^*$; $O^* \oslash O^* = E^*$.

For the brevity of presentation, in the process of theoretical derivation thereafter, we shall not consider the particular case where $t = 0$ in subtraction operation and $t = 1$ in division operation. It is noted that the HFS encompasses the IFS as a particular case; thus, the subtraction and division operations over HFEs should be equivalent to the subtraction and division operations over IFNs when not considering the nonmembership degree of each IFN. Comparing Definitions 1.10 and 1.11, we can see that this requirement is met. The following theorems show that the subtraction and division operations over HFEs in Definition 1.11 are convincing and they satisfy some basic properties:

Theorem 1.2 (Liao and Xu 2014a). *Let h_1 and h_2 be two HFEs, then*

(1) $(h_1 \ominus h_2) \oplus h_2 = h_1$, if $\gamma_1 \geq \gamma_2$, $\gamma_2 \neq 1$.
(2) $(h_1 \oslash h_2) \otimes h_2 = h_1$, if $\gamma_1 \leq \gamma_2$, $\gamma_2 \neq 0$.

Theorem 1.3 (Liao and Xu 2014a). *Let h_1 and h_2 be two HFEs, $\lambda > 0$, then*

(1) $\lambda(h_1 \ominus h_2) = \lambda h_1 \ominus \lambda h_2$, if $\gamma_1 \geq \gamma_2, \gamma_2 \neq 1$.
(2) $(h_1 \oslash h_2)^\lambda = h_1^\lambda \oslash h_2^\lambda$, if $\gamma_1 \leq \gamma_2, \gamma_2 \neq 0$.

Theorem 1.4 (Liao and Xu 2014a). *Let* $h = \cup_{\gamma \in h}\{\gamma\}$ *be a HFE, and* $\lambda_1 \geq \lambda_2 > 0$, *then*

(1) $\lambda_1 h \ominus \lambda_2 h = (\lambda_1 - \lambda_2)h$, if $\gamma \neq 1$.
(2) $h^{\lambda_1} \oslash h^{\lambda_2} = h^{(\lambda_1 - \lambda_2)}$, if $\gamma \neq 0$.

Theorem 1.5 (Liao and Xu 2014a). *For three HFEs* h_1, h_2, *and* h_3, *the following conclusions are valid:*

(1) $h_1 \ominus h_2 \ominus h_3 = h_1 \ominus h_3 \ominus h_2$, if
$\gamma_1 \geq \gamma_2, \gamma_1 \geq \gamma_3, \gamma_2 \neq 1, \gamma_3 \neq 1, \gamma_1 - \gamma_2 - \gamma_3 + \gamma_2 \gamma_3 \geq 0.$
(2) $h_1 \oslash h_2 \oslash h_3 = h_1 \oslash h_3 \oslash h_2$, if $\gamma_1 \leq \gamma_2 \gamma_3, \gamma_2 \neq 0, \gamma_3 \neq 0.$

Theorem 1.6 (Liao and Xu 2014a). *For three HFEs* h_1, h_2, *and* h_3, *the following conclusions are valid:*

(1) $h_1 \ominus h_2 \ominus h_3 = h_1 \ominus (h_2 \oplus h_3)$, if
$\gamma_1 \geq \gamma_2, \gamma_1 \geq \gamma_3, \gamma_2 \neq 1, \gamma_3 \neq 1, \gamma_1 - \gamma_2 - \gamma_3 + \gamma_2 \gamma_3 \geq 0.$
(2) $h_1 \oslash h_2 \oslash h_3 = h_1 \oslash (h_2 \otimes h_3)$, if $\gamma_1 \leq \gamma_2 \gamma_3, \gamma_2 \neq 0, \gamma_3 \neq 0.$

It should be noted that in the above theorems, the equations hold only under the given precondition. Moreover, the relationship between IFNs and HFEs can be further verified in terms of these two operations:

Theorem 1.7 (Liao and Xu 2014a). *Let* h_1 *and* h_2 *be two HFEs, then*

(1) $A_{env}(h_1 \ominus h_2) = A_{env}(h_1) \ominus A_{env}(h_2).$
(2) $A_{env}(h_1 \oslash h_2) = A_{env}(h_1) \oslash A_{env}(h_2).$

Theorem 1.7 further reveals that the subtraction and division operations defined for HFEs are consistent with the ones for IFNs. The following theorem reveals the relationship between these two operations:

Theorem 1.8 (Liao and Xu 2014a). *For two HFEs* h_1 *and* h_2, *the following conclusions are valid:*

(1) $h_1^c \ominus h_2^c = (h_1 \oslash h_2)^c.$
(2) $h_1^c \oslash h_2^c = (h_1 \ominus h_2)^c.$

Example 1.4 (Liao and Xu 2014a). Consider two HFEs $h_1 = \{0.3, 0.2\}$ and $h_2 = \{0.1, 0.2\}$. According to Definition 1.11, we have

$$h_1 \ominus h_2 = \left\{ \frac{0.3 - 0.1}{1 - 0.1}, \frac{0.3 - 0.2}{1 - 0.2}, \frac{0.2 - 0.1}{1 - 0.1}, \frac{0.2 - 0.2}{1 - 0.2} \right\} = \left\{ \frac{2}{9}, \frac{1}{8}, \frac{1}{9}, 0 \right\}$$

In addition, as $h_1^c = \{0.7, 0.8\}$, and $h_2^c = \{0.9, 0.8\}$, by Definition 1.11, we obtain

$$h_1^c \oslash h_2^c = \left\{ \frac{0.7}{0.9}, \frac{0.8}{0.9}, \frac{0.7}{0.8}, \frac{0.8}{0.8} \right\} = \left\{ \frac{7}{9}, \frac{8}{9}, \frac{7}{8}, 1 \right\}$$

Since

$$(h_1 \ominus h_2)^c = \left\{ 1 - \frac{2}{9}, 1 - \frac{1}{8}, 1 - \frac{1}{9}, 1 - 0 \right\} = \left\{ \frac{7}{9}, \frac{8}{9}, \frac{7}{8}, 1 \right\}$$

Then, $(h_1 \ominus h_2)^c = h_1^c \oslash h_2^c$, which verifies (2) of Theorem 1.8. In analogous, (1) of Theorem 1.8 can also be verified.

The subtraction and division operations are significantly important in forming the integral theoretical framework of HFS. Meanwhile, it is also critical in developing some well-known decision making method such as PROMETHEE (Behzadian et al. 2010) with hesitant fuzzy information. The operations of HFEs can be immediately extended into interval-valued HFEs and dual HFEs.

1.1.3 Comparison Laws of Hesitant Fuzzy Elements

It is noted that the number of values in different HFEs may be different. Let l_{h_j} be the number of values in h_j. For two HFEs h_1 and h_2, let $l = \max\{l_{h_1}, l_{h_2}\}$. To operate correctly, Xu and Xia (2011a) gave the following regulation, which is based on the assumption that all the decision makers are pessimistic: If $l_{h_1} < l_{h_2}$, then h_1 should be extended by adding the minimum value in it until it has the same length with h_2; If $l_{h_1} > l_{h_2}$, then h_2 should be extended by adding the minimum value in it until it has the same length with h_1. For example, let $h_1 = \{0.1, 0.2, 0.3\}$, $h_2 = \{0.4, 0.5\}$. To operate correctly, we should extend h_2 until it has the same length with h_1. The pessimist may extend it as $h_2 = \{0.4, 0.4, 0.5\}$, and the optimist may extend h_2 as $h_2 = \{0.4, 0.5, 0.5\}$ which adds the maximum value instead. The results may be different if we extend the shorter one by adding different values. It is reasonable because the decision makers' risk preferences can directly influence the final decision. As to the situation where the decision makers are neither pessimistic nor optimistic, then the added value should be the mean value of the shorter HFE. We can also extend the shorter one by adding the value of 0.5 in it. In such a case, we assume that the decision makers have uncertain information.

Xia and Xu (2011a) defined the score function of a HFE:

Definition 1.12 (*Xia and Xu* 2011a). For a HFE h,

$$s(h) = \frac{1}{l_h} \sum_{\gamma \in h} \gamma \tag{1.17}$$

is called the score function of h, where l_h is the number of values in h. For two HFEs h_1 and h_2, if $s(h_1) > s(h_2)$, then $h_1 > h_2$; if $s(h_1) = s(h_2)$, then $h_1 = h_2$.

However, in some special cases, this comparison law cannot be used to distinguish two HFEs:

Example 1.5 (Liao et al. 2014a). Let $h_1 = (0.1, 0.2, 0.6)$ and $h_2 = (0.2, 0.4)$ be two HFEs, then by (1.17), we have

$$s(h_1) = \frac{0.1 + 0.2 + 06}{3} = 0.3, \quad s(h_2) = \frac{0.2 + 0.4}{2} = 0.3.$$

Since $s(h_1) = s(h_2)$, we cannot tell the difference between h_1 and h_2 by only using Definition 1.12. Actually, such a case is common in practice. Hence, Liao et al. (2014a) introduced the variance function of HFE.

Definition 1.13 (*Liao et al.* 2014a). For a HFE h,

$$v_1(h) = \frac{1}{l_h} \sqrt{\sum_{\gamma_i, \gamma_j \in h} (\gamma_i - \gamma_j)^2} \tag{1.18}$$

is called the variance function of h, where l_h is the number of values in h, and $v_1(h)$ is called the variance degree of h. For two HFEs h_1 and h_2, if $v_1(h_1) > v_1(h_2)$, then $h_1 < h_2$; if $v_1(h_1) = v_1(h_2)$, then $h_1 = h_2$.

Example 1.6 (Liao et al. 2014a). According to Eq. (1.18), in Example 1.5, we have

$$v_1(h_1) = \frac{\sqrt{0.2^2 + 0.4^2 + 0.5^2}}{3} = 0.2160, v_1(h_2) = \frac{\sqrt{0.2^2}}{2} = 0.1$$

Then, $v_1(h_1) > v_1(h_2)$, i.e., the variance degree of h_1 is higher than that of h_2. Thus, $h_1 < h_2$.

From the above analysis, we can see that the relationship between the score function and the variance function is similar to the relationship between mean and variance in statistics. It is noted that recently, Liao and Xu (2015c) modified the variance function into the following form:

$$v_2(h) = \frac{2}{l_h(l_h - 1)} \sqrt{\sum_{\gamma_i, \gamma_j \in h} (\gamma_i - \gamma_j)^2} \tag{1.19}$$

where l_h in the coefficient of Eq. (1.18) is replaced by $C_{l_h}^2 = \frac{l_h(l_h - 1)}{2}$.

In addition, Chen et al. (2015) introduced the deviation function of a HFE:

Definition 1.14 (*Chen et al.* 2015). For a HFE h, we define the deviation degree $v_3(h)$ of h as:

$$v_3(h) = \sqrt{\frac{1}{l_h} \sum_{\gamma \in h} (\gamma - s(h))^2} \tag{1.20}$$

As it can be seen that $v_3(h)$ is just conventional standard variance in statistics, which reflects the deviation degree between all values in the HFE h and their mean value.

Based on the score function $s(h)$ and the variance function $v_q(h)(q = 1, 2, 3)$, a comparison scheme can be developed to rank any HFEs (Liao et al. 2014a):

- If $s(h_1) < s(h_2)$, then $h_1 < h_2$;
- If $s(h_1) = s(h_2)$, then

  - If $v_q(h_1) < v_q(h_2)$, then $h_1 > h_2$.
  - If $v_q(h_1) = v_q(h_2)$, then $h_1 = h_2$.

Note that we cannot claim that "For two HFEs h_1 and h_2, if $v(h_1) > v(h_2)$, then $h_1 < h_2$; If $v(h_1) = v(h_2)$, then $h_1 = h_2$" due to the fact that sometimes variance is bad, while sometimes variance is good. This assentation holds only under the precondition that $s(h_1) = s(h_2)$. It is well known that an efficient estimator is a measure of the variance of an estimate's sampling distribution in statistics. Hence, under the condition that the score values are equal, which implies that the average values are the same in statistics, it is appropriate to stipulate that the smaller the variance, the more stable the HFE, and thus, the greater the HFE. Similar schemes can be seen in the process of comparing two vague sets (Hong and Choi 2000), and also the comparison between two IFNs (Xu and Yager 2006).

1.2 Extensions of Hesitant Fuzzy Set

1.2.1 Interval-Valued Hesitant Fuzzy Set

In many decision making problems, due to the insufficiency of available information, it may be difficult for decision makers (or experts) to exactly quantify the membership degrees of an element to a set by crisp numbers but by interval-valued numbers within [0, 1]. Consequently, it is necessary to introduce the concept of interval-valued hesitant fuzzy set (IVHFS), which permits the membership degree of an element to a given set to have a few different interval values. The situation is similar to that encounters in intuitionistic fuzzy environment where the concept of IFS has been extended to interval-valued IFS (Atanassov and Gargov 1989).

Definition 1.15 (*Chen et al.* 2013b). Let X be a reference set, and $D[0, 1]$ be the set of all closed subintervals of [0, 1]. An IVHFS on X is

$$\tilde{H} = \{ <x, \tilde{h}_A(x) > | x \in X \} \tag{1.21}$$

where $\tilde{h}_A(x) : X \to D[0,1]$ denotes all possible interval-valued membership degrees of the element $x \in X$ to the set $A \subset X$. For convenience, we call $\tilde{h}_A(x)$ an interval-valued hesitant fuzzy element (IVHFE), which reads

$$\tilde{h}_A(x) = \{\tilde{\gamma}|\tilde{\gamma} \in \tilde{h}_A(x)\} \tag{1.22}$$

Here $\tilde{\gamma} = [\tilde{\gamma}^L, \tilde{\gamma}^U]$ is an interval-valued number. $\tilde{\gamma}^L = \inf \tilde{\gamma}$ and $\tilde{\gamma}^U = \sup \tilde{\gamma}$ represent the lower and upper limits of $\tilde{\gamma}$, respectively.

The IVHFE is the basic unit of the IVHFS. It can be considered as a special case of the IVHFS. The relationship between IVHFE and IVHFS is similar to that between the interval-valued fuzzy number and interval-valued fuzzy set (Zadeh 1975).

Example 1.7 (Chen et al. 2013b). Let $X = \{x_1, x_2\}$ be a reference set, and the IVHFEs $h_A(x_1) = \{[0.1, \ 0.3], [\ 0.4, 0.5]\}$ and $h_A(x_2) = \{[0.1, 0.2], [0.3, 0.5], [0.7, 0.9]\}$ denote the membership degrees of $x_i (i = 1, 2)$ to a set $A \subset X$ respectively. We call $\tilde{H}$ an IVHFS, where

$$\tilde{H} = \{<x_1, \{[0.1, \ 0.3], \ [0.4, 0.5]\} >, \ <x_2, \{[0.1, 0.2], [0.3, 0.5], [0.7, 0.9]\}\}$$

When a decision making problem needs to be characterized by interval-valued numbers rather than crisp numbers, the IVHFS is a preferable choice because it has a great ability in handling imprecise and ambiguous information. For example, supposing two decision makers (or experts) discuss the membership degree of an element x to a set A, one wants to assign [0.3, 0.5] and the other wants to assign [0.6, 0.7]. They cannot reach consensus. In such a circumstance, the degree can be represented by an IVHFE $\{[0.3, \ 0.5], \ [0.6, \ 0.7]\}$. Furthermore, in a usual interval-valued fuzzy logic, it is common to average these interval membership degrees or take the smallest interval that contains all these interval degrees. However, the IVHFE can keep all interval values proposed by the decision makers (or experts). That is to say, potentially, it keeps more information about the decision makers' (or experts') opinions, the information that is normally dismissed. It therefore can give a better result in information aggregation.

It should be noted that when the upper and lower bounds of the interval values are identical, the IVHFS becomes the HFS, indicating that the HFS is a special case of the IVHFS. Moreover, when the membership degree of each element belonging to a given set only has an interval value, the IVHFE reduces to the interval-valued fuzzy number and the IVHFS becomes the interval-valued fuzzy set. We can introduce some special IVHFEs, such as:

(1) Empty set: $\tilde{O}^* = \{<x, \ \tilde{h}^\circ(x) > |x \in X\}$, where $\tilde{h}^\circ(x) = \{[0,0]\}$, $\forall x \in X$.
(2) Full set: $\tilde{E}^* = \{<x, \ \tilde{h}^*(x) > |x \in X\}$, where $\tilde{h}^*(x) = \{[1,1]\}$, $\forall x \in X$.
(3) Complete ignorance (all is possible): $\tilde{U}^* = \{<x, \ \tilde{h}(x) > |x \in X\}$, where $\tilde{h}(x) = \{[0,1]\}$, $\forall x \in X$.
(4) Nonsense set: $\tilde{\oslash}^* = \{<x, \ \tilde{h}(x) > |x \in X\}$, where $\tilde{h}(x) = \varnothing$, $\forall x \in X$.

Chen et al. (2013b) defined some operations on IVHFEs through the connection between IVHFEs and HFEs.

Definition 1.16 (*Chen et al.* 2013b). Let $\tilde{h}$, $\tilde{h}_1$ and $\tilde{h}_2$ be three IVHFEs, then

(1) $\tilde{h}^c = \left\{[1 - \tilde{\gamma}^U, 1 - \tilde{\gamma}^L] | \tilde{\gamma} \in \tilde{h}\right\}$.

(2) $\tilde{h}_1 \cup \tilde{h}_2 = \left\{[\max(\tilde{\gamma}_1^L, \tilde{\gamma}_2^L), \max(\tilde{\gamma}_1^U, \tilde{\gamma}_2^U)] | \tilde{\gamma}_1 \in \tilde{h}_1, \tilde{\gamma}_2 \in \tilde{h}_2\right\}$.

(3) $\tilde{h}_1 \cap \tilde{h}_2 = \left\{[\min(\tilde{\gamma}_1^L, \tilde{\gamma}_2^L), \min(\tilde{\gamma}_1^U, \tilde{\gamma}_2^U)] | \tilde{\gamma}_1 \in \tilde{h}_1, \tilde{\gamma}_2 \in \tilde{h}_2\right\}$.

(4) $\tilde{h}^\lambda = \left\{[(\tilde{\gamma}^L)^\lambda, (\tilde{\gamma}^U)^\lambda] | \tilde{\gamma} \in \tilde{h}\right\}$, $\lambda > 0$.

(5) $\lambda\tilde{h} = \left\{[1 - (1 - \tilde{\gamma}^L)^\lambda, 1 - (1 - \tilde{\gamma}^U)^\lambda] | \tilde{\gamma} \in \tilde{h}\right\}$, $\lambda > 0$.

(6) $\tilde{h}_1 \oplus \tilde{h}_2 = \left\{[\tilde{\gamma}_1^L + \tilde{\gamma}_2^L - \tilde{\gamma}_1^L \cdot \tilde{\gamma}_2^L, \tilde{\gamma}_1^U + \tilde{\gamma}_2^U - \tilde{\gamma}_1^U \cdot \tilde{\gamma}_2^U] | \tilde{\gamma}_1 \in \tilde{h}_1, \tilde{\gamma}_2 \in \tilde{h}_2\right\}$.

(7) $\tilde{h}_1 \otimes \tilde{h}_2 = \left\{[\tilde{\gamma}_1^L \cdot \tilde{\gamma}_2^L, \tilde{\gamma}_1^U \cdot \tilde{\gamma}_2^U] | \tilde{\gamma}_1 \in \tilde{h}_1, \tilde{\gamma}_2 \in \tilde{h}_2\right\}$.

Example 1.8 (Chen and Xu 2014). Suppose there are three IVHFEs $\tilde{h}_1 = \{[0.4, 0.6]\}$, $\tilde{h}_2 = \{[0.2, 0.3], [0.5, 0.7], [0.6, 0.8]\}$, $\tilde{h}_3 = \{[0.3, 0.4], [0.7, 0.8]\}$. Let $\lambda = 2$, then we have

$$
\begin{aligned}
(1) \quad \tilde{h}_3^c &= \left\{[1 - \tilde{\gamma}_3^U, 1 - \tilde{\gamma}_3^L] | \tilde{\gamma}_3 \in \tilde{h}_3\right\} \\
&= \{[1 - 0.8, 1 - 0.7], [1 - 0.4, 1 - 0.3]\} = \{[0.2, 0.3], [0.6, 0.7]\}.
\end{aligned}
$$

$$
\begin{aligned}
(2) \quad \tilde{h}_1 \cup \tilde{h}_2 &= \left\{[\max(\tilde{\gamma}_1^L, \tilde{\gamma}_2^L), \max(\tilde{\gamma}_1^U, \tilde{\gamma}_2^U)] | \tilde{\gamma}_1 \in \tilde{h}_1, \tilde{\gamma}_2 \in \tilde{h}_2\right\} \\
&= \{[\max(0.4, 0.2), \max(0.6, 0.3)], [\max(0.4, 0.5), \max(0.6, 0.7)], \\
&\quad [\max(0.4, 0.6), \max(0.6, 0.8)]\} \\
&= \{[0.4, 0.6], [0.5, 0.7], [0.6, 0.8]\}.
\end{aligned}
$$

$$
\begin{aligned}
(3) \quad \tilde{h}_1 \cap \tilde{h}_2 &= \left\{[\min(\tilde{\gamma}_1^L, \tilde{\gamma}_2^L), \min(\tilde{\gamma}_1^U, \tilde{\gamma}_2^U)] | \tilde{\gamma}_1 \in \tilde{h}_1, \tilde{\gamma}_2 \in \tilde{h}_2\right\} \\
&= \{[\min(0.4, 0.2), \min(0.6, 0.3)], [\min(0.4, 0.5), \min(0.6, 0.7)], \\
&\quad [\min(0.4, 0.6), \min(0.6, 0.8)] \\
&= \{[0.2, 0.3], [0.4, 0.6]\}
\end{aligned}
$$

Noted that the symbol "{}" means the set of interval-valued numbers. Considering that any two elements in a set must be different, the repeated elements are thus deleted.

$$
\begin{aligned}
(4) \quad \tilde{h}_1 \oplus \tilde{h}_2 &= \left\{[\tilde{\gamma}_1^L + \tilde{\gamma}_2^L - \tilde{\gamma}_1^L \cdot \tilde{\gamma}_2^L, \tilde{\gamma}_1^U + \tilde{\gamma}_2^U - \tilde{\gamma}_1^U \cdot \tilde{\gamma}_2^U] | \tilde{\gamma}_1 \in \tilde{h}_1, \tilde{\gamma}_2 \in \tilde{h}_2\right\} \\
&= \{[0.4 + 0.2 - 0.4 \cdot 0.2, 0.6 + 0.3 - 0.6 \cdot 0.3], [0.4 + 0.5 - 0.4 \cdot 0.5, 0.6 + 0.7 - 0.6 \cdot 0.7], \\
&\quad [0.4 + 0.6 - 0.4 \cdot 0.6, 0.6 + 0.8 - 0.6 \cdot 0.8]\} \\
&= \{[0.52, 0.72], [0.7, 0.88], [0.76, 0.92]\}.
\end{aligned}
$$

(5) $\quad \tilde{h}_1 \otimes \tilde{h}_2 = \left\{ \left[\tilde{\gamma}_1^L \cdot \tilde{\gamma}_2^L, \ \tilde{\gamma}_1^U \cdot \tilde{\gamma}_2^U \right] \big| \tilde{\gamma}_1 \in \tilde{h}_1, \tilde{\gamma}_2 \in \tilde{h}_2 \right\}$

$\qquad\quad = \{ [0.4 \cdot 0.2, \ 0.6 \cdot 0.3], [0.4 \cdot 0.5, \ 0.6 \cdot 0.7], [0.4 \cdot 0.6, \ 0.6 \cdot 0.8] \}$

$\qquad\quad = \{ [0.08, \ 0.18], [0.2, \ 0.42], [0.24, \ 0.48] \}.$

(6) $\quad \lambda \tilde{h}_3 = \left\{ \left[1 - (1 - \tilde{\gamma}_3^L)^2, 1 - (1 - \tilde{\gamma}_3^U)^2 \right] \big| \tilde{\gamma}_3 \in \tilde{h}_3 \right\}$

$\qquad\quad = \left\{ \left[1 - (1 - 0.3)^2, 1 - (1 - 0.4)^2 \right], \left[1 - (1 - 0.7)^2, 1 - (1 - 0.8)^2 \right] \right\}$

$\qquad\quad = \{ [0.51, 0.64], [0.91, 0.96] \}.$

(7) $\quad \tilde{h}_3^2 = \left\{ \left[(\tilde{\gamma}_3^L)^2, (\tilde{\gamma}_3^U)^2 \right] \big| \tilde{\gamma}_3 \in \tilde{h}_3 \right\}$

$\qquad\quad = \left\{ \left[(0.3)^2, (0.4)^2 \right], \left[(0.7)^2, (0.8)^2 \right] \right\} = \{ [0.09, 0.16], [0.49, 0.64] \}.$

It is pointed out that if $\tilde{\gamma}^L = \tilde{\gamma}^U$, then the operations in Definition 1.16 reduce to those of HFEs.

Theorem 1.9 (Chen and Xu 2014). *Let $\tilde{h}$ be an IVHFE and $\lambda, \lambda_1, \lambda_2 > 0$, then*

(1) $\tilde{h} \cup \tilde{h} = \tilde{h}, \ \tilde{h} \cap \tilde{h} = \tilde{h}.$

(2) $\tilde{h} \cup \tilde{h}^\circ = \tilde{h}, \ \tilde{h} \cap \tilde{h}^\circ = \tilde{h}^\circ.$

(3) $\tilde{h} \cup \tilde{h}^* = \tilde{h}^*, \ \tilde{h} \cap \tilde{h}^* = \tilde{h}.$

(4) $\tilde{h} \oplus \tilde{h}^\circ = \tilde{h}, \ \tilde{h} \otimes \tilde{h}^\circ = \tilde{h}^\circ.$

(5) $\tilde{h} \oplus \tilde{h}^* = \tilde{h}^*, \ \tilde{h} \otimes \tilde{h}^* = \tilde{h}.$

(6) $\lambda \tilde{h}^\circ = \tilde{h}^\circ, \ \lambda \tilde{h}^* = \tilde{h}^*.$

(7) $\left(\tilde{h}^\circ \right)^\lambda = \tilde{h}^\circ, \ \left(\tilde{h}^* \right)^\lambda = \tilde{h}^*.$

(8) $\left(\tilde{h}^{\lambda_1} \right)^{\lambda_2} = \left(\tilde{h}^{\lambda_2} \right)^{\lambda_1} = \tilde{h}^{\lambda_1 \lambda_2}, \ \lambda_2 \left(\lambda_1 \tilde{h} \right) = \lambda_1 \left(\lambda_2 \tilde{h} \right) = (\lambda_1 \lambda_2) \tilde{h}.$

Theorem 1.10 (Chen and Xu 2014). *Let $\tilde{h}_1, \tilde{h}_2$ and $\tilde{h}_3$ be three IVHFEs, then*

(1) $\tilde{h}_1 \cup \tilde{h}_2 = \tilde{h}_2 \cup \tilde{h}_1.$

(2) $\tilde{h}_1 \cap \tilde{h}_2 = \tilde{h}_2 \cap \tilde{h}_1.$

(3) $\tilde{h}_1 \cap (\tilde{h}_2 \cap \tilde{h}_3) = (\tilde{h}_1 \cap \tilde{h}_2) \cap \tilde{h}_3.$

(4) $\tilde{h}_1 \cup (\tilde{h}_2 \cup \tilde{h}_3) = (\tilde{h}_1 \cup \tilde{h}_2) \cup \tilde{h}_3.$

(5) $\tilde{h}_1 \oplus (\tilde{h}_2 \oplus \tilde{h}_3) = (\tilde{h}_1 \oplus \tilde{h}_2) \oplus \tilde{h}_3.$

(6) $\tilde{h}_1 \otimes (\tilde{h}_2 \otimes \tilde{h}_3) = (\tilde{h}_1 \otimes \tilde{h}_2) \otimes \tilde{h}_3.$

(7) $\tilde{h}_1 \cap (\tilde{h}_2 \cup \tilde{h}_3) = (\tilde{h}_1 \cap \tilde{h}_2) \cup (\tilde{h}_1 \cap \tilde{h}_3).$

(8) $\tilde{h}_1 \cup (\tilde{h}_2 \cap \tilde{h}_3) = (\tilde{h}_1 \cup \tilde{h}_2) \cap (\tilde{h}_1 \cup \tilde{h}_3).$

Theorem 1.11 (Chen and Xu 2014). *Let $\tilde{h}_1$ and $\tilde{h}_2$ be two IVHFEs, then*

(1) $\tilde{h}_1 \cap (\tilde{h}_1 \cup \tilde{h}_2) = \tilde{h}_1$.
(2) $\tilde{h}_1 \cup (\tilde{h}_1 \cap \tilde{h}_2) = \tilde{h}_1$.

Theorem 1.12 (Chen and Xu 2014). *Let $\tilde{h}_1$ and $\tilde{h}_2$ be two IVHFEs and $\lambda > 0$, then*

(1) $\lambda(\tilde{h}_1 \cup \tilde{h}_2) = \lambda\tilde{h}_1 \cup \lambda\tilde{h}_2$.
(2) $\lambda(\tilde{h}_1 \cap \tilde{h}_2) = \lambda\tilde{h}_1 \cap \lambda\tilde{h}_2$.
(3) $(\tilde{h}_1 \cup \tilde{h}_2)^\lambda = \tilde{h}_1^\lambda \cup \tilde{h}_2^\lambda$.
(4) $(\tilde{h}_1 \cap \tilde{h}_2)^\lambda = \tilde{h}_1^\lambda \cap \tilde{h}_2^\lambda$.

Theorem 1.13 (Chen et al. 2013b). *Let $\tilde{h}$, $\tilde{h}_1$ and $\tilde{h}_2$ be three IVHFEs, we have*

(1) $\tilde{h}_1 \oplus \tilde{h}_2 = \tilde{h}_2 \oplus \tilde{h}_1$.
(2) $\tilde{h}_1 \otimes \tilde{h}_2 = \tilde{h}_2 \otimes \tilde{h}_1$.
(3) $\lambda(\tilde{h}_1 \oplus \tilde{h}_2) = \lambda\tilde{h}_1 \oplus \lambda\tilde{h}_2$, $\lambda > 0$.
(4) $(\tilde{h}_1 \otimes \tilde{h}_2)^\lambda = \tilde{h}_1^\lambda \otimes \tilde{h}_2^\lambda$, $\lambda > 0$.
(5) $\lambda_1\tilde{h} \oplus \lambda_2\tilde{h} = (\lambda_1 + \lambda_2)\tilde{h}$, $\lambda_1, \lambda_2 > 0$.
(6) $\tilde{h}^{\lambda_1} \otimes \tilde{h}^{\lambda_2} = \tilde{h}^{(\lambda_1 + \lambda_2)}$, $\lambda_1, \lambda_2 > 0$.

Proof For three IVHFEs $\tilde{h}$, $\tilde{h}_1$ and $\tilde{h}_2$, we have

(1) $$\tilde{h}_1 \oplus \tilde{h}_2 = \left\{ [\tilde{\gamma}_1^L + \tilde{\gamma}_2^L - \tilde{\gamma}_1^L \cdot \tilde{\gamma}_2^L,\ \tilde{\gamma}_1^U + \tilde{\gamma}_2^U - \tilde{\gamma}_1^U \cdot \tilde{\gamma}_2^U] \big| \tilde{\gamma}_1 \in \tilde{h}_1, \tilde{\gamma}_2 \in \tilde{h}_2 \right\} \cdot$$
$$= \left\{ [\tilde{\gamma}_2^L + \tilde{\gamma}_1^L - \tilde{\gamma}_2^L \cdot \tilde{\gamma}_1^L,\ \tilde{\gamma}_2^U + \tilde{\gamma}_1^U - \tilde{\gamma}_2^U \cdot \tilde{\gamma}_1^U] \big| \tilde{\gamma}_1 \in \tilde{h}_1, \tilde{\gamma}_2 \in \tilde{h}_2 \right\}$$
$$= \tilde{h}_2 \oplus \tilde{h}_1$$

(2) $$\tilde{h}_1 \otimes \tilde{h}_2 = \left\{ [\tilde{\gamma}_1^L \cdot \tilde{\gamma}_1^L,\ \tilde{\gamma}_1^U \cdot \tilde{\gamma}_1^U] \big| \tilde{\gamma}_1 \in \tilde{h}_1, \tilde{\gamma}_2 \in \tilde{h}_2 \right\}$$
$$= \left\{ [\tilde{\gamma}_2^L \cdot \tilde{\gamma}_1^L,\ \tilde{\gamma}_2^U \cdot \tilde{\gamma}_2^U] \big| \tilde{\gamma}_1 \in \tilde{h}_1, \tilde{\gamma}_2 \in \tilde{h}_2 \right\} = \tilde{h}_2 \otimes \tilde{h}_1.$$

(3) $$\lambda(\tilde{h}_1 \oplus \tilde{h}_2) = \left\{ [1 - (1 - (\tilde{\gamma}_1^L + \tilde{\gamma}_2^L - \tilde{\gamma}_1^L \cdot \tilde{\gamma}_2^L))^\lambda, 1 - (1 - (\tilde{\gamma}_1^U + \tilde{\gamma}_2^U - \tilde{\gamma}_1^U \cdot \tilde{\gamma}_2^U)^\lambda] \big| \tilde{\gamma}_1 \in \tilde{h}_1, \tilde{\gamma}_2 \in \tilde{h}_2 \right\}$$
$$= \left\{ [1 - (1 - \tilde{\gamma}_1^L)^\lambda (1 - \tilde{\gamma}_2^L)^\lambda, 1 - (1 - \tilde{\gamma}_1^U)^\lambda (1 - \tilde{\gamma}_2^U)^\lambda] \big| \tilde{\gamma}_1 \in \tilde{h}_1, \tilde{\gamma}_2 \in \tilde{h}_2 \right\}$$
$$= \left\{ [1 - (1 - \tilde{\gamma}_1^L)^\lambda + 1 - (1 - \tilde{\gamma}_2^L)^\lambda - (1 - (1 - \tilde{\gamma}_1^L)^\lambda)(1 - (1 - \tilde{\gamma}_2^L)^\lambda), \right.$$
$$\left. 1 - (1 - \tilde{\gamma}_1^U)^\lambda + 1 - (1 - \tilde{\gamma}_2^U)^\lambda - (1 - (1 - \tilde{\gamma}_1^U)^\lambda)(1 - (1 - \tilde{\gamma}_2^U)^\lambda)] \big| \tilde{\gamma}_1 \in \tilde{h}_1, \tilde{\gamma}_2 \in \tilde{h}_2 \right\}$$
$$= \lambda\tilde{h}_1 \oplus \lambda\tilde{h}_2.$$

(4) $\quad (\tilde{h}_1 \otimes \tilde{h}_2)^\lambda = \left\{ [(\tilde{\gamma}_1^L \cdot \tilde{\gamma}_2^L)^\lambda, (\tilde{\gamma}_1^U \cdot \tilde{\gamma}_2^U)^\lambda] | \tilde{\gamma}_1 \in \tilde{h}_1, \tilde{\gamma}_2 \in \tilde{h}_2 \right\}$

$\qquad = \left\{ [(\tilde{\gamma}_1^L)^\lambda \cdot (\tilde{\gamma}_2^L)^\lambda, (\tilde{\gamma}_1^U)^\lambda \cdot (\tilde{\gamma}_2^U)^\lambda] | \tilde{\gamma}_1 \in \tilde{h}_1, \tilde{\gamma}_2 \in \tilde{h}_2 \right\}$

$\qquad = \tilde{h}_1^\lambda \otimes \tilde{h}_2^\lambda.$

(5) $\quad \lambda_1 \tilde{h} \oplus \lambda_2 \tilde{h} = \left\{ [1 - (1 - \tilde{\gamma}^L)^{\lambda_1} + 1 - (1 - \tilde{\gamma}^L)^{\lambda_2} - (1 - (1 - \tilde{\gamma}^L)^{\lambda_1})(1 - (1 - \tilde{\gamma}^L)^{\lambda_2}), \right.$

$\qquad \left. 1 - (1 - \tilde{\gamma}^U)^{\lambda_1} + 1 - (1 - \tilde{\gamma}^U)^{\lambda_2} - (1 - (1 - \tilde{\gamma}^U)^{\lambda_1})(1 - (1 - \tilde{\gamma}^U)^{\lambda_2})] | \tilde{\gamma} \in \tilde{h} \right\}$

$\qquad = \left\{ [1 - (1 - \tilde{\gamma}^L)^{\lambda_1}(1 - \tilde{\gamma}^L)^{\lambda_2}, 1 - (1 - \tilde{\gamma}^U)^{\lambda_1}(1 - \tilde{\gamma}^U)^{\lambda_2}] | \tilde{\gamma} \in \tilde{h} \right\}$

$\qquad = \left\{ [1 - (1 - \tilde{\gamma}^L)^{\lambda_1 + \lambda_2}, 1 - (1 - \tilde{\gamma}^U)^{\lambda_1 + \lambda_2}] | \tilde{\gamma} \in \tilde{h} \right\}$

$\qquad = (\lambda_1 + \lambda_2)\tilde{h}.$

(6) $\quad \tilde{h}^{\lambda_1} \otimes \tilde{h}^{\lambda_2} = \left\{ [(\tilde{\gamma}^L)^{\lambda_1} \cdot (\tilde{\gamma}^L)^{\lambda_2}, (\tilde{\gamma}^U)^{\lambda_1} \cdot (\tilde{\gamma}^U)^{\lambda_2}] | \tilde{\gamma} \in \tilde{h} \right\}$

$\qquad = \left\{ [(\tilde{\gamma}^L)^{\lambda_1 + \lambda_2}, (\tilde{\gamma}^U)^{\lambda_1 + \lambda_2}] | \tilde{\gamma} \in \tilde{h} \right\} = \tilde{h}^{(\lambda_1 + \lambda_2)}. \quad \square$

This completes the proof.

The relationships between the defined operations on IVHFEs are given in Theorem 1.14.

Theorem 1.14 (Chen et al. 2013b). *For three IVHFEs $\tilde{h}$, $\tilde{h}_1$ and $\tilde{h}_2$, we have*

(1) $\tilde{h}_1^c \cup \tilde{h}_2^c = (\tilde{h}_1 \cap \tilde{h}_2)^c.$

(2) $\tilde{h}_1^c \cap \tilde{h}_2^c = (\tilde{h}_1 \cup \tilde{h}_2)^c.$

(3) $(\tilde{h}^c)^\lambda = (\lambda\tilde{h})^c.$

(4) $\lambda(\tilde{h}^c) = (\tilde{h}^\lambda)^c.$

(5) $\tilde{h}_1^c \oplus \tilde{h}_2^c = (\tilde{h}_1 \otimes \tilde{h}_2)^c.$

(6) $\tilde{h}_1^c \otimes \tilde{h}_2^c = (\tilde{h}_1 \oplus \tilde{h}_2)^c.$

Proof For three IVHFEs $\tilde{h}$, $\tilde{h}_1$ and $\tilde{h}_2$, we have

(1) $\quad \tilde{h}_1^c \cup \tilde{h}_2^c = \left\{ [\max(1 - \tilde{\gamma}_1^U, 1 - \tilde{\gamma}_2^U), \max(1 - \tilde{\gamma}_1^L, 1 - \tilde{\gamma}_2^L)] | \tilde{\gamma}_1 \in h_1, \tilde{\gamma}_2 \in h_2 \right\}$

$\qquad = \left\{ [1 - \min(\tilde{\gamma}_1^U, \tilde{\gamma}_2^U), 1 - \min(\tilde{\gamma}_1^L, \tilde{\gamma}_2^L)] | \tilde{\gamma}_1 \in h_1, \tilde{\gamma}_2 \in h_2 \right\} = (\tilde{h}_1 \cap \tilde{h}_2)^c.$

(2) $\quad \tilde{h}_1^c \cap \tilde{h}_2^c = \left\{ [\min(1 - \tilde{\gamma}_1^U, 1 - \tilde{\gamma}_2^U), \min(1 - \tilde{\gamma}_1^L, 1 - \tilde{\gamma}_2^L)] | \tilde{\gamma}_1 \in h_1, \tilde{\gamma}_2 \in h_2 \right\}$

$\qquad = \left\{ [1 - \max(\tilde{\gamma}_1^U, \tilde{\gamma}_2^U), 1 - \max(\tilde{\gamma}_1^L, \tilde{\gamma}_1^L)] | \tilde{\gamma}_1 \in h_1, \tilde{\gamma}_2 \in h_2 \right\} = (\tilde{h}_1 \cup \tilde{h}_2)^c.$

(3) $\quad (\tilde{h}^c)^\lambda = \left\{ [(1 - \tilde{\gamma}^U)^\lambda, (1 - \tilde{\gamma}^L)^\lambda] | \tilde{\gamma} \in \tilde{h} \right\}$

$\qquad = \left\{ [1 - (1 - \tilde{\gamma}^L)^\lambda, 1 - (1 - \tilde{\gamma}^U)^\lambda] | \tilde{\gamma} \in \tilde{h} \right\}^c = (\lambda\tilde{h})^c.$

(4) $\lambda \tilde{h}^c = \lambda \left\{ \left[1 - \tilde{\gamma}^U, 1 - \tilde{\gamma}^L \right] | \tilde{\gamma} \in \tilde{h} \right\} = \left\{ \left[1 - (1 - (1 - \tilde{\gamma}^U))^\lambda, 1 - (1 - (1 - \tilde{\gamma}^L))^\lambda \right] | \tilde{\gamma} \in \tilde{h} \right\}$

$\qquad = \left\{ \left[1 - (\tilde{\gamma}^U)^\lambda, 1 - (\tilde{\gamma}^L)^\lambda \right] | \tilde{\gamma} \in \tilde{h} \right\} = (\tilde{h}^\lambda)^c.$

(5) $\tilde{h}_1^c \oplus \tilde{h}_2^c = \left\{ \left[(1 - \tilde{\gamma}_1^U) + (1 - \tilde{\gamma}_2^U) - (1 - \tilde{\gamma}_1^U)(1 - \tilde{\gamma}_2^U), \right. \right.$

$\qquad\qquad \left. \left. (1 - \tilde{\gamma}_1^L) + (1 - \tilde{\gamma}_2^L) - (1 - \tilde{\gamma}_1^L)(1 - \tilde{\gamma}_2^L) \right] | \tilde{\gamma}_1 \in \tilde{h}_1, \tilde{\gamma}_2 \in \tilde{h}_2 \right\}$

$\qquad\qquad = \left\{ \left[1 - \tilde{\gamma}_1^U \cdot \tilde{\gamma}_2^U, 1 - \tilde{\gamma}_1^L \cdot \tilde{\gamma}_2^L \right] | \tilde{\gamma}_1 \in \tilde{h}_1, \tilde{\gamma}_2 \in \tilde{h}_2 \right\} = \left(\tilde{h}_1 \otimes \tilde{h}_2 \right)^c.$

(6) $\tilde{h}_1^c \otimes \tilde{h}_2^c = \left\{ \left[(1 - \tilde{\gamma}_1^U)(1 - \tilde{\gamma}_2^U), (1 - \tilde{\gamma}_1^L)(1 - \tilde{\gamma}_2^L) \right] | \tilde{\gamma}_1 \in \tilde{h}_1, \tilde{\gamma}_2 \in \tilde{h}_2 \right\}$

$\qquad\qquad = \left\{ \left[1 - (\tilde{\gamma}_1^U + \tilde{\gamma}_2^U - \tilde{\gamma}_1^U \cdot \tilde{\gamma}_2^U), 1 - (\tilde{\gamma}_1^L + \tilde{\gamma}_2^L - \tilde{\gamma}_1^L \cdot \tilde{\gamma}_2^L) \right] | \tilde{\gamma}_1 \in \tilde{h}_1, \tilde{\gamma}_2 \in \tilde{h}_2 \right\} = \left(\tilde{h}_1 \oplus \tilde{h}_2 \right)^c. \square$

This completes the proof.

Since the number of interval values for different IVHFEs could be different and the interval values are usually out of order, we arrange them in any order using Eq. (1.23). To facilitate the calculation between two IVHFEs, we let $l = \max\{l_{\tilde{\alpha}}, l_{\tilde{\beta}}\}$ with $l_{\tilde{\alpha}}$ and $l_{\tilde{\beta}}$ being the number of intervals in the IVHFEs $\tilde{\alpha}$ and $\tilde{\beta}$. To operate correctly, we give the following regulation: when $l_{\tilde{\alpha}} \neq l_{\tilde{\beta}}$, we can make them equivalent through adding elements to the IVHFE that has a less number of elements. In terms of pessimistic principles, the smallest element can be added while the opposite case will be adopted following optimistic principles. In this study we adopt the latter. Specifically, if $l_{\tilde{\alpha}} < l_{\tilde{\beta}}$, then $\tilde{\alpha}$ should be extended by adding the maximum value in it until it has the same length as $\tilde{\beta}$; if $l_{\tilde{\alpha}} > l_{\tilde{\beta}}$, then $\tilde{\beta}$ should be extended by adding the maximum value in it until it has the same length as $\tilde{\alpha}$.

Definition 1.17 (*Xu and Da* 2002). Let $\tilde{a} = [\tilde{a}^L, \tilde{a}^U]$ and $\tilde{b} = [\tilde{b}^L, \tilde{b}^U]$ be two interval numbers, and $\lambda \geq 0$, then

(1) $\tilde{a} = \tilde{b} \Leftrightarrow \tilde{a}^L = \tilde{b}^L$ and $\tilde{a}^U = \tilde{b}^U$.
(2) $\tilde{a} + \tilde{b} = [\tilde{a}^L + \tilde{b}^L, \tilde{a}^U + \tilde{b}^U]$.
(3) $\lambda \tilde{a} = [\lambda \tilde{a}^L, \lambda \tilde{a}^U]$, especially, $\lambda \tilde{a} = 0$, if $\lambda = 0$.

The possibility degree is proposed to compare two interval numbers:

Definition 1.18 (*Xu and Da* 2002). Let $\tilde{a} = [\tilde{a}^L, \tilde{a}^U]$ and $\tilde{b} = [\tilde{b}^L, \tilde{b}^U]$, and let $l_{\tilde{a}} = \tilde{a}^U - \tilde{a}^L$ and $l_{\tilde{b}} = \tilde{b}^U - \tilde{b}^L$. Then the degree of possibility of $\tilde{a} \geq \tilde{b}$ is formulated by

$$p(\tilde{a} \geq \tilde{b}) = \max \left\{ 1 - \max \left(\frac{\tilde{b}^U - \tilde{a}^L}{l_{\tilde{a}} + l_{\tilde{b}}}, 0 \right), 0 \right\} \qquad (1.23)$$

The score function for IVHFEs is defined as follows:

Definition 1.19 (*Chen et al.* 2013b). For an IVHFE $\tilde{h}$,

$$s(\tilde{h}) = \frac{1}{l_{\tilde{h}}} \sum_{\tilde{\gamma} \in \tilde{h}} \tilde{\gamma} \tag{1.24}$$

is called the score function of $\tilde{h}$ with $l_{\tilde{h}}$ being the number of interval values in $\tilde{h}$, and $s(\tilde{h})$ is an interval value belonging to [0, 1]. For two IVHFEs $\tilde{h}_1$ and $\tilde{h}_2$, if $s(\tilde{h}_1) \geq s(\tilde{h}_2)$, then $\tilde{h}_1 \geq \tilde{h}_2$.

Note that we can compare two scores using Eq. (1.23). Moreover, with Definition 1.19, we can compare two IVHFEs.

1.2.2 Dual Hesitant Fuzzy Set

Zhu et al. (2012) defined the dual hesitant fuzzy set in terms of two functions that return two sets of membership values and nonmembership values respectively for each element in the domain:

Definition 1.20 (*Zhu et al.* 2012). Let X be a fixed set, then a dual hesitant fuzzy set (DHFS) D on X is described as:

$$D = \{ <x, \, h_A(x), g_A(x) > |x \in X \} \tag{1.25}$$

in which $h_A(x)$ and $g_A(x)$ are two sets of some values in [0, 1], denoting the possible membership degrees and nonmembership degrees of the element $x \in X$ to the set $A \subset X$ respectively, with the conditions:

$$0 \leq \gamma, \eta \leq 1, \, 0 \leq \gamma^+ + \eta^+ \leq 1, \tag{1.26}$$

where $\gamma \in h_A(x)$, $\eta \in g_A(x)$, $\gamma^+ \in h^+(x) = \cup_{\gamma \in h_A(x)} \max\{\gamma\}$, and $\eta^+ \in g^+(x) = \cup_{\eta \in g_A(x)} \max\{\eta\}$ for all $x \in X$. For convenience, the pair $d_A(x) = (h_A(x), g_A(x))$ is called a dual hesitant fuzzy element (DHFE) denoted by $d = (h, g)$, with the conditions: $\gamma \in h$, $\eta \in g$, $\gamma^+ \in h^+ = \cup_{\gamma \in h} \max\{\gamma\}$, $\eta^+ \in g^+ = \cup_{\eta \in g} \max\{\eta\}$, $0 \leq \gamma, \eta \leq 1$ and $0 \leq \gamma^+ + \eta^+ \leq 1$.

There are some special DHFEs:

(1) Complete uncertainty: $d = (\{0\}, \{1\})$.
(2) Complete certainty: $d = (\{1\}, \{0\})$.
(3) Complete ill-known (all is possible): $d = [0, 1]$.
(4) Nonsense element: $d = \oslash$, i.e., $h = \oslash, g = \oslash$.

For a given $d \neq \oslash$, if h and g have only one value γ and η respectively, and $\gamma + \eta < 1$, then the DHFS reduces to an IFS. If h and g have only one value γ and η respectively, and $\gamma + \eta = 1$, or h owns one value, and $g = \oslash$, then the DHFS reduces to a fuzzy set. If $g = \oslash$ and $h \neq \oslash$, then the DHFS reduces to a HFS.

Hence, the DHFS encompasses the fuzzy set, the IFS, and the HFS as special cases. DHFS consists of two parts, i.e., the membership hesitancy function and the non-membership hesitancy function, which confront several different possible values indicating the cognitive degrees whether certainty or uncertainty. As we all know, when the decision makers provide their judgments over the objects, the more the information they take into account, the more the values we will obtain from the decision makers. As the DHFS can reflect the original information given by the decision makers as much as possible, it can be regarded as a more comprehensive set supporting a more flexible approach.

For simplicity, let $\gamma^- \in h^- = \cup_{\gamma \in h(x)} \min\{\gamma\}$, $\eta^- \in g^- = \cup_{\eta \in g(x)} \min\{\eta\}$. γ^+ and η^+ are defined as above. For a typical DHFS, h and g can be represented by two intervals as:

$$h = [\gamma^-, \gamma^+], \ g = [\eta^-, \eta^+] \tag{1.27}$$

Based on Definition 1.4, there is a transformation between IFN and HFE, we can also transform g to the second HFE $h^2(x) = [1 - \eta^+, 1 - \eta^-]$ denoting the possible membership degrees of the element $x \in X$ to the set $A \subset X$. In this way, both h and h^2 indicate the membership degrees. As such, we can use a "nested interval" to represent $d(x)$ as:

$$d = [[\gamma^-, \gamma^+], [1 - \eta^+, 1 - \eta^-]] \tag{1.28}$$

The common ground of these sets is to reflect fuzzy degrees to an object, according to either fuzzy numbers or interval-valued fuzzy numbers. Therefore, we use nonempty closed interval as a uniform framework to represent a DHFE d, which is divided into different cases as follows:

$$d = \begin{cases} \oslash, \text{ if } g = \oslash \text{ and } h = \oslash \\ \langle (\gamma) \begin{cases} \text{if } g = \oslash \text{ and } h \neq \oslash, \gamma^- = \gamma^+ = \gamma \\ \text{if } g \neq \oslash \text{ and } h \neq \oslash, \gamma^- = \gamma^+ = \gamma = 1 - \eta^- = 1 - \eta^+ = 1 - \eta \end{cases} \\ (1 - \eta), \text{ if } g \neq \oslash \text{ and } h = \oslash, \eta^- = \eta^+ = \eta \\ \langle [\gamma^-, \gamma^+], \text{ if } g = \oslash \text{ and } h \neq \oslash, \gamma^- \neq \gamma^+ \\ \langle [1 - \eta^+, 1 - \eta^-], \text{ if } g \neq \oslash \text{ and } h = \oslash, \eta^- \neq \eta^+ \\ \langle [\gamma, [1 - \eta^+, 1 - \eta^-]], \text{ if } g \neq \oslash \text{ and } h \neq \oslash, \eta^- \neq \eta^+, \gamma^- = \gamma^+ = \gamma \\ [[\gamma^-, \gamma^+], \eta], \text{ if } g \neq \oslash \text{ and } h \neq \oslash, \gamma^- \neq \gamma^+, \eta^- = \eta^+ = \eta \\ [[\gamma^-, \gamma^+], [1 - \eta^+, 1 - \eta^-]], \text{ if } g \neq \oslash \text{ and } h \neq \oslash, \eta^- \neq \eta^+, \gamma^- \neq \gamma^+ \end{cases}$$

$$\tag{1.29}$$

Equation (1.29) reflects the connections between DHFS and other types of fuzzy set extensions. The merit of DHFS is more flexible to be valued in multifold ways according to the practical demands than the existing sets, taking into account much more information given by decision makers.

The complement of the DHFS can be defined regarding to different situations.

Definition 1.21 (*Zhu et al.* 2012). Given a DHFE represented by the function d, and $d \neq \oslash$, its complement is defined as:

$$d^c = \begin{cases} (\cup_{\eta \in g}\{\eta\}, \cup_{\gamma \in h}\{\gamma\}), & \text{if } g \neq \oslash \text{ and } h \neq \oslash \\ (\cup_{\gamma \in h}\{1 - \gamma\}, \{\oslash\}), & \text{if } g = \oslash \text{ and } h \neq \oslash \\ (\{\oslash\}, \cup_{\eta \in g}\{1 - \eta\}), & \text{if } h = \oslash \text{ and } g \neq \oslash \end{cases} \quad (1.30)$$

Apparently, the complement can be correspondingly represented as $(d^c)^c = d$.

For two DHFSs d_1 and d_2, the corresponding lower and upper bounds to h and g are h^-, h^+, g^- and g^+, respectively, where $h^- = \cup_{\gamma \in h}\min\{\gamma\}$, $h^+ = \cup_{\gamma \in h}\max\{\gamma\}$, $g^- = \cup_{\eta \in g}\min\{\eta\}$, and $g^+ = \cup_{\eta \in g}\max\{\eta\}$. Then the union and intersection of DHFSs can be defined as follows:

Definition 1.22 (*Zhu et al.* 2012). Let d_1 and d_2 be two DHFEs. Then,

(1) $d_1 \cup d_2 = (\{h \in (h_1 \cup h_2) | h \geq \max(h_1^-, h_2^-)\}, \{g \in (g_1 \cap g_2) | g \leq \min(g_1^+, g_2^+)\})$.

(2) $d_1 \cap d_2 = (\{h \in (h_1 \cap h_2) | h \leq \min(h_1^+, h_2^+)\}, \{g \in (g_1 \cup g_2) | g \geq \max(g_1^-, g_2^-)\})$.

Example 1.9 (Zhu et al. 2012). Let $d_1 = (\{0.1, 0.3, 0.4\}, \{0.3, 0.5\})$ and $d_2 = (\{0.2, 0.5\}, \{0.1, 0.2, 0.4\})$ be two DHFEs, then we have

(1) Complement: $d_1^c = (\{0.3, 0.5\}, \{0.1, 0.3, 0.4\})$.

(2) Union: $d_1 \cup d_2 = (\{0, 2, 0.3, 0.4, 0.5\}, \{0.1, 0.2, 0.3, 0.4\})$.

(3) Intersection: $d_1 \cap d_2 = (\{0, 1, 0.2, 0.3, 0.4\}, \{0.3, 0.4, 0.5\})$.

Definition 1.23 (*Zhu et al.* 2012). For two DHFEs d_1 and d_2, let n be a positive integer, then the following operations are valid:

(1) $$d_1 \oplus d_2 = (h_{d_1} \oplus h_{d_2}, g_{d_1} \otimes g_{d_2})$$
$$= (\cup_{\gamma_{d_1} \in h_{d_1}, \gamma_{d_2} \in h_{d_2}}\{\gamma_{d_1} + \gamma_{d_2} - \gamma_{d_1}\gamma_{d_2}\}, \cup_{\eta_{d_1} \in g_{d_1}, \eta_{d_2} \in g_{d_2}}\{\eta_{d_1}\eta_{d_2}\}).$$

(2) $$d_1 \otimes d_2 = (h_{d_1} \otimes h_{d_2}, g_{d_1} \oplus g_{d_2})$$
$$= (\cup_{\gamma_{d_1} \in h_{d_1}, \gamma_{d_2} \in h_{d_2}}\{\gamma_{d_1}\gamma_{d_2}\}, \cup_{\eta_{d_1} \in g_{d_1}, \eta_{d_2} \in g_{d_2}}\{\eta_{d_1} + \eta_{d_2} - \eta_{d_1}\eta_{d_2}\}).$$

(3) $nd = (\cup_{\gamma_d \in h_d}\{1 - (1 - \gamma_d)^n\}, \cup_{\eta_d \in g_d}\{(\eta_d)^n\})$.

(4) $d^n = (\cup_{\gamma_d \in h_d}\{(\gamma_d)^n\}, \cup_{\eta_d \in g_d}\{1 - (1 - \eta_d)^n\})$.

Theorem 1.15 (Zhu et al. 2012). *Let d, d_1 and d_2 be any three DHFEs, $\lambda \geq 0$, then*

(1) $d_1 \oplus d_2 = d_2 \oplus d_1$.

(2) $d_1 \otimes d_2 = d_2 \otimes d_1$.

(3) $\lambda(d_1 \otimes d_2) = \lambda d_1 \otimes \lambda d_2$.

(4) $(d_1 \otimes d_2)^\lambda = d_1^\lambda \otimes d_2^\lambda$.

It is noted that the above operations for DHFEs are based on the Algebraic t-conorm and t-norm. In fact, there are various types of t-conorm and t-norm. If we replace the Algebraic t-conorm and t-norm in the above operations for DHFEs with other forms of t-conorm and t-norm, we shall get more operational methods for DHFEs. For example, the Einstein t-conorm and t-norm are given as:

$$S^E(x, y) = \frac{x + y}{1 + xy}, \quad T^E(x, y) = \frac{xy}{1 + (1 - x)(1 - y)} \tag{1.31}$$

Based on the Einstein t-conorm and t-norm, Zhao et al. (2015) defined the Einstein sum and the Einstein product of DHFEs as follows:

Definition 1.24 (*Zhao et al.* 2016a). For any two DHFEs $d_1 = (h_1, g_1)$ and $d_2 = (h_2, g_2)$, we have

(1) $d_1 \dot{\oplus} d_2 = (\cup_{\gamma_1 \in h_1, \gamma_2 \in h_2} \{\frac{\gamma_1 + \gamma_2}{1 + \gamma_1 \gamma_2}\}, \cup_{\eta_1 \in g_1, \eta_2 \in g_2} \{\frac{\eta_1 \eta_2}{1 + (1 - \eta_1)(1 - \eta_2)}\}).$

(2) $d_1 \dot{\otimes} d_2 = (\cup_{\gamma_1 \in h_1, \gamma_2 \in h_2} \{\frac{\gamma_1 \gamma_2}{1 + (1 - \gamma_1)(1 - \gamma_2)}\}, \cup_{\eta_1 \in g_1, \eta_2 \in g_2} \{\frac{\eta_1 + \eta_2}{1 + \eta_1 \eta_2}\}).$

To get the Einstein scalar multiplication and the Einstein power for DHFEs, the following theorems are introduced:

Theorem 1.16 (Zhao et al. 2016a). *Let $d = (h, g)$ be a DHFE, and n be any positive real number, then*

$$nd = (\cup_{\gamma \in h} \{\frac{(1 + \gamma)^n - (1 - \gamma)^n}{(1 + \gamma)^n + (1 - \gamma)^n}\}, \cup_{\eta \in g} \{\frac{2\eta^n}{(2 - \eta)^n + \eta^n}\}) \tag{1.32}$$

where $nd = \overbrace{d \dot{\oplus} d \dot{\oplus} \cdots \dot{\oplus} d}^{n}$. *Moreover, nd is a DHFE.*

Proof We use mathematical induction to prove that Eq. (1.32) holds for the positive integer n.

(1) For $n = 1$, it is obvious that Eq. (1.32) holds.
(2) Assume Eq. (1.32) holds for $n = k$. Then for $n = k + 1$, we have

$$(k + 1)d = kd \dot{\oplus} d = (\cup_{\gamma \in h} \{\frac{(1 + \gamma)^k - (1 - \gamma)^k}{(1 + \gamma)^k + (1 - \gamma)^k}\}, \cup_{\eta \in g} \{\frac{2\eta^k}{(2 - \eta)^k + \eta^k}\}) \dot{\oplus} (\cup_{\gamma \in h} \{\gamma\}, \cup_{\eta \in g} \{\eta\})$$

$$= (\cup_{\gamma \in h} \{\frac{\frac{(1 + \gamma)^k - (1 - \gamma)^k}{(1 + \gamma)^k + (1 - \gamma)^k} + \gamma}{1 + \frac{(1 + \gamma)^k - (1 - \gamma)^k}{(1 + \gamma)^k + (1 - \gamma)^k} \gamma}\}, \cup_{\eta \in g} \{\frac{\frac{2\eta^k}{(2 - \eta)^k + \eta^k} \eta}{1 + (1 - \frac{2\eta^k}{(2 - \eta)^k + \eta^k})(1 - \eta)}\})$$

$$= (\cup_{\gamma \in h} \{\frac{(1 + \gamma)^{k + 1} - (1 - \gamma)^{k + 1}}{(1 + \gamma)^{k + 1} + (1 - \gamma)^{k + 1}}\}, \cup_{\eta \in g} \{\frac{2\eta^{k + 1}}{(2 - \eta)^{k + 1} + \eta^{k + 1}}\})$$

Thus, Eq. (1.32) holds for $n = k + 1$.

In the following, we prove that Theorem 1.16 holds when n is a positive real number.

Since $0 \leq \gamma \leq 1$, $0 \leq \eta \leq 1$, $1 \leq 2 - \eta \leq 2$, and $1 - \gamma \geq \eta \geq 0$, $1 - \eta \geq \gamma \geq 0$, obviously, we have

$$0 \leq \frac{(1+\gamma)^n - (1-\gamma)^n}{(1+\gamma)^n + (1-\gamma)^n} \leq 1 \qquad (1.33)$$

$$0 \leq \frac{2\eta^n}{(2-\eta)^n + \eta^n} \leq 1 \qquad (1.34)$$

$$0 \leq \frac{(1+\gamma)^n - (1-\gamma)^n}{(1+\gamma)^n + (1-\gamma)^n} \leq \frac{(1+\gamma)^n - (1-\gamma)^n}{(1+\gamma)^n + \eta^n} \leq \frac{(1+\gamma)^n - \eta^n}{(1+\gamma)^n + \eta^n} \qquad (1.35)$$

$$0 \leq \frac{2\eta^n}{(2-\eta)^n + \eta^n} = \frac{2\eta^n}{(1+(1-\eta))^n + \eta^n} \leq \frac{2\eta^n}{(1+(1-\eta))^n + \eta^n} \leq \frac{2\eta^n}{(1+\gamma)^n + \eta^n} \qquad (1.36)$$

From Eqs. (1.35) and (1.36), we have

$$0 \leq \frac{(1+\gamma)^n - (1-\gamma)^n}{(1+\gamma)^n + (1-\gamma)^n} + \frac{2\eta^n}{(2-\eta)^n + \eta^n} \leq 1 \qquad (1.37)$$

Combining Eqs. (1.33), (1.34) and (1.37), we know that the DHFE nd is a DHFE for any positive real number n. This completes the proof of Theorem 1.16.

Theorem 1.17 (Zhao et al. 2016a). *Let $d = (h, g)$ be a DHFE, and n be any positive real number, then*

$$d^n = (\cup_{\gamma \in h}\{\frac{2\gamma^n}{(2-\gamma)^n + \gamma^n}\}, \ \cup_{\eta \in g}\{\frac{(1+\eta)^n - (1-\eta)^n}{(1+\eta)^n + (1-\eta)^n}\}) \qquad (1.38)$$

where $d^n = \overbrace{d \dot{\otimes} d \dot{\otimes} \cdots \otimes d}^{n}$, and d^n is a DHFE.

Based on Theorems 1.16 and 1.17, the Einstein scalar multiplication and the Einstein power of DHFE can be defined:

Definition 1.25 (*Zhao et al. 2016a*). Let $d = (h, g)$ be a DHFE, $\lambda > 0$, then

(1) $\lambda d = (\cup_{\gamma \in h}\{\frac{(1+\gamma)^\lambda - (1-\gamma)^\lambda}{(1+\gamma)^\lambda + (1-\gamma)^\lambda}\}, \ \cup_{\eta \in g}\{\frac{2\eta^\lambda}{(2-\eta)^\lambda + \eta^\lambda}\})$.

(2) $d^\lambda = (\cup_{\gamma \in h}\{\frac{2\gamma^\lambda}{(2-\gamma)^\lambda + \gamma^\lambda}\}, \ \cup_{\eta \in g}\{\frac{(1+\eta)^\lambda - (1-\eta)^\lambda}{(1+\eta)^\lambda + (1-\eta)^\lambda}\})$.

Theorem 1.18 (Zhao et al. 2016a). *Let d, d_1 and d_2 be any three DHFEs, $\lambda > 0$, then*

(1) $d_1 \dot{\oplus} d_2 = d_2 \dot{\oplus} d_1$.
(2) $d_1 \dot{\otimes} d_2 = d_2 \dot{\otimes} d_1$.
(3) $\lambda(d_1 \dot{\oplus} d_2) = \lambda d_1 \dot{\oplus} \lambda d_2$.
(4) $(d_1 \dot{\otimes} d_2)^\lambda = d_1^\lambda \dot{\otimes} d_2^\lambda$.
(5) $\lambda_1 d \dot{\oplus} \lambda_2 d = (\lambda_1 \dot{\oplus} \lambda_2) d$.
(6) $d^{\lambda_1} \dot{\otimes} d^{\lambda_2} = d^{\lambda_1 + \lambda_2}$.

Proof (1) and (2) are obvious. We prove (3) and (5), while (4) and (6) can be proven similarly.

(3) $\lambda(d_1 \dot{\oplus} d_2) = \lambda\Big(\cup_{\gamma_1 \in h_1, \gamma_2 \in h_2}\Big\{ \frac{\gamma_1 + \gamma_2}{1 + \gamma_1 \gamma_2}\Big\}, \cup_{\eta_1 \in g_1, \eta_2 \in g_2}\Big\{ \frac{\eta_1 \eta_2}{1 + (1 - \eta_1)(1 - \eta_2)}\Big\}\Big)$

$$= \Big(\cup_{\gamma_1 \in h_1, \gamma_2 \in h_2}\Big\{ \frac{(1 + \frac{\gamma_1 + \gamma_2}{1 + \gamma_1 \gamma_2})^\lambda - (1 - \frac{\gamma_1 + \gamma_2}{1 + \gamma_1 \gamma_2})^\lambda}{(1 + \frac{\gamma_1 + \gamma_2}{1 + \gamma_1 \gamma_2})^\lambda + (1 - \frac{\gamma_1 + \gamma_2}{1 + \gamma_1 \gamma_2})^\lambda}\Big\},$$

$$\cup_{\eta_1 \in g_1, \eta_2 \in g_2}\Big\{ \frac{2(\frac{\eta_1 \eta_2}{1 + (1 - \eta_1)(1 - \eta_2)})^\lambda}{(2 - \frac{\eta_1 \eta_2}{1 + (1 - \eta_1)(1 - \eta_2)})^\lambda + (\frac{\eta_1 \eta_2}{1 + (1 - \eta_1)(1 - \eta_2)})^\lambda}\Big\}\Big)$$

$$= \Big(\cup_{\gamma_1 \in h_1, \gamma_2 \in h_2}\Big\{ \frac{(1 + \gamma_1)^\lambda(1 + \gamma_2)^\lambda - (1 - \gamma_1)^\lambda(1 - \gamma_2)^\lambda}{(1 + \gamma_1)^\lambda(1 + \gamma_2)^\lambda + (1 - \gamma_1)^\lambda(1 - \gamma_2)^\lambda}\Big\},$$

$$\cup_{\eta_1 \in g_1, \eta_2 \in g_2}\Big\{ \frac{2(\eta_1 \eta_2)^\lambda}{(2 - \eta_1)^\lambda(2 - \eta_2)^\lambda + (\eta_1 \eta_2)^\lambda}\Big\}\Big)$$

$$\lambda d_1 \dot{\oplus} \lambda d_2 = \Big(\cup_{\gamma_1 \in h_1}\Big\{ \frac{(1 + \gamma_1)^\lambda - (1 - \gamma_1)^\lambda}{(1 + \gamma_1)^\lambda + (1 - \gamma_1)^\lambda}\Big\}, \cup_{\eta_1 \in g_1}\Big\{ \frac{2\eta_1^\lambda}{(2 - \eta_1)^\lambda + \eta_1^\lambda}\Big\}\Big)$$

$$\dot{\oplus}\Big(\cup_{\gamma_2 \in h_2}\Big\{ \frac{(1 + \gamma_2)^\lambda - (1 - \gamma_2)^\lambda}{(1 + \gamma_2)^\lambda + (1 - \gamma_2)^\lambda}\Big\}, \cup_{\eta_2 \in g_2}\Big\{ \frac{2\eta_2^\lambda}{(2 - \eta_2)^\lambda + \eta_2^\lambda}\Big\}\Big)$$

$$= \Big(\cup_{\gamma_1 \in h_1, \gamma_2 \in h_2}\Big\{ \frac{\frac{(1 + \gamma_1)^\lambda - (1 - \gamma_1)^\lambda}{(1 + \gamma_1)^\lambda + (1 - \gamma_1)^\lambda} + \frac{(1 + \gamma_2)^\lambda - (1 - \gamma_2)^\lambda}{(1 + \gamma_2)^\lambda + (1 - \gamma_2)^\lambda}}{1 + \frac{(1 + \gamma_1)^\lambda - (1 - \gamma_1)^\lambda}{(1 + \gamma_1)^\lambda + (1 - \gamma_1)^\lambda} \frac{(1 + \gamma_2)^\lambda - (1 - \gamma_2)^\lambda}{(1 + \gamma_2)^\lambda + (1 - \gamma_2)^\lambda}}\Big\},$$

$$\cup_{\eta_1 \in g_1, \eta_2 \in g_2}\Big\{ \frac{\frac{2\eta_1^\lambda}{(2 - \eta_1)^\lambda + \eta_1^\lambda} \frac{2\eta_2^\lambda}{(2 - \eta_2)^\lambda + \eta_2^\lambda}}{1 + (1 - \frac{2\eta_1^\lambda}{(2 - \eta_1)^\lambda + \eta_1^\lambda})(1 - \frac{2\eta_2^\lambda}{(2 - \eta_2)^\lambda + \eta_2^\lambda})}\Big\}\Big)$$

$$= \Big(\cup_{\gamma_1 \in h_1, \gamma_2 \in h_2}\Big\{ \frac{(1 + \gamma_1)^\lambda(1 + \gamma_2)^\lambda - (1 - \gamma_1)^\lambda(1 - \gamma_2)^\lambda}{(1 + \gamma_1)^\lambda(1 + \gamma_2)^\lambda + (1 - \gamma_1)^\lambda(1 - \gamma_2)^\lambda}\Big\},$$

$$\cup_{\eta_1 \in g_1, \eta_2 \in g_2}\Big\{ \frac{2(\eta_1 \eta_2)^\lambda}{(2 - \eta_1)^\lambda(2 - \eta_2)^\lambda + (\eta_1 \eta_2)^\lambda}\Big\}\Big).$$

Thus, $\lambda(d_1 \dot{\oplus} d_2) = \lambda d_1 \dot{\oplus} \lambda d_2$.

(5) $\lambda_1 d \dot{\oplus} \lambda_2 d = (\cup_{\gamma \in h} \{ \dfrac{(1+\gamma)^{\lambda_1} - (1-\gamma)^{\lambda_1}}{(1+\gamma)^{\lambda_1} + (1-\gamma)^{\lambda_1}} \}, \cup_{\eta \in g} \{ \dfrac{2\eta^{\lambda_1}}{(2-\eta)^{\lambda_1} + \eta^{\lambda_1}} \})$

$\dot{\oplus} (\cup_{\gamma \in h} \{ \dfrac{(1+\gamma)^{\lambda_2} - (1-\gamma)^{\lambda_2}}{(1+\gamma)^{\lambda_2} + (1-\gamma)^{\lambda_2}} \}, \cup_{\eta \in g} \{ \dfrac{2\eta^{\lambda_2}}{(2-\eta)^{\lambda_2} + \eta^{\lambda_2}} \})$

$= (\cup_{\gamma \in h} \{ \dfrac{\frac{(1+\gamma)^{\lambda_1} - (1-\gamma)^{\lambda_1}}{(1+\gamma)^{\lambda_1} + (1-\gamma)^{\lambda_1}} + \frac{(1+\gamma)^{\lambda_2} - (1-\gamma)^{\lambda_2}}{(1+\gamma)^{\lambda_2} + (1-\gamma)^{\lambda_2}}}{1 + \frac{(1+\gamma)^{\lambda_1} - (1-\gamma)^{\lambda_1}}{(1+\gamma)^{\lambda_1} + (1-\gamma)^{\lambda_1}} \frac{(1+\gamma)^{\lambda_2} - (1-\gamma)^{\lambda_2}}{(1+\gamma)^{\lambda_2} + (1-\gamma)^{\lambda_2}}} \},$

$\cup_{\eta \in g} \{ \dfrac{\frac{2\eta^{\lambda_1}}{(2-\eta)^{\lambda_1} + \eta^{\lambda_1}} \frac{2\eta^{\lambda_2}}{(2-\eta)^{\lambda_2} + \eta^{\lambda_2}}}{1 + (1 - \frac{2\eta^{\lambda_1}}{(2-\eta)^{\lambda_1} + \eta^{\lambda_1}})(1 - \frac{2\eta^{\lambda_2}}{(2-\eta)^{\lambda_2} + \eta^{\lambda_2}})} \})$

$= (\cup_{\gamma \in h} \{ \dfrac{(1+\gamma)^{\lambda_1 + \lambda_2} - (1-\gamma)^{\lambda_1 + \lambda_2}}{(1+\gamma)^{\lambda_1 + \lambda_2} + (1-\gamma)^{\lambda_1 + \lambda_2}} \},$

$\cup_{\eta \in g} \{ \dfrac{2\eta^{\lambda_1 + \lambda_2}}{(2-\eta)^{\lambda_1 + \lambda_2} + \eta^{\lambda_1 + \lambda_2}} \}) = (\lambda_1 \dot{\oplus} \lambda_2)d.$

Thus, $\lambda_1 d \dot{\oplus} \lambda_2 d = (\lambda_1 \dot{\oplus} \lambda_2)d$. This completes the proof.

To compare the DHFEs, inspired by the comparison method of HFEs, the following definition is given:

Definition 1.26 (*Zhu et al.* 2012). Let $d = \{h, g\}$ be any two DHFEs,

$$s(d) = \frac{1}{l_h} \sum_{\gamma \in h} \gamma - \frac{1}{l_g} \sum_{\eta \in g} \eta \qquad (1.39)$$

is called the score function of d, and

$$p(d) = \frac{1}{l_h} \sum_{\gamma \in h} \gamma + \frac{1}{l_g} \sum_{\eta \in g} \eta \qquad (1.40)$$

is called the accuracy function of d, where l_h and l_g are the numbers of the elements in h and g, respectively.

Based on the score function and accuracy function of DHFEs, the following scheme is proposed to compare any two DHFEs d_1 and d_2:

(1) If $s(d_1) > s(d_2)$, then d_1 is superior to d_2, denoted by $d_1 \succ d_2$.
(2) If $s(d_1) = s(d_2)$, then

   (a) if $p(d_1) = p(d_2)$, then d_1 is equivalent to d_2, denoted by $d_1 \sim d_2$.
   (b) If $p(d_1) > p(d_2)$, then d_1 is superior than d_2, denoted by $d_1 \succ d_2$.

Example 1.10 (Zhu et al. 2012). Let $d_1 = (\{0.1, 0.3\}, \{0.3, 0.5\})$ and $d_2 = (\{0.2, 0.4\}, \{0.4, 0.6\})$ be two DHFEs, then based on Definition 1.26, we obtain $s(d_1) = s(d_2) = 0$, $p(d_2)(= 0.8) > p(d_1)(= 0.6)$. Thus, $d_2 \succ d_1$.

1.2.3   Hesitant Fuzzy Linguistic Term Set

It is noted that the above mentioned different forms of fuzzy sets suit the problems that are defined as quantitative situations. However, in real world decision making problems, many aspects of different activities cannot be assessed in a quantitative form, but rather in a qualitative one. Using linguistic information to express experts' opinions is suitable and straightforward because it is very close to human's cognitive processes. A common approach to model linguistic information is the fuzzy linguistic approach proposed by Zadeh (1975), which represents qualitative information as linguistic variables. Although it is less precise than a number, the linguistic variable, defined as "*a variable whose values are not numbers but words or sentences in a natural or artificial language*", enhances the flexibility and reliability of decision making models and provides good results in different fields. Nevertheless, similar to fuzzy sets, the fuzzy linguistic approach has some limitations and thus different linguistic representation models have been introduced, such as the 2-tuple fuzzy linguistic representation model (Herrera and Martínez 2000), the linguistic model based on type-2 fuzzy set (Türkşen 2002), the virtual linguistic model (Xu 2004a), the proportional 2-tuple model (Wang and Hao 2006), and so on. However, all these extended models are still very limited due to the fact that they are based on the elicitation of single or simple terms that should encompass and describe the information provided by decision makers (or experts) regarding to a linguistic variable. When the experts hesitate among different linguistic terms and need to use a more complex linguistic term that is not usually defined in the linguistic term set to depict their assessments, the above mentioned fuzzy linguistic approaches are out of use. Thus, motivated by the HFS, Rodríguez et al. (2012) proposed the concept of hesitant fuzzy linguistic term set (HFLTS), which provides a different and great flexible form to represent the assessments of decision makers.

In fuzzy linguistic approach, the decision makers' opinions are taken as the values of a linguistic variable which is established by linguistic descriptors and their corresponding semantics (Herrera and Herrera-Viedma 2000b). Once the experts provide the linguistic evaluation information, the following step is to translate these linguistic inputs into a machine manipulative format in which the computation can be carried out. Such translation is conducted by some fuzzy tools. Meanwhile, the outputs of the computing with words (CWW) model should also be easy to be converted into the linguistic information. To do so, Xu (2005b) proposed the subscript-symmetric additive linguistic term set, shown as

$$S = \{s_t | t = -\tau, \ldots, -1, 0, 1, \ldots, \tau\} \tag{1.41}$$

where the mid linguistic label s_0 represents an assessment of "indifference", and the rest of them are placed symmetrically around it. In particular, $s_{-\tau}$ and s_τ are the lower and upper bounds of the linguistic labels used by the decision makers in practical applications. τ is a positive integer, and S satisfies the following conditions:

(1) If $\alpha > \beta$, then $s_\alpha > s_\beta$;
(2) The negation operator is defined: neg $(s_\alpha) = s_{-\alpha}$, especially, neg $(s_0) = s_0$.

The linguistic term set S is a discrete linguistic term set and thus is not convenient for calculation and analysis. To preserve all given linguistic information, Xu (2005b) extended the discrete linguistic term set to a continuous linguistic term set $\bar{S} = \{s_\alpha | \alpha \in [-q, q]\}$, where $q(q > \tau)$ is a sufficiently large positive integer. In general, the linguistic term $s_\alpha(s_\alpha \in S)$ is determined by the decision makers, and the virtual linguistic term $\bar{s}_\alpha(\bar{s}_\alpha \in \bar{S})$ only appears in computation. The virtual linguistic term provides a tool to compute with the linguistic terms. The mapping between virtual linguistic terms and their corresponding semantics is easy to build, shown as Fig. 1.1 (Liao et al. 2014b).

As traditional fuzzy linguistic approach can only use single linguistic term, such as *"medium"*, *"high"* or *"a little high"*, to represent the value of a linguistic variable but cannot express complicated linguistic expressions such as *"between medium and high"*, *"at least a little high"*, Rodríguez et al. (2012) introduced the concept of HFLTS, which can be used to elicit several linguistic terms or linguistic expression for a linguistic variable.

Definition 1.27 (*Rodríguez et al.* 2012). Let $S = \{s_0, \ldots, s_\tau\}$ be a linguistic term set. A HFLTS, H_S, is an ordered finite subset of the consecutive linguistic terms of S.

Since Definition 1.27 does not give any mathematical form for HFLTS, Liao et al. (2015a) redefined the HFLTS mathematically as follows, which is much easier to be understood. Liao and Xu (2015c) also replaced the linguistic term set by the subscript-symmetric linguistic term set $S = \{s_t | t = -\tau, \ldots, -1, 0, 1, \ldots, \tau\}$.

Fig. 1.1 Semantics of virtual linguistic terms

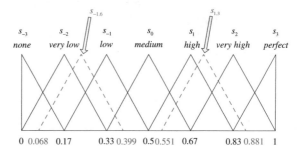

Definition 1.28 (*Liao et al.* 2015a). Let $x \in X$, be fixed and $S = \{s_t | t = -\tau, \ldots, -1, 0, 1, \ldots, \tau\}$ be a linguistic term set. A HFLTS in X, H_S, is in mathematical terms of

$$H_S = \{ <x, h_S(x) > | x \in X \} \tag{1.42}$$

where the function $h_S(x) : X \to S$ defines the possible membership grades of the element $x \in X$ to the set $A \subset X$ and for every $x \in X$, the value of $h_S(x)$ is represented by a set of some values in the linguistic term set S and can be expressed as $h_S(x) = \{s_{\varphi_l}(x) | s_{\varphi_l}(x) \in S, l = 1, \ldots, L(x)\}$ with $\varphi_l \in \{-\tau, \ldots, -1, 0, 1, \ldots, \tau\}$ being the subscript of a linguistic term $s_{\varphi_l}(x)$ and $L(x)$ being the number of linguistic terms in $h_S(x)$.

For convenience, Liao et al. (2015a) called $h_S(x)$ the hesitant fuzzy linguistic element (HFLE) and let $\mathbb{H}_S$ be the set of all HFLEs on S. For simplicity, $h_S(x)$, $s_{\varphi_l}(x)$ and $L(x)$ can be written respectively as h_S, s_{φ_l} and L for short. There are several special HFLEs, such as:

(1)   empty HFLE: $h_S = \{\}$.
(2)   full HFLE: $h_S = S$.
(3)   the complement of HFLE h_S: $h_S^c = S - h_S = \{s_{\varphi_l} | s_{\varphi_l} \in S \text{ and } s_{\varphi_l} \notin h_S\}$.

Although the HFLTS can be used to elicit several linguistic values for a linguistic variable, it is still not similar to the human way of thinking and reasoning. Thus, Rodríguez et al. (2012) further proposed a context-free grammar to generate simple but elaborated linguistic expressions ll that are more similar to the human expressions and can be easily represented by means of HFLTS. The grammar G_H is a 4-tuple (V_N, V_T, I, P) where V_N is a set of nonterminal symbols, V_T is the set of terminals' symbols, I is the starting symbols, and P is the production rules.

Definition 1.29 (*Rodríguez et al.* 2012). Let S be a linguistic term set, and G_H be a context-free grammar. The elements of $G_H = (V_N, V_T, I, P)$ are defined as:

$$V_N = \{ <primary\ term>, <composite\ term>,$$
$$<unary\ relation>, <binary\ relation>, <conjunction> \}$$
$$V_T = \{lower\ than, greater\ than, at\ least, at\ most, between,$$
$$I \in V_N;$$

$P = \{I ::= \ <primary\ term> \ | <composite\ term>$

$<composite\ term> \ ::= \ <unary\ relation> \ <primary\ term> \ |$

$<binary\ relation> \ <conjunction> \ <primary\ term>$

$<primary\ term> \ ::= s_{-\tau}|\cdots|s_{-1}|s_0|s_1|\cdots|s_\tau$

$<unary\ relation> \ ::= lower\ than|greater\ than$

$<binary\ relation> \ ::= between$

$<conjunction> \ ::= and\}.$

Note: In the above definition, the brackets enclose optional elements and the symbol "|" indicates alternative elements.

The expressions ll generated by the context-free grammar G_H may be either single valued linguistic terms $s_t \in S$ or linguistic expressions. The transformation function E_{G_H} can be used to transform the expressions ll that are produced by G_H into HFLTS.

Definition 1.30 (*Rodríguez et al.* 2012). Let E_{G_H} be a function that transforms linguistic expressions $ll \in S_{ll}$, obtained by using G_H, into the HFLTS H_S. S is the linguistic term set used by G_H, and S_{ll} is the expression domain generated by G_H:

$$E_{G_H} : S_{ll} \to H_S \tag{1.43}$$

The linguistic expression generated by G_H using the production rules are converted into HFLTS by means of the following transformations:

- $E_{G_H}(s_t) = \{s_t|s_t \in S\};$
- $E_{G_H}(at\ most\ s_m) = \{s_t|s_t \in S\ and\ s_t \leq s_m\};$
- $E_{G_H}(lower\ than\ s_m) = \{s_t|s_t \in S\ and\ s_t < s_m\};$
- $E_{G_H}(at\ least\ s_m) = \{s_t|s_t \in S\ and\ s_t \geq s_m\};$
- $E_{G_H}(great\ than\ s_m) = \{s_t|s_t \in S\ and\ s_t > s_m\};$
- $E_{G_H}(between\ s_m\ and\ s_n) = \{s_t|s_t \in S\ and\ s_m \leq s_t \leq s_n\}.$

With the transformation function E_{G_H} defined as Definition 1.30, it is easy to transform the initial linguistic expressions into HFLTS. Liao et al. (2015a) used a figure (see Fig. 1.2) to show the relationships among the context-free grammar G_H, the linguistic expression ll and the HFLTS H_S.

Example 1.11 (Liao et al. 2015a). Quality management is more and more popular in our daily life. In the process of quality management, many aspects of certain products cannot be measured as crisp values but only qualitative values. Here we

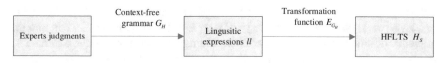

Fig. 1.2 The way to obtain a HFLTS

just consider a simple example that an expert evaluates the operational complexity of three automatic systems, represented as x_1, x_2 and x_3. Since this criterion is qualitative, it is impossible to give crisp values but only linguistic terms. The operational complexity of these automatic systems can be taken as a linguistic variable. The linguistic term set for the operational complexity can be set up as:

$$S = \{s_{-3} = very\ complex,\ s_{-2} = complex, s_{-1} = a\ little\ complex,\ s_0 = medium,$$
$$s_1 = a\ little\ easy,\ s_2 = easy,\ s_3 = very\ easy\}$$

With the linguistic term set and also the context-free grammar, the expert determines his/her judgments over these three automatic systems with linguistic expressions, which are $ll_1 = at\ least\ a\ little\ easy$, $ll_2 = between\ complex\ and\ medium$ and $ll_3 = great\ than\ easy$. These linguistic expressions are similar to human way of thinking and they can reflect the expert's hesitant cognition intuitively. Using the transformation function E_{G_H}, a HFLTS can be yielded as $H_S(x) = \{<x_1, h_S(x_1)>, <x_2, h_S(x_2)>, <x_3, h_S(x_3)>\}$ with $h_S(x_1) = \{s_1, s_2, s_3\}$, $h_S(x_2) = \{s_{-2}, s_{-1}, s_0\}$, and $h_S(x_3) = \{s_3\}$ being three HFLEs.

Example 1.12 (Liao et al. 2015a). Consider a simple example that a Chief Information Officer (CIO) of a company evaluates the candidate ERP system in terms of three criteria, i.e., x_1 (potential cost), x_2 (function), and x_3 (operation complexity). Since the three criteria are qualitative, the CIO gives his evaluation values in linguistic expressions. Different criteria are associated with different linguistic term sets and different semantics. The linguistic term sets for these three criteria are set up as:

$$S_1 = \{s_{-3} = very\ expensive,\ s_{-2} = expensive,\ s_{-1} = a\ little\ expensive,\ s_0 = medium,$$
$$s_1 = a\ little\ cheap,\ s_2 = cheap,\ s_3 = very\ cheap\}$$

$$S_2 = \{s_{-3} = none,\ s_{-2} = very\ low,\ s_{-1} = low, s_0 = medium,$$
$$s_1 = high, s_2 = very\ high,\ s_3 = perfect\}$$

$$S_3 = \{s_{-3} = too\ complex,\ s_{-2} = complex, s_{-1} = a\ little\ complex,\ s_0 = medium,$$
$$s_1 = a\ little\ easy,\ s_2 = easy,\ s_3 = every\ easy\}$$

respectively. With these linguistic term sets and also the context-free grammar, the CIO provides his evaluation values in linguistic expressions for a ERP system as: $ll_1 = between\ cheap\ and\ very\ cheap$, $ll_2 = at\ least\ high$, $ll_3 = great\ than\ easy$. Using the transformation function E_{G_H}, a HFLTS is obtained as $H(x) = \{<x_1, h_{S_1}(x_1)>, <x_2, h_{S_2}(x_2)>, <x_3, h_{S_3}(x_3)>\}$ with $h_{S_1}(x_1) = \{s_2, s_3 | s_2, s_3 \in S_1\}$, $h_{S_2}(x_2) = \{s_1, s_2, s_3 | s_1, s_2, s_3 \in S_2\}$ and $h_{S_3}(x_3) = \{s_3 | s_3 \in S_3\}$. Furthermore, if we ignore the influence of different semantics over different linguistic term sets on the criteria, i.e., let $S = \{s_{-3}, s_{-2}, s_{-1}, s_0, s_1, s_2, s_3\}$, then the HFLTS $H(x)$ can be rewritten as

$H_S(x) = \{ <x_1, h_S(x_1)> , <x_2, h_S(x_2)> , <x_3, h_S(x_3)> \}$ with $h_S(x_1) = \{s_2, s_3\}$, $h_S(x_2) = \{s_1, s_2, s_3\}$ and $h_S(x_3) = \{s_3\}$.

From Examples 1.11 and 1.12, we can find that X, in Definition 1.28, could be either a set of objects on a linguistic variable or a set of linguistic variables of an object (in this case, the influence of different semantics over different linguistic term sets on different linguistic variables should be ignored).

Rodríguez et al. (2012) defined the complement, union and intersection of HFLTSs:

Definition 1.31 (*Rodríguez et al.* 2012). For three HFLEs h_S, h_S^1 and h_S^2, the following operations are defined:

(1) Lower bound: $h_S^- = \min(s_t) = s_k$, $s_t \in h_S$ and $s_t \geq s_k$, $\forall t$.
(2) Upper bound: $h_S^+ = \max(s_t) = s_k$, $s_t \in h_S$ and $s_t \leq s_k$, $\forall t$.
(3) $h_S^c = S - h_S = \{s_t | s_t \in S \text{ and } s_t \notin h_S\}$.
(4) $h_S^1 \cup h_S^2 = \{s_t | s_t \in h_S^1 \text{ or } s_t \in h_S^2\}$.
(5) $h_S^1 \cap h_S^2 = \{s_t | s_t \in h_S^1 \text{ and } s_t \in h_S^2\}$.

For a HFLE $h_S = \{s_{\varphi_l} | l = 1, 2, \ldots, L\}$, the linguistic terms in it might be out of order. To simplify the computation, we can arrange the linguistic terms s_{φ_l} $(l = 1, \ldots, L)$ in any of the following orders (Liao and Xu 2015c): ① ascending order $\delta : (1, 2, \ldots, n) \rightarrow (1, 2, \ldots, n)$ is a permutation satisfying $\delta_l \leq \delta_{l+1}$, $l = 1, \ldots, L$; ② descending order $\eta : (1, 2, \ldots, n) \rightarrow (1, 2, \ldots, n)$ is a permutation satisfying $\eta_l \geq \eta_{l+1}$, $l = 1, \ldots, L$. In addition, considering that different HFLEs may have different numbers of linguistic terms, we can extend the short HFLEs by adding some linguistic terms in it till they have same length. Liao et al. (2014b) introduced a method to add linguistic terms in a HFLE. For a HFLE $h_S = \{s_{\varphi_l} | l = 1, 2, \ldots, L\}$, let s^+ and s^- be the maximal and minimal linguistic terms in the HFLE h_S, defined as $s^+ = \max_{\varphi_l} \{s_{\varphi_l} | l = 1, 2, \ldots, L\}$ and $s^- = \min_{\varphi_l} \{s_{\varphi_l} | l = 1, 2, \ldots, L\}$, respectively, and ξ $(0 \leq \xi \leq 1)$ be an optimized parameter, then we can add the linguistic term:

$$\bar{s} = \xi s^+ \oplus (1 - \xi) s^- \tag{1.44}$$

to the HFLE. The optimized parameter, which is used to reflect the decision makers' risk preferences, is provided by the decision makers.

Motivated by the score function and the variance function of HFS, Liao et al. (2015c) introduced the score function and the variance function for HFLE.

Definition 1.32 (*Liao et al.* 2015c). For a HFLE $h_S = \cup_{s_{\delta_l} \in h_S} \{s_{\delta_l} | l = 1, \ldots, L\}$ where L is the number of linguistic terms in h_S, $\rho(h_S) = \frac{1}{L} \sum_{s_{\delta_l} \in h_S} s_{\delta_l} = s_{\frac{1}{L} \sum_{l=1}^{L} \delta_l}$ is called the score function of h_S.

Definition 1.33 (*Liao et al.* 2015c). For a HFLE $h_S = \cup_{s_{\delta_l} \in h_S} \{ s_{\delta_l} | l = 1, \ldots, L \}$ where L is the number of linguistic terms in h_S, $\sigma(h_S) = \frac{1}{L} \sqrt{\sum_{s_{\delta_l}, s_{\delta_k} \in h_S} (s_{\delta_l} - s_{\delta_k})^2}$ $= s_{\frac{1}{L} \sqrt{\sum_{s_{\delta_l}, s_{\delta_k} \in h_S} (\delta_l - \delta_k)^2}}$ is called the variance function of h_S.

The relationship between the score function and the variance function of HFLE is similar to the relationship between mean and variance in statistics. Thus, for two HFLEs h_S^1 and h_S^2, the following approach can be used to compare any two HFLEs:

- If $\rho(h_S^1) > \rho(h_S^2)$, then $h_S^1 > h_S^2$.
- Else if $\rho(h_S^1) = \rho(h_S^2)$, then,

    - if $\sigma(h_S^1) < \sigma(h_S^2)$, then $h_S^1 > h_S^2$;
    - else if $\sigma(h_S^1) = \sigma(h_S^2)$, then $h_S^1 = h_S^2$.

Example 1.13 (Liao et al. 2015c). Let $S = \{ s_{-3} = none, \ s_{-2} = very \ low, \ s_{-1} = low, \ s_0 = medium, \ s_1 = high, \ s_2 = very \ high, \ s_3 = perfect \}$ be a linguistic term set. The linguistic information obtained by means of the context-free grammar is $\phi_1 = high$, $\phi_2 = lower \ than \ medium$, $\phi_3 = greater \ than \ high$, and $\phi_4 = between \ medium \ and \ very \ high$. With the transformation function, the above linguistic information can be represented as $H_S = \{ h_S^1, h_S^2, h_S^3, h_S^4 \}$ with $h_S^1 = \{ s_1 \}$, $h_S^2 = \{ s_{-3}, s_{-2}, s_{-1}, s_0 \}$, $h_S^3 = \{ s_1, s_2, s_3 \}$ and $h_S^4 = \{ s_0, s_1, s_2 \}$. Then, we have $\rho(h_S^1) = s_1$, $\rho(h_S^2) = s_{-1.5}$, $\rho(h_S^3) = s_2$, $\rho(h_S^4) = s_1$. Since $\rho(h_S^3) > \rho(h_S^1) = \rho(h_S^4) > \rho(h_S^2)$, it yields that $MAX(H_S) = h_S^3$, $MIN(H_S) = h_S^2$.

Calculating the variance functions of h_S^1 and h_S^4, we have $\sigma(h_S^1) = s_0$, $\sigma(h_S^4) = s_{0.8165}$. Since $\sigma(h_S^1) < \sigma(h_S^4)$, then we get $h_S^1 > h_S^4$. Hence, the rank of these four HFLEs is $h_S^3 > h_S^1 > h_S^4 > h_S^2$.

References

Atanassov KT (1983) Intuitionistic fuzzy sets. In: VII ITKR's Session, Sgurev V (Ed.), Sofia.

Atanassov KT (1986) Intuitionistic fuzzy sets. Fuzzy Sets Syst 20:87–96

Atanassov KT (2009) Remark on operations "subtraction" over intuitionistic fuzzy sets. Notes Intuitionistic Fuzzy Sets 15(3):20–24

Atanassov KT (2012) On intuitionistic fuzzy sets theory. Springer, Berlin

Atanassov KT, Gargov G (1989) Interval-valued intuitionistic fuzzy sets. Fuzzy Sets Syst 31: 343–349

Atanassov KT, Riečan B (2006) On two operations over intuitionistic fuzzy sets. J Appl Math Stat Inform 2(2):145–148

Behzadian M, Kazemzadeh RB, Albadvi A, Aghdasi M (2010) PROMETHEE: a comprehensive literature review on methodologies and applications. Eur J Oper Res 200:198–215

Bezdek JC (1993) Fuzzy models-what are they, and why? IEEE Trans Fuzzy Syst 1(1):1–6

Chen TY (2007) Remarks on the subtraction and division operations over intuitionistic fuzzy sets and interval-valued fuzzy sets. J Intell Fuzzy Syst 9(3):169–172

Chen N, Xu ZS (2014) Properties of interval-valued hesitant fuzzy sets. J Intell Fuzzy Syst 27(1): 143–158

Chen N, Xu ZS, Xia MM (2013) Interval-valued hesitant preference relations and their applications to group decision making. Knowl-Based Syst 37:528–540

Chen N, Xu ZS, Xia MM (2015) The ELECTRE I multi-criteria decision making method based on hesitant fuzzy sets. Int J Inf Technol Decis Making 14(3):621–657

Dubois D, Prade H (1980) Fuzzy sets and systems: theory and applications. Academic Press, London

Herrera F, Herrera-Viedma E (2000) Linguistic decision analysis: steps for solving decision problems under linguistic information. Fuzzy Sets Syst 115:67–82

Herrera F, Martínez L (2000) A 2-tuple fuzzy linguistic representation model for computing with words. IEEE Trans Fuzzy Syst 8:746–752

Hong DH, Choi CH (2000) Multicriteria fuzzy decision-making problems based on vague set theory. Fuzzy Sets Syst 114:103–113

Liao HC, Xu ZS (2013) A VIKOR-based method for hesitant fuzzy multi-criteria decision making. Fuzzy Optim Decis Making 12(4):373–392

Liao HC, Xu ZS (2014a) Subtraction and division operations over hesitant fuzzy sets. J Intell Fuzzy Syst 27(1):65–72

Liao HC, Xu ZS (2014b) Satisfaction degree based interactive decision making method under hesitant fuzzy environment with incomplete weights. Int J Uncertainty Fuzziness Knowl-Based Syst 22(4):553–572

Liao HC, Xu ZS (2015a) Extended hesitant fuzzy hybrid weighted aggregation operators and their application in decision making. Soft Comput 19(9):2551–2564

Liao HC, Xu ZS (2015b) Approaches to manage hesitant fuzzy linguistic information based on the cosine distance and similarity measures for HFLTSs and their application in qualitative decision making. Expert Syst Appl 42:5328–5336

Liao HC, Xu ZS, Xia MM (2014a) Multiplicative consistency of hesitant fuzzy preference relation and its application in group decision making. Int J Inf Technol Decis Making 13(1):47–76

Liao HC, Xu ZS, Zeng XJ (2014b) Distance and similarity measures for hesitant fuzzy linguistic term sets and their application in multi-criteria decision making. Inf Sci 271:125–142

Liao HC, Xu ZS, Zeng XJ, Merigó JM (2015a) Qualitative decision making with correlation coefficients of hesitant fuzzy linguistic term sets. Knowl-Based Syst 76:127–138

Liao HC, Xu ZS, Zeng XJ (2015b) Novel correlation coefficients between hesitant fuzzy sets and their application in decision making. Knowl-Based Syst 82:115–127

Liao HC, Xu ZS, Zeng XJ (2015c) Hesitant fuzzy linguistic VIKOR method and its application in qualitative multiple criteria decision making. IEEE Trans Fuzzy Syst 23(5):1343–1355

Mizumoto M, Tanaka K (1976) Some properties of fuzzy sets of type 2. Inf Control 31:312–340

Rodríguez RM, Martínez L, Herrera F (2012) Hesitant fuzzy linguistic terms sets for decision making. IEEE Trans Fuzzy Syst 20(1):109–119

Torra V (2010) Hesitant fuzzy sets. Int J Intell Syst 25:529–539

Torra V, Narukawa Y (2009) On hesitant fuzzy sets and decision. The 18th IEEE international conference on fuzzy systems (FS'09). Jeju Island, Korea, pp 1378–1382

Türkşen IB (2002) Type 2 representation and reasoning for CWW. Fuzzy Sets Syst 127:17–36

Wang JH, Hao J (2006) A new version of 2-tuple fuzzy linguistic representation model for computing with words. IEEE Trans Fuzzy Syst 14:435–445

Xia MM, Xu ZS (2011a) Hesitant fuzzy information aggregation in decision making. Int J Approximate Reasoning 52(3):395–407

Xia MM, Xu ZS (2011b) Methods for fuzzy complementary preference relations based on multiplicative consistency. Comput Ind Eng 61(4):930–935

Xu ZS (2004) A method based on linguistic aggregation operators for group decision making with linguistic preference relations. Inf Sci 166:19–30

Xu ZS (2005) Deviation measures of linguistic preference relations in group decision making. Omega-Int J Manage Sci 33:249–254

Xu ZS (2007) Intuitionistic fuzzy aggregation operators. IEEE Trans Fuzzy Syst 15:1179–1187

Xu ZS, Da QL (2002) The uncertain OWA operator. Int J Intell Syst 17:569–575

Xu ZS, Xia MM (2011a) Distance and similarity measures for hesitant fuzzy sets. Inf Sci 181:2128–2138

Xu ZS, Xia MM (2011b) On distance and correlation measures of hesitant fuzzy information. Int J Intell Syst 26:410–425

Xu ZS, Yager RR (2006) Some geometric aggregation operators based on intuitionistic fuzzy sets. Int J Gen Syst 35:417–433

Yager RR (1986) On the theory of bags. Int J Gen Syst 13:23–37

Yager RR (2014) Pythagorean membership grades in multicriteria decision making. IEEE Trans Fuzzy Syst 22(4):958–965

Yu PL (1973) A class of solutions for group decision problems. Manage Sci 19:936–946

Zadeh LA (1965) Fuzzy sets. Inf Control 8:338–353

Zadeh LA (1975) The concept of a linguistic variable and its application to approximate reasoning-I. Inf Sci 8:199–249

Zhao N, Xu ZS, Liu FJ (2015) Uncertainty measures for hesitant fuzzy information. Int J Intell Syst 30(7):818–836

Zhao H, Xu ZS, Liu SS (2016) Dual hesitant fuzzy information aggregation with Einstein t-conorm and t-norm. J Syst Sci Syst Eng. doi:10.1007/s11518-015-5289-6

Zhu B, Xu ZS, Xia MM (2012) Dual hesitant fuzzy set. J Appl Math 879629, 13 p. doi:10.1155/2012/879629

Chapter 2
Novel Correlation and Entropy Measures of Hesitant Fuzzy Sets

Correlation is one of the most widely used indices in data analysis, pattern recognition, machine learning, decision making, etc. It measures how well two variables move together in a linear fashion. The correlation coefficient, which was originally appeared in Karl Pearson's proposal related to statistics, has been extended into different fuzzy circumstances. Different forms of fuzzy correlation coefficients have been proposed, such as the fuzzy correlation coefficients, the intuitionistic fuzzy correlation coefficients, and the hesitant fuzzy correlation coefficients. Xu and Xia (2011b) defined several correlation coefficients for HFEs. Afterwards, Chen et al. (2013a) proposed a formula to calculate the correlation coefficient between two HFSs. In this chapter, we first point out the weaknesses of the existing correlation coefficients between HFSs, and then introduce some novel correlation coefficient formulas for HFSs. Some new concepts, such as the mean of a HFS, the variance of a HFS and the correlation between two HFSs are defined. Based on these concepts, a novel correlation coefficient formula between two HFSs is introduced. Afterwards, the upper and lower bounds of the correlation coefficient are defined. A theorem is given to determine these two bounds. It is stated that the correlation coefficient between two HFSs should also be hesitant, and thus, the upper and lower bounds can further help to identify the correlation coefficient between HFSs. The significant characteristic of the introduced correlation coefficient is that it lies in the interval $[-1, 1]$, which is in accordance with the classical correlation coefficient in statistics, whereas all the old correlation coefficients between HFSs in the literature are within the unit interval $[0, 1]$. The weighted correlation coefficient is also proposed to make it more applicable. In order to show the efficiency of the proposed correlation coefficients, they are implemented in medical diagnosis and cluster analysis. Some numerical examples are given in this chapter to illustrate the applicability and efficiency of the proposed correlation coefficient between HFSs.

Entropy is another important index for fuzzy information, which measures the degree of uncertainty of a fuzzy set. Usually, there are two aspects of uncertainty associated with a fuzzy set. One is related to fuzziness, which results from the lack

© Springer Nature Singapore Pte Ltd. 2017
H. Liao and Z. Xu, *Hesitant Fuzzy Decision Making Methodologies and Applications*, Uncertainty and Operations Research,
DOI 10.1007/978-981-10-3265-3_2

of clear discrimination between the elements belonging or not belonging to a set. For classical fuzzy set, Zadeh (1965) first defined the entropy to measure the fuzziness of a fuzzy set and then many scholars developed different kinds of entropy formulas for fuzzy set (De Luca and Termini 1972; Kaufmann and Swanson 1975; Yager 1979; Parkash et al. 2008) and IFS (Burillo and Bustince 1996; Szmidt and Kacprzyk 2001; Wei et al. 2012). The other aspect of uncertainty associated with a fuzzy set is related to the lack of specificity. Specificity measures the amount of information contained in a fuzzy set. Yager (1992, 1998) put forward several specificity measures to quantify the degree that a fuzzy set contains just one element. Based on three t-norms and a negation, Garmendia et al. (2003) gave a general expression for specificity measures of a fuzzy set. Later, Yager (2008c) studied the formula of specificity measures in continuous domain. Pal et al. (2013) pointed out that there are two types of uncertainty for an IFS, i.e., the fuzzy-type uncertainty and the non-specificity type uncertainty. As to the entropy measure of HFS, Xu and Xia (2012) gave the axiomatic definition of entropy for HFEs and developed several entropy formulas to measure the degree of fuzziness of a HFE. Later, Farhadinia (2013) also proposed some entropy measures to quantify the degree of fuzziness of a HFS. In the second subsection of this chapter, we review the existing entropy measures for HFEs and demonstrate that the existing entropy measures for HFEs fail to effectively distinguish some apparently different HFEs in some cases. Then, we give a new axiomatic framework of entropy measures for HFEs by taking into account two facets of uncertainty associated with a HFE (i.e., fuzziness and non-specificity). We adopt a two-tuple entropy model to represent the two types of uncertainty associated with a HFE. Additionally, we discuss how to formulate each kind of uncertainty. Several examples are given to illustrate each method, and the comparisons with the existing entropy measures are also offered.

2.1 Novel Correlation Measures of Hesitant Fuzzy Sets

2.1.1 The Existing Correlation Measures of Hesitant Fuzzy Sets

As the correlation measure is one of the most important indices in measuring the relationship between two sets, it has been investigated in-depth within in the context of fuzzy sets and their extensions. As a representation, in the following, we just review the advances in correlation coefficient related to fuzzy sets and IFSs.

After discussing various properties which are attributed to "correlation" in statistics, Murthy et al. (1985) first introduced the correlation coefficient $\rho(\mu_1, \mu_2)$, similar to the correlation coefficient in statistics, between two fuzzy membership functions. It was proven that the correlation coefficient they defined satisfies many good properties, including $\rho(\mu_1, \mu_2) \in [-1, 1]$. In the case that the elements of fuzzy sets are ranked in terms of memberships, Chaudhuri and Bhattacharya (2001) proposed a rank correlation coefficient for fuzzy sets and then compared it with Murthy et al. (1985)'s correlation coefficient formula. Also adopting the concepts from conventional statistics, Chiang and Lin (1999) derived another formula of correlation coefficient in the domain of fuzzy sets. All these three kinds of correlation coefficients over fuzzy sets lie in the interval $[-1, 1]$ and have similar meaning as that in conventional statistics. On the other hand, Yu (1993) introduced quite different concepts of correlation and correlation coefficient to measure the interrelation between fuzzy numbers. The value of correlation coefficient he introduced is within the interval $[0, 1]$. That is to say, the correlation coefficient he proposed can only represent the strength of relationship between fuzzy sets, but cannot manifest the positive or negative correlation.

It is stated that all the above achievements calculate the correlation coefficient between fuzzy sets as a crisp number. By using the sup-min convolution, Liu and Kao (2002) proposed a mathematical programming approach to calculate the correlation coefficient as a fuzzy number. After that, by applying the T_w-based extension principle, Hong (2006) gave an exact solution of a fuzzy correlation coefficient without relying on programming.

Regarding to IFSs, many different forms of correlation measures have also been investigated. Hung (2001) proposed the correlation coefficient for IFSs from statistics point of view by considering the membership degree and non-membership degree as two separate fuzzy sets. After that, Szmidt and Kacprzyk (2010) extended his formula by taking the hesitant degrees of IFSs into account. Mitchell (2004) also proposed an improved version of correlation coefficient, in which he interpreted the IFSs as the ensembles of ordinary membership functions. As these correlation coefficients are motivated from traditional statistics, the correlation coefficients of IFSs they developed are within the interval $[-1, 1]$. On the other side, motivated by the information energy of a fuzzy set, Gerstenkorn and Manko (1991) developed a quite different form of correlation coefficient for IFSs. Further, Hong and Hwang (1995) extended this type of correlation coefficient into possibility space in which the set $\{x_i\}$ is an infinite universe of discourse. Moreover, Hung and Wu (2002) improved the correlation coefficient and introduced the so-called centroid-method-based correlation coefficient for IFSs. As these correlation coefficients cannot guarantee that the correlation coefficient between any two IFSs equals to one if and only if these two IFSs are the same, Xu (2006b) proposed a new form of correlation coefficient for IFSs and circumvented this weakness. It should be stated that all the correlation coefficients proposed in Gerstenkorn and Manko (1991), Hong and Hwang (1995), Huang and Wu (2002), and Xu (2006b) lie in the unit interval $[0, 1]$.

Some scholars also proposed distinct correlation measures within the context of HFSs. Xu and Xia (2011b) proposed several correlation coefficients from the point of HFEs. For two HFEs $h_A = \{\gamma_{A1}, \gamma_{A2}, \ldots, \gamma_{Al_A}\}$ and $h_B = \{\gamma_{B1}, \gamma_{B2}, \ldots, \gamma_{Bl_B}\}$, it is possible that the values in h_A and h_B are out of order. In addition, the number of values, l_A and l_B, in different HFEs may be different. Thus, to introduce the definition of correlation coefficient between two HFEs, Xu and Xia (2011b) firstly supposed that the values in different HFEs were arranged in ascending order;

meanwhile, they also assumed that the HFEs have the same length l. Based on these two assumptions, they proposed five different kinds of correlation coefficients for HFEs. Here we just set out one as a representation (for more others, readers can refer to Xu and Xia (2011b)):

$$\rho(h_A, h_B) = \frac{\sum_{k=1}^{l} \gamma_{A\sigma(k)} \gamma_{B\sigma(k)}}{\left(\sum_{k=1}^{l} \gamma_{A\sigma(k)}^2 \sum_{k=1}^{l} \gamma_{B\sigma(k)}^2\right)^{1/2}} \tag{2.1}$$

where $\sigma : (1, 2, \ldots, l) \to (1, 2, \ldots, l)$ is a permutation satisfying $\gamma_{\sigma(k)} \leq \gamma_{\sigma(k+1)}$, $k = 1, 2, \ldots, l-1$. Although Xu and Xia (2011b) stated that $|c(h_A, h_B)| \leq 1$, it is obvious that $c(h_A, h_B) \geq 0$ as $\gamma \in [0, 1]$, which means $c(h_A, h_B) \in [0, 1]$.

Chen et al. (2013a) proposed a formula to calculate the correlation coefficient between two HFSs. Let X be a reference set, $A = \{h_A(x_i)\}$ and $B = \{h_B(x_i)\}$ $(i = 1, 2, \ldots, n)$ be two HFSs. As the values in HFEs are out of order, and the number of values in different HFEs may be different, in order to introduce the correlation coefficient between two HFSs, the following assumptions are given in advance:

- The values in a HFE are arranged in ascending order.
- The lengths of different HFEs are assumed to have equal length.

The first assumption is easy to be satisfied. For the second one, sometimes the cardinality of two HFEs are different. In such case, as to Chen et al. (2013a)'s method, we need to make the lengths of the two HFEs be the same. There are many different regulations to extend the shorter HFE to the same length as the longer one. The most representative regulations are the pessimistic principle and the optimistic principle. For two HFEs h_A and h_B, let $l = \max\{l_{h_A}, l_{h_B}\}$ where l_{h_A} and l_{h_B} are the number of values in h_A and h_B, respectively. When $l_{h_A} \neq l_{h_B}$, one can extend the short HFE by adding some values in it until it has the same length with the other. In terms of the pessimistic principle, the short HFE is extended by adding the minimum value in it until it has the same length with the other HFE; while as to the optimistic principle, the maximum value of the short HFE should be added till the HFE has the same length as the longer one. In Chen et al. (2013a)'s definition, they used the former case and thus the correlation coefficient between two HFSs was defined as:

$$\rho_*(A, B) = \frac{\sum_{i=1}^{n} \left(\frac{1}{l_i} \sum_{k=1}^{l_i} \gamma_{A\sigma(k)}(x_i) \cdot \gamma_{B\sigma(k)}(x_i)\right)}{\left[\sum_{i=1}^{n} \left(\frac{1}{l_i} \sum_{k=1}^{l_i} \gamma_{A\sigma(k)}^2(x_i)\right)\right]^{1/2} \cdot \left[\sum_{i=1}^{n} \left(\frac{1}{l_i} \sum_{k=1}^{l_i} \gamma_{B\sigma(k)}^2(x_i)\right)\right]^{1/2}} \tag{2.2}$$

where $\gamma_{A\sigma(k)}(x_i)$ and $\gamma_{B\sigma(k)}(x_i)$ are the kth value in $h_A(x_i)$ and $h_B(x_i)$.

In summary, the correlation coefficients defined as Eqs. (2.1) and (2.2) have a few weaknesses:

(1) In Eq. (2.1), the HFEs were assumed to have equal length. This is not in accordance with real cases because it is impossible to make sure that all HFEs have equal length. As to Eq. (2.2), the pessimistic (or optimistic) principle was applied to fill the short HFE with some artificial values. It should be pointed out that filling some artificial values into a HFE would change its original information.

The following two examples show that the extensional regulations used in Xu and Xia (2011b) and Chen et al. (2013a) in the process of defining correlation coefficients between two HFEs or HFSs are not reasonable:

Example 2.1 (*Liao et al.* 2015b) For two HFEs $h_1 = \{0.1, 0.3\}$ and $h_2 = \{0.1, 0.3, 0.8\}$. In order to calculate the correlation coefficient between h_1 and h_2 by Eq. (2.1), according to Xu and Xia (2011b)'s assumptions, we should firstly extend h_1 to make it have equal length with h_2. Suppose that the pessimistic principle is applied to h_1, i.e., the minimum element in h_1 should be added to h_1. Then, h_1 is modified as $h_1' = \{0.1, 0.1, 0.3\}$. By Eqs. (1.7) and (1.8), we have $\bar{h}_1 = 0.2$, $\varphi_{h_1} = 0.1$, and $\bar{h}_1' = 0.1667$, $\varphi_{h_1'} = 0.0943$. It is obvious that the revised HFE h_1' is quite different from the original HFE h_1.

Example 2.2 (*Liao et al.* 2015b) Suppose that we are going to measure the correlation coefficient between two HFEs $h_1 = \{0.1, 0.8\}$ and $h_2 = \{0.1, 0.2, 0.3, 0.4, 0.5, 0.6, 0.7, 0.8\}$. Then, according to the pessimistic principle, h_1 should be modified as $h_1' = \{0.1, 0.1, 0.1, 0.1, 0.1, 0.1, 0.1, 0.8\}$. Via Eqs. (1.7) and (1.8), we have $\bar{h}_1 = 0.45$, $\varphi_{h_1} = 0.35$, and $\bar{h}_1' = 0.1875$, $\varphi_{h_1'} = 0.2315$. It should be noted that the mean of the revised HFE h_1' is more than two times smaller than that of the original HFE h_1; meanwhile, the hesitant degree also changes apparently.

Examples 2.1 and 2.2 reveal that adding some artificial values in a HFE, no matter by pessimistic principle or by optimistic principle, would change the information of the original one. Thus, it is not very reasonable to measure the correlation coefficient between HFEs or HFSs by Eq. (2.1) or Eq. (2.2), and some new correlation coefficients need to be proposed for HFSs.

(2) The correlation coefficient defined in Xu and Xia (2011b) and Chen et al. (2013a) is always positive but this ignores the negative situation. In traditional random variable case, the correlation coefficient lies in $[-1, 1]$. For those correlation coefficients defined over fuzzy sets or IFSs, the correlation coefficients also lie in the interval $[-1, 1]$. Hence, it is not adequate to use the always positive variable to denote the correlation degree between two HFSs. The positive correlation coefficient can only demonstrate the strength of the relationship between HFSs, but cannot manifest the positive or negative correlation.

(3) It is not a best choice to use just one crisp number to represent the correlation degree between two HFEs or HFSs as the HFEs or HFSs per se are hesitant but

not precise. In other words, the correlation coefficient for HFSs should have certain degree of hesitance rather than just a fixed value.

2.1.2 Novel Correlation Measures of Hesitant Fuzzy Sets

This subsection introduces some novel correlation coefficients for HFSs. As to HFS, the following definitions are given:

Definition 2.1 (*Liao et al.* 2015b). For a reference set X, let $A = \{<x, h_A(x_i) > |x_i \in X\}$ be a HFS on X with $h_A(x_i) = \{\gamma_{Ai1}, \gamma_{Ai2}, \ldots, \gamma_{Ail_{Ai}}\}$, $i = 1, 2, \ldots, n$. The mean of the HFS A is defined as:

$$\bar{A} = E(A) = \frac{1}{n}\sum_{i=1}^{n}\bar{h}_A(x_i) = \frac{1}{n}\sum_{i=1}^{n}\left(\frac{1}{l_{Ai}}\sum_{k=1}^{l_{Ai}}\gamma_{Aik}\right) \tag{2.3}$$

Definition 2.2 (*Liao et al.* 2015b). For a reference set X, let $A = \{<x, h_A(x_i) > |x_i \in X\}$ be a HFS on X with $h_A(x_i) = \{\gamma_{Ai1}, \gamma_{Ai2}, \ldots, \gamma_{Ail_{Ai}}\}$, $i = 1, 2, \ldots, n$. The variance of the HFS A is defined as:

$$Var(A) = \frac{1}{n}\sum_{i=1}^{n}(\bar{h}_A(x_i) - \bar{A}) \tag{2.4}$$

Definition 2.3 (*Liao et al.* 2015b). For a reference set X, let $A = \{h_A(x_i)\}$ and $B = \{h_B(x_i)\}$ be two HFSs on X, where $h_A(x_i) = \{\gamma_{Ai1}, \gamma_{Ai2}, \ldots, \gamma_{Ail_{Ai}}\}$, $h_B(x_i) = \{\gamma_{Bi1}, \gamma_{Bi2}, \ldots, \gamma_{Bil_{Bi}}\}$, $i = 1, 2, \ldots, n$. Then the correlation between HFSs A and B is defined as:

$$C(A, B) = \frac{1}{n}\sum_{i=1}^{n}[\bar{h}_A(x_i) - \bar{A}] \cdot [\bar{h}_B(x_i) - \bar{B}]$$

$$= \frac{1}{n}\sum_{i=1}^{n}\left[\bar{h}_A(x_i) - \frac{1}{n}\sum_{i=1}^{n}\bar{h}_A(x_i)\right] \cdot \left[\bar{h}_B(x_i) - \frac{1}{n}\sum_{i=1}^{n}\bar{h}_B(x_i)\right] \tag{2.5}$$

where

$$\bar{h}_A(x_i) = \frac{1}{l_{Ai}}\sum_{k=1}^{l_{Ai}}\gamma_{Aik}, \quad \bar{h}_B(x_i) = \frac{1}{l_{Bi}}\sum_{k=1}^{l_{Bi}}\gamma_{Bik}, \quad i = 1, 2, \ldots, n \tag{2.6}$$

Note 2.1 In Definition 2.3, we do not need the HFSs A and B to have the same length, that is to say, $l_{Ai} \neq l_{Bi}$ is acceptable.

Note 2.2 The correlation above can be positive or negative. So the negative correlation can be modeled too.

Based on Definitions 2.2 and 2.3, it is easy to verify that the correlation coefficient between HFSs satisfies the following theorem:

Theorem 2.1 (Liao et al. 2015b). *For a HFS $A = \{<x, h(x_i)> |x_i \in X\}$ on X with $h(x_i) = \{\gamma_{i1}, \gamma_{i2}, \ldots, \gamma_{il_i}\}$, $i = 1, 2, \ldots, n$, the following equation holds:*

$$C(A,A) = Var(A) \tag{2.7}$$

Now we can define the correlation coefficient for HFSs:

Definition 2.4 (*Liao et al.* 2015b). For a reference set X, let $A = \{h_A(x_i)\}$ and $B = \{h_B(x_i)\}$ be two HFSs on X, where $h_A(x_i) = \{\gamma_{Ai1}, \gamma_{Ai2}, \ldots, \gamma_{Ail_{Ai}}\}$ and $h_B(x_i) = \{\gamma_{Bi1}, \gamma_{Bi2}, \ldots, \gamma_{Bil_{Bi}}\}$, $i = 1, 2, \ldots, n$. Then the correlation coefficient between the HFSs A and B is defined as:

$$\rho(A,B) = \frac{C(A,B)}{[C(A,A) \cdot C(B,B)]^{1/2}} \tag{2.8}$$

Theorem 2.2 reveals the fundamental properties of correlation coefficient between HFSs:

Theorem 2.2 (Liao et al. 2015b) *The correlation coefficient $\rho(A, B)$ between HFSs A and B satisfies the following properties:*

(1) $\rho(A,B) = \rho(B,A)$.
(2) $\rho(A,A) = 1$.
(3) $\rho(A,A^c) = -1$, where A^c is defined as $A^c = \{<x, h^c(x_i)> |x_i \in X\}$ with $h^c(x_i) = \{1 - \gamma_{i1}, 1 - \gamma_{i2}, \ldots, 1 - \gamma_{il_i}\}, i = 1, 2, \ldots, n$.
(4) $-1 \leq \rho(A,B) \leq 1$.

Proof The proofs of (1), (2) and (3) are obvious according to Definition 2.4.
(4) According to Eq. (2.5), we have

$$|C(A,B)| = \left| \frac{1}{n} \sum_{i=1}^{n} [\bar{h}_A(x_i) - \bar{A}] \cdot [\bar{h}_B(x_i) - \bar{B}] \right| \leq \frac{1}{n} \sum_{i=1}^{n} |\bar{h}_A(x_i) - \bar{A}| \cdot |\bar{h}_B(x_i) - \bar{B}|$$

Using the Cauchy-Schwarz inequality:

$$(a_1 b_1 + a_2 b_2 + \cdots + a_n b_n)^2 \leq (a_1^2 + a_2^2 + \cdots + a_n^2) \cdot (b_1^2 + b_2^2 + \cdots + b_n^2)$$

where $a_i, b_i \in R, i = 1, 2, \ldots, N$, it follows that

$$|C(A,B)| \leq \sqrt{\frac{1}{n}\sum_{i=1}^{n}\left|\bar{h}_A(x_i) - \bar{A}\right|^2} \cdot \sqrt{\frac{1}{n}\sum_{i=1}^{n}\left|\bar{h}_B(x_i) - \bar{B}\right|^2}$$
$$= [C(A,A)]^{1/2} \cdot [C(B,B)]^{1/2}$$

Thus, we have

$$|\rho(A,B)| = \frac{|C(A,B)|}{[C(A,A)]^{1/2} \cdot [C(B,B)]^{1/2}} \leq 1$$

Hence, $-1 \leq \rho(A,B) \leq 1$, which ends the proof.

In the following, we discuss how to measure the hesitant degree of the correlation coefficient $\rho(A,B)$ between two HFSs A and B.

We first rewrite the definition of correlation coefficient $\rho(A,B)$ between two HFSs A and B as:

$$\rho(A,B) = \frac{\sum_{i=1}^{n}\left(\bar{h}_A(x_i) - \bar{A}\right) \cdot \left(\bar{h}_B(x_i) - \bar{B}\right)}{\sqrt{\sum_{i=1}^{n}\left(\bar{h}_A(x_i) - \bar{A}\right)^2} \cdot \sqrt{\sum_{i=1}^{n}\left(\bar{h}_B(x_i) - \bar{B}\right)^2}} \tag{2.9}$$

Then we define the upper and lower bounds of the correlation coefficient $\rho(A,B)$ between A and B as follows:

$$\rho^U(A,B) = \max_{\substack{u_i \in h_A(x_i) \\ v_i \in h_B(x_i) \\ i = 1,2,\ldots,n}} \frac{\sum_{i=1}^{n}\left(u_i - \bar{A}\right) \cdot \left(v_i - \bar{B}\right)}{\sqrt{\sum_{i=1}^{n}\left(u_i - \bar{A}\right)^2} \cdot \sqrt{\sum_{i=1}^{n}\left(v_i - \bar{B}\right)^2}} \tag{2.10}$$

$$\rho^L(A,B) = \min_{\substack{u_i \in h_A(x_i) \\ v_i \in h_B(x_i) \\ i = 1,2,\ldots,n}} \frac{\sum_{i=1}^{n}\left(u_i - \bar{A}\right) \cdot \left(v_i - \bar{B}\right)}{\sqrt{\sum_{i=1}^{n}\left(u_i - \bar{A}\right)^2} \cdot \sqrt{\sum_{i=1}^{n}\left(v_i - \bar{B}\right)^2}} \tag{2.11}$$

Theorem 2.3 (Liao et al. 2015b). *For two HFSs $A = \{h_A(x_i)\}$ and $B = \{h_B(x_i)\}$ on X with $h_A(x_i) = \left\{\gamma_{Ai1}, \gamma_{Ai2}, \ldots, \gamma_{Ai1_{Ai}}\right\}$, $h_B(x_i) = \left\{\gamma_{Bi1}, \gamma_{Bi2}, \ldots, \gamma_{Bi1_{Bi}}\right\}$, $i = 1, 2, \ldots, n$, let $h_A^U(x_i) = \max\left\{\gamma_{Ai1}, \gamma_{Ai2}, \ldots, \gamma_{Ail_{Ai}}\right\}$, $h_A^L(x_i) = \min\left\{\gamma_{Ai1}, \gamma_{Ai2}, \ldots, \gamma_{Ail_{Ai}}\right\}$, $h_B^U(x_i) = \max\left\{\gamma_{Bi1}, \gamma_{Bi2}, \ldots, \gamma_{Bil_{Bi}}\right\}$ and $h_B^L(x_i) = \min\left\{\gamma_{Bi1}, \gamma_{Bi2}, \ldots, \gamma_{Bil_{Bi}}\right\}$. Then,*

(1) $\rho^L(A,B) \le \rho(A,B) \le \rho^U(A,B)$.

(2) $\rho^U(A,B) = \dfrac{\sum_{i=1}^{n}\left(h_A^U(x_i)-\bar{A}\right)\cdot\left(h_B^U(x_i)-\bar{B}\right)}{\sqrt{\sum_{i=1}^{n}\left(h_A^U(x_i)-\bar{A}\right)^2}\cdot\sqrt{\sum_{i=1}^{n}\left(h_B^U(x_i)-\bar{B}\right)^2}}$.

(3) $\rho^L(A,B) = \dfrac{\sum_{i=1}^{n}\left(h_A^L(x_i)-\bar{A}\right)\cdot\left(h_B^L(x_i)-\bar{B}\right)}{\sqrt{\sum_{i=1}^{n}\left(h_A^L(x_i)-\bar{A}\right)^2}\cdot\sqrt{\sum_{i=1}^{n}\left(h_B^L(x_i)-\bar{B}\right)^2}}$.

Note 2.3 Once we know $\rho^U(A,B)$ and $\rho^L(A,B)$, the difference

$$\Delta\rho(A,B) = \rho^U(A,B) - \rho^L(A,B) \tag{2.12}$$

can be used as an indicator how hesitant the correlation relationship is. The lager $\Delta\rho(A,B)$ is, the more hesitant the decision maker should be.

To prove the above theorem, a lemma is given first:

Lemma 2.1 (*Liao et al.* 2015b). Let x be any real number, $f(x) = \frac{x}{\sqrt{x^2+a}}$ with $a > 0$. Then, the function $f(x)$ is monotonically increasing.

Proof To prove the above, we only need to prove that $f'(x) > 0$. Since

$$f'(x) = \frac{\sqrt{x^2+a} - x\cdot\frac{2x}{2\sqrt{x^2+a}}}{x^2+a} = \frac{x^2+a-x^2}{(x^2+a)\cdot\sqrt{x^2+a}} = \frac{a}{(x^2+a)\cdot\sqrt{x^2+a}} > 0$$

then $f(x)$ is monotonically increasing, which competes the proof.

Based on the above lemma, we know for any real numbers x and y, if $x \ge y$, then

$$\frac{x}{\sqrt{x^2+a}} \ge \frac{y}{\sqrt{y^2+a}}, \quad a > 0 \tag{2.13}$$

In the following, we give the proof of Theorem 2.3.

Proof We only prove the case of $\rho^U(A,B)$. The case for $\rho^L(A,B)$ can be proven similarly.

Let $p_i = h_A^U(x_i) - \bar{A}$ and $q_i = u_i - \bar{A}$. Since $h_A^U(x_i) = \max\{\gamma_{Ai1}, \gamma_{Ai2}, \ldots, \gamma_{Ail_{Ai}}\}$ and $u_i \in h_A(x_i) = \{\gamma_{Ai1}, \gamma_{Ai2}, \ldots, \gamma_{Ail_{Ai}}\}$, then we get $h_A^U(x_i) \ge u_i$. Thus, $p_i \ge q_i$. Furthermore, we let $a_i = a_i(u_1, u_2, \ldots, u_{i-1}, u_{i+1}, \ldots, u_n) \triangleq \sum_{\substack{j=1 \\ j\neq i}}^{n}\left(u_j - \bar{A}\right)^2$. Obviously, $a_i > 0$.

Analogously, for the HFS B, let $s_i = h_B^U(x_i) - \bar{B}$, $t_i = v_i - \bar{B}$, $b_i = b_i(v_1, v_2, \ldots, v_{i-1}, v_{i+1}, \ldots, v_n) \triangleq \sum_{\substack{j=1 \\ j\neq i}}^{n}\left(v_j - \bar{B}\right)^2$. We can also obtain $s_i \ge t_i$ as well as $b_i > 0$.

Based on the above transformation, the following equation holds:

$$\frac{\sum_{i=1}^n \left(h_A^U(x_i) - \bar{A}\right) \cdot \left(h_B^U(x_i) - \bar{B}\right)}{\sqrt{\sum_{\substack{j=1 \\ j \neq i}}^n \left(u_j - \bar{A}\right)^2 + \left(h_A^U(x_i) - \bar{A}\right)^2} \cdot \sqrt{\sum_{\substack{j=1 \\ j \neq i}}^n \left(v_j - \bar{B}\right)^2 + \left(h_B^U(x_i) - \bar{B}\right)^2}}$$

$$= \frac{\sum_{i=1}^n p_i \cdot s_i}{\sqrt{a_i + p_i^2} \cdot \sqrt{b_i + s_i^2}}$$

According to Eq. (2.13), it follows

$$\frac{\sum_{i=1}^n p_i \cdot s_i}{\sqrt{a_i + p_i^2} \cdot \sqrt{b_i + s_i^2}} = \sum_{i=1}^n \frac{p_i}{\sqrt{a_i + p_i^2}} \cdot \frac{s_i}{\sqrt{b_i + s_i^2}}$$

$$\geq \sum_{i=1}^n \frac{q_i}{\sqrt{a_i + q_i^2}} \cdot \frac{t_i}{\sqrt{b_i + t_i^2}} = \frac{\sum_{i=1}^n q_i \cdot t_i}{\sqrt{a_i + q_i^2} \cdot \sqrt{b_i + t_i^2}}$$

$$= \frac{\sum_{i=1}^n (u_i - \bar{A}) \cdot (v_i - \bar{B})}{\sqrt{\sum_{\substack{j=1 \\ j \neq i}}^n (u_j - \bar{A})^2 + (u_i - \bar{A})^2} \cdot \sqrt{\sum_{\substack{j=1 \\ j \neq i}}^n (v_j - \bar{B})^2 + (v_i - \bar{B})^2}} = \frac{\sum_{i=1}^n (u_i - \bar{A}) \cdot (v_i - \bar{B})}{\sqrt{\sum_{i=1}^n (u_i - \bar{A})^2} \cdot \sqrt{\sum_{i=1}^n (v_i - \bar{B})^2}}$$

Therefore, we can obtain

$$\frac{\sum_{i=1}^n \left(h_A^U(x_i) - \bar{A}\right) \cdot \left(h_B^U(x_i) - \bar{B}\right)}{\sqrt{\sum_{\substack{j=1 \\ j \neq i}}^n \left(u_j - \bar{A}\right)^2 + \left(h_A^U(x_i) - \bar{A}\right)^2} \cdot \sqrt{\sum_{\substack{j=1 \\ j \neq i}}^n \left(v_j - \bar{B}\right)^2 + \left(h_B^U(x_i) - \bar{B}\right)^2}}$$

$$\geq \frac{\sum_{i=1}^n (u_i - \bar{A}) \cdot (v_i - \bar{B})}{\sqrt{\sum_{i=1}^n (u_i - \bar{A})^2} \cdot \sqrt{\sum_{i=1}^n (v_i - \bar{B})^2}}, \text{ for all } u_i \in \xi(x_i),\, v_i \in \xi(y_i),\, i = 1, 2, \cdots, n \tag{2.14}$$

In the left side of Eq. (2.14), we set $v_i = h_B^U(x_i)$. Then, it comes

$$\frac{\sum_{i=1}^n \left(h_A^U(x_i) - \bar{A}\right) \cdot \left(h_B^U(x_i) - \bar{B}\right)}{\sqrt{\sum_{\substack{j=1 \\ j \neq i}}^n \left(u_j - \bar{A}\right)^2 + \left(h_A^U(x_i) - \bar{A}\right)^2} \cdot \sqrt{\sum_{\substack{j=1 \\ j \neq i}}^n \left(v_j - \bar{B}\right)^2 + \left(h_B^U(x_i) - \bar{B}\right)^2}}$$

$$= \frac{\sum_{i=1}^n \left(h_A^U(x_i) - \bar{A}\right) \cdot \left(h_B^U(x_i) - \bar{B}\right)}{\sqrt{\sum_{i=1}^n \left(h_A^U(x_i) - \bar{A}\right)^2} \cdot \sqrt{\sum_{i=1}^n \left(h_B^U(x_i) - \bar{B}\right)^2}} \geq \frac{\sum_{i=1}^n (u_i - \bar{A}) \cdot (v_i - \bar{B})}{\sqrt{\sum_{i=1}^n (u_i - \bar{A})^2} \cdot \sqrt{\sum_{i=1}^n (v_i - \bar{B})^2}},$$

$$\text{for all } u_i \in h_A(x_i),\, v_i \in h_B(x_i),\, i = 1, 2, \ldots, n \tag{2.15}$$

Combining Eqs. (2.9) and (2.15), we can obtain

$$\rho^U(A,B) = \frac{\sum_{i=1}^n \left(h_A^U(x_i) - \bar{A}\right) \cdot \left(h_B^U(x_i) - \bar{B}\right)}{\sqrt{\sum_{i=1}^n \left(h_A^U(x_i) - \bar{A}\right)^2} \cdot \sqrt{\sum_{i=1}^n \left(h_B^U(x_i) - \bar{B}\right)^2}}$$

which is to say, (2) in Theorem 2.3 holds.

Additionally, in the right side of the inequality (2.15), let $v_i = \bar{h}_B(x_i)$, then it follows

$$\frac{\sum_{i=1}^n \left(h_A^U(x_i) - \bar{A}\right) \cdot \left(h_B^U(x_i) - \bar{B}\right)}{\sqrt{\sum_{i=1}^n \left(h_A^U(x_i) - \bar{A}\right)^2} \cdot \sqrt{\sum_{i=1}^n \left(h_B^U(x_i) - \bar{B}\right)^2}} \geq \frac{\sum_{i=1}^n \left(\bar{h}_A(x_i) - \bar{A}\right) \cdot \left(\bar{h}_B(x_i) - \bar{B}\right)}{\sqrt{\sum_{i=1}^n \left(\bar{h}_A(x_i) - \bar{A}\right)^2} \cdot \sqrt{\sum_{i=1}^n \left(\bar{h}_B(x_i) - \bar{B}\right)^2}}$$

i.e.,

$$\rho^U(A,B) \geq \rho(A,B)$$

This completes the proof.

Definition 2.5 (*Liao et al.* 2015b). For a reference set X, let $A = \{h_A(x_i)\}$ and $B = \{h_B(x_i)\}$ be two HFSs on X, where $h_A(x_i) = \{\gamma_{Ai1}, \gamma_{Ai2}, \ldots, \gamma_{Ail_{Ai}}\}$, and $h_B(x_i) = \{\gamma_{Bi1}, \gamma_{Bi2}, \ldots, \gamma_{Bil_{Bi}}\}$, $i = 1, 2, \ldots, n$. Given the correlation coefficient $\rho(A,B)$ between A and B being defined as Eq. (2.9), then the hesitant degree of $\rho(A,B)$ is measured in terms of

$$\varphi_{(A,B)} = \rho^U(A,B) - \rho^L(A,B) \tag{2.16}$$

where $\rho^U(A,B)$ and $\rho^L(A,B)$ are the upper and lower bounds of the correlation coefficient $\rho(A,B)$.

It is noted that the value of correlation coefficient $\rho(A,B)$ between the HFSs A and B defined as Eq. (2.8) is also a crisp value. However, as both A and B are HFSs, it is not adequate to use just a crisp value to represent their relationship. The correlation coefficient defined as Eq. (2.8) can only be taken as the expected (or averaging) correlation coefficient between the HFSs A and B. In order to describe the correlation coefficient between HFSs more objectively, we can also use the upper bound $\rho^U(A,B)$, the lower bound $\rho^L(A,B)$, or the hesitant degree $\varphi_{(A,B)}$ to better identify the correlation coefficient between two HFSs A and B.

Consider that in some cases, the objects $x_i \in X$ ($i = 1, 2, \ldots, n$) may be assigned different weights. Liao et al. (2015b) proposed the weighted form of the correlation coefficient for HFSs.

Definition 2.6 (*Liao et al.* 2015b). Let $w = (w_1, w_2, \ldots, w_n)$ be the weight vector of x_i ($i = 1, 2, \ldots, n$) with $w_i \in [0, 1]$, ($i = 1, 2, \ldots, n$) and $\sum_{i=1}^n w_i = 1$. For two HFSs $A = \{h_A(x_i)\}$ and $B = \{h_B(x_i)\}$ with $h_A(x_i) = \{\gamma_{Ai1}, \gamma_{Ai2}, \ldots, \gamma_{Ail_{Ai}}\}$, and

$h_B(x_i) = \{\gamma_{Bi1}, \gamma_{Bi2}, \ldots, \gamma_{Bil_{Bi}}\}$, $i = 1, 2, \ldots, n$, the following definition can be developed:

(1) The weighted mean of the HFS A is defined as:

$$\bar{A}_w = \frac{1}{n}\sum_{i=1}^{n} w_i \bar{h}_A(x_i) = \frac{1}{n}\sum_{i=1}^{n}\left(\frac{w_i}{l_{Ai}}\sum_{k=1}^{l_{Ai}}\gamma_{Aik}\right) \tag{2.17}$$

(2) The weighted variance of the HFS A is defined as:

$$Var_w(A) = \frac{1}{n}\sum_{i=1}^{n}\left(w_i \bar{h}_A(x_i) - \bar{A}_w\right) \tag{2.18}$$

(3) The weighted correlation between the HFSs A and B is defined as:

$$
\begin{aligned}
C_w(A, B) &= \frac{1}{n}\sum_{i=1}^{n}\left[w_i \bar{h}_A(x_i) - \bar{A}_w\right]\cdot\left[w_i \bar{h}_B(x_i) - \bar{B}_w\right] \\
&= \frac{1}{n}\sum_{i=1}^{n}\left[w_i \bar{h}_A(x_i) - \frac{1}{n}\sum_{i=1}^{n} w_i \bar{h}_A(x_i)\right]\cdot\left[w_i \bar{h}_B(x_i) - \frac{1}{n}\sum_{i=1}^{n} w_i \bar{h}_B(x_i)\right]
\end{aligned}
\tag{2.19}
$$

where

$$\bar{h}_A(x_i) = \frac{1}{l_{Ai}}\sum_{k=1}^{l_{Ai}}\gamma_{Aik}, \quad \bar{h}_B(x_i) = \frac{1}{l_{Bi}}\sum_{k=1}^{l_{Bi}}\gamma_{Bik}, \quad i = 1, 2, \ldots, n$$

(4) The weighted correlation coefficient between the HFSs A and B is defined as:

$$
\begin{aligned}
\rho_w(A, B) &= \frac{C_w(A, B)}{[C_w(A, A)\cdot C_w(B, B)]^{1/2}} \\
&= \frac{\sum_{i=1}^{n}\left(w_i \bar{h}_A(x_i) - \bar{A}_w\right)\cdot\left(w_i \bar{h}_B(x_i) - \bar{B}_w\right)}{\sqrt{\sum_{i=1}^{n}\left(w_i \bar{h}_A(x_i) - \bar{A}_w\right)^2}\cdot\sqrt{\sum_{i=1}^{n}\left(w_i \bar{h}_B(x_i) - \bar{B}_w\right)^2}}
\end{aligned}
\tag{2.20}
$$

The weighted correlation coefficient $\rho_w(A, B)$ between the HFSs A and B also satisfies the following properties:

(a) $\rho_w(A, B) = \rho_w(B, A)$;
(b) $\rho_w(A, A) = 1$;
(c) $\rho_w(A, A^c) = -1$, where A^c is defined as $A^c = \{ <x, h^c(x_i) > |x_i \in X\}$ with $h^c(x_i) = \{1 - \gamma_{i1}, 1 - \gamma_{i2}, \ldots, 1 - \gamma_{il_i}\}$, $i = 1, 2, \ldots, n$;
(d) $-1 \leq \rho_w(A, B) \leq 1$.

(5) The upper and lower bounds of the weighted correlation coefficient $\rho_w(A, B)$ are defined as:

$$\rho_w^U(A, B) = \max_{\substack{u_i \in h_A(x_i) \\ v_i \in h_B(x_i) \\ i = 1, 2, \ldots, n}} \frac{\sum_{i=1}^n (w_i u_i - \bar{A}_w) \cdot (w_i v_i - \bar{B}_w)}{\sqrt{\sum_{i=1}^n (w_i u_i - \bar{A}_w)^2} \cdot \sqrt{\sum_{i=1}^n (w_i v_i - \bar{B}_w)^2}} \quad (2.21)$$

$$\rho_w^L(A, B) = \min_{\substack{u_i \in h_A(x_i) \\ v_i \in h_B(x_i) \\ i = 1, 2, \ldots, n}} \frac{\sum_{i=1}^n (w_i u_i - \bar{A}_w) \cdot (w_i v_i - \bar{B}_w)}{\sqrt{\sum_{i=1}^n (w_i u_i - \bar{A}_w)^2} \cdot \sqrt{\sum_{i=1}^n (w_i v_i - \bar{B}_w)^2}} \quad (2.22)$$

The following properties hold as well:

(a) $$\rho_w^L(A, B) \leq \rho_w(A, B) \leq \rho_w^U(A, B).$$

(b) $$\rho_w^U(A, B) = \frac{\sum_{i=1}^n \left(w_i h_A^U(x_i) - \bar{A}_w\right) \cdot \left(w_i h_B^U(x_i) - \bar{B}_w\right)}{\sqrt{\sum_{i=1}^n \left(w_i h_A^U(x_i) - \bar{A}_w\right)^2} \cdot \sqrt{\sum_{i=1}^n \left(w_i h_B^U(x_i) - \bar{B}_w\right)^2}} \cdot$$

(c) $$\rho_w^L(A, B) = \frac{\sum_{i=1}^n \left(w_i h_A^L(x_i) - \bar{A}_w\right) \cdot \left(w_i h_B^L(x_i) - \bar{B}_w\right)}{\sqrt{\sum_{i=1}^n \left(w_i h_A^L(x_i) - \bar{A}_w\right)^2} \cdot \sqrt{\sum_{i=1}^n \left(w_i h_B^L(x_i) - \bar{B}_w\right)^2}}, \quad \text{where}$$

$$h_A^U(x_i) = \max\{\gamma_{Ai1}, \gamma_{Ai2}, \ldots, \gamma_{Ail_{Ai}}\}, \quad h_A^L(x_i) = \min\{\gamma_{Ai1}, \gamma_{Ai2}, \ldots, \gamma_{Ail_{Ai}}\}$$
$$h_B^U(x_i) = \max\{\gamma_{Bi1}, \gamma_{Bi2}, \ldots, \gamma_{Bil_{Bi}}\}, \quad h_B^L(x_i) = \min\{\gamma_{Bi1}, \gamma_{Bi2}, \ldots, \gamma_{Bil_{Bi}}\}$$

(6) The hesitant degree of $\rho_w(A, B)$ is measured in terms of

$$\varphi_{(A,B)_w} = \rho_w^U(A, B) - \rho_w^L(A, B) \quad (2.23)$$

2.1.3 Applications of the Correlation Measures of Hesitant Fuzzy Sets

(1) The application of the correlation coefficients in medical diagnosis

The correlation coefficient can be implemented into many practical applications. The first case given below is related to medical diagnosis.

Example 2.3 (Liao et al. 2015b). Suppose that a doctor wants to make a proper diagnosis D = {Viral fever, Malaria, Typhoid, Stomach problem, Chest problem} for a set of patients P = {Al, Bob, Joe, Ted} with the values of symptoms

Table 2.1 Symptom characteristics for the considered diagnoses in terms of HFSs

	Temperature	Headache	Cough	Stomach pain	Stomach pain
Viral fever	$\{0.6, 0.4, 0.3\}$	$\{0.7, 0.5, 0.3, 0.2\}$	$\{0.5, 0.3\}$	$\{0.5, 0.4, 0.3, 0.2, 0.1\}$	$\{0.5, 0.4, 0.2, 0.1\}$
Malaria	$\{0.9, 0.8, 0.7\}$	$\{0.5, 0.3, 0.2, 0.1\}$	$\{0.2, 0.1\}$	$\{0.6, 0.5, 0.3, 0.2, 0.1\}$	$\{0.4, 0.3, 0.2, 0.1\}$
Typhoid	$\{0.6, 0.3, 0.1\}$	$\{0.9, 0.8, 0.7, 0.6\}$	$\{0.5, 0.3\}$	$\{0.5, 0.4, 0.3, 0.2, 0.1\}$	$\{0.6, 0.4, 0.3, 0.1\}$
Stomach problem	$\{0.5, 0.4, 0.2\}$	$\{0.4, 0.3, 0.2, 0.1\}$	$\{0.4, 0.3\}$	$\{0.9, 0.8, 0.7, 0.6, 0.5\}$	$\{0.5, 0.4, 0.2, 0.1\}$
Chest problem	$\{0.3, 0.2, 0.1\}$	$\{0.5, 0.3, 0.2, 0.1\}$	$\{0.3, 0.2\}$	$\{0.7, 0.6, 0.5, 0.3, 0.2\}$	$\{0.9, 0.8, 0.7, 0.6\}$

Table 2.2 Symptom characteristics for the considered patients in terms of HFSs

	Temperature	Headache	Cough	Stomach pain	Chester pain
Al	$\{0.9, 0.7, 0.5\}$	$\{0.4, 0.3, 0.2, 0.1\}$	$\{0.4, 0.3\}$	$\{0.6, 0.5, 0.4, 0.2, 0.1\}$	$\{0.4, 0.3, 0.2, 0.1\}$
Bob	$\{0.5, 0.4, 0.2\}$	$\{0.5, 0.4, 0.3, 0.1\}$	$\{0.2, 0.1\}$	$\{0.9, 0.8, 0.6, 0.5, 0.4\}$	$\{0.5, 0.4, 0.3, 0.2\}$
Joe	$\{0.9, 0.7, 0.6\}$	$\{0.7, 0.4, 0.3, 0.1\}$	$\{0.3, 0.2\}$	$\{0.6, 0.4, 0.3, 0.2, 0.1\}$	$\{0.6, 0.3, 0.2, 0.1\}$
Ted	$\{0.8, 0.7, 0.5\}$	$\{0.6, 0.5, 0.4, 0.2\}$	$\{0.4, 0.3\}$	$\{0.6, 0.4, 0.3, 0.2, 0.1\}$	$\{0.5, 0.4, 0.2, 0.1\}$

$V = \{$temperature, headache, cough, stomach pain, chest pain$\}$. As in many cases such as in traditional Chinese medical diagnosis or in emergency case that crisp measuring instruments cannot be obtained, it is impossible to get the crisp values of the symptoms but only vague information, which is described in terms of HFEs. Before starting the diagnosis, a medical knowledge-based data set involving symptom characteristic of the considered diagnoses is necessary to be constructed (see Table 2.1). The symptoms of the patients are given in Table 2.2.

To derive a diagnosis for each patient, we can calculate the correlation coefficient between the symptom characteristic of each diagnose and that of each patient. Using the correlation coefficient formula shown as Eq. (2.9), the correlation coefficient values are obtained, shown in Table 2.3 and Fig. 2.1. From Table 2.3 and Fig. 2.1, it is clear to see that Al, Joe and Ted suffer from Malaria, but Bob suffers from Stomach problem.

Meanwhile, Xu and Xia (2011b) utilized the correlation formula Eq. (2.1) to calculate the correlation coefficients and yielded their results, illustrated in Table 2.4 and Fig. 2.2. Table 2.4 and Fig. 2.2 imply that Al and Ted suffer from viral fever; Bob suffer from stomach problem; Joe suffer from malaria.

Table 2.3 Correlation coefficient values for each considered patient to the set of possible diagnoses by using our approach

	Viral fever	Malaria	Typhoid	Stomach problem	Chest problem
Al	0.4597	**0.9187**	−0.4288	0.1323	−0.5372
Bob	−0.5715	0.2546	−0.3166	**0.8074**	0.3042
Joe	0.5395	**0.9803**	−0.1217	−0.1017	−0.4636
Ted	0.7330	**0.9082**	0.0210	−0.2230	−0.6506

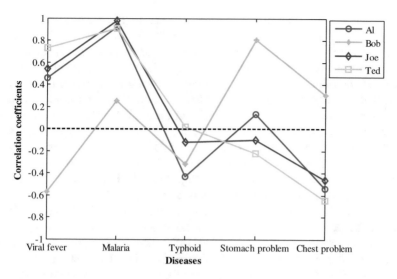

Fig. 2.1   Correlation coefficient values by using our approach

Table 2.4   Correlation coefficient values for each considered patient to the set of possible diagnoses by using Xu and Xia (2011b)'s approach

	Viral fever	Malaria	Typhoid	Stomach problem	Chest Problem
Al	**0.9969**	0.9929	0.9800	0.9902	0.9878
Bob	0.9900	0.9862	0.9792	**0.9921**	0.9909
Joe	0.9927	**0.9929**	0.9677	0.9817	0.9750
Ted	**0.9942**	0.9899	0.9787	0.9879	0.9772

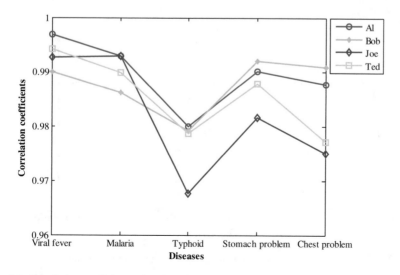

Fig. 2.2   Correlation coefficient values by using Xu and Xia (2011b)'s approach

Comparing the results in Table 2.3 with those in Table 2.4, some interesting findings can be derived. Firstly, we can find that all the values in Table 2.4 are positive values within the unit interval [0, 1], but in Table 2.3, there are some negative values. It is quite strange that all the HFEs are positively correlated even though all the values over different symptom characteristics are quite different. This is the first weakness of Xu and Xia (2011b)'s method. For example, let us look into the symptom characteristics of Typhoid and those of Al. It is obvious that Al's symptoms are negative correlated to those of Typhoid. However, according to Eq. (2.1), the correlation between Typhoid and Al is 0.9800, which implies that it is highly probable that Al suffers from Typhoid. This is definitely wrong.

In addition, comparing Table 2.3 (or Fig. 2.1) with Table 2.4 (or Fig. 2.2), we can find that all the correlation coefficients shown in Table 2.4 are quite close and vary from 0.9677 to 0.9969. These similar values cannot clearly distinguish the different between different diagnoses. Actually, if we draw a new figure (see Fig. 2.3) according to Xu and Xia (2011b)'s results but restrict the correlation coefficient values vary within the same domain as in Fig. 2.1, then it is very hard or even impossible for us to distinguish the diagnoses. In other words, the results derived from Table 2.4 are not very convincing (or at least not applicable) especially when the number of objects is a little large. However, Table 2.3 presents a striking contrast to Table 2.4 as all the values in it lies between −0.4288 and 0.9803, which shows the differences among the diagnoses significantly. All these above points imply that the correlation coefficient proposed in this chapter is much more convincing in medical diagnosis.

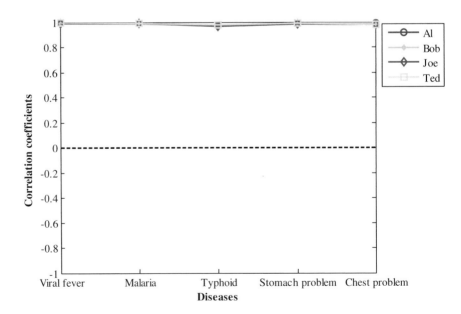

Fig. 2.3 Correlation coefficient values by using Xu and Xia (2011b)'s approach

It is stated that in this example, we just use Eq. (2.9) as a representation to describe the correlation coefficient between HFSs and illustrate its advantages over the existing correlation coefficients for HFSs. In fact, we can also use the upper bound $\rho^U(A, B)$, the lower bound $\rho^L(A, B)$, or the hesitant degree $\varphi_{(A,B)}$ to better identify the correlation coefficients in the above example.

(2) The application of the correlation coefficients in cluster analysis

To better understand the strength of the novel correlation coefficients, in the following, we show the applicability of the correlation coefficient between HFSs in the process of clustering. Cluster analysis, or clustering, is defined as the unsupervised process of group a set of data objects in such a way that objects in the same group (called a cluster) are somehow more similar to each other than to those in other groups (clusters) (Jain et al. 1999). It can be applied either as an exploratory tool (to discover previously unknown pattern in data), or as an input to a decision making process (Friedman et al. 2007). There are many algorithms for clustering, which differ significantly in their notion of what constitutes a cluster and how to efficiently find them. Within the context of hesitant fuzzy information, Chen et al. (2013a) proposed an algorithm to cluster hesitant fuzzy data into different clusters. In that algorithm, the correlation coefficient defined as Eq. (2.2) is used to measure the relationship between different objects. In the following, we do not intend to propose new clustering algorithm but use that algorithm to illustrate the efficiency of our proposed correlation coefficient. The algorithm proposed by Chen et al. (2013a) is described below:

Algorithm 2.1

Step 1. Let $\{A_1, A_2, \ldots, A_m\}$ be a set of HFSs on X. We construct a correlation matrix $C = (\rho_{ij})_{m \times m}$ where $\rho_{ij} = \rho(A_i, A_j)$ and can be calculated via Eq. (2.1) or Eq. (2.9) or Eq. (2.20).

Step 2. Check whether the correlation matrix satisfies $C^2 \subseteq C$, where $C^2 = C \circ C = (\rho'_{ij})_{m \times m}$, and $\rho'_{ij} = \max_k \{\min\{\rho_{ik}, \rho_{kj}\}\}$, $i, j = 1, 2, \ldots, m$. If it does not hold, then we construct the equivalent correlation matrix $C^{2^k} : C \to C^2 \to C^4 \to \cdots \to C^{2^k} \to \cdots$ until $C^{2^k} = C^{2^{(k+1)}}$.

Step 3. For a given confidence level $\lambda \in [0, 1]$, we construct a λ-cutting matrix $C_\lambda = (\rho_{ij}^\lambda)_{m \times m}$ where ρ_{ij}^λ is defined as:

$$\rho_{ij}^\lambda = \begin{cases} 0 & \text{if } \rho_{ij} < \lambda \\ 1 & \text{if } \rho_{ij} \geq \lambda \end{cases} \quad i, j = 1, 2, \ldots, m \quad (2.24)$$

Step 4. Classify the HFEs by the principle: if all elements of the ith line in C_λ are the same as the corresponding elements of the jth line, then the HFEs A_i and A_j are supposed as the same type.

Step 5. End.

An application example concerning the assessment of business failure risk is utilized to validate the above algorithm and our proposed correlation coefficient for HFSs. In this example, the weighted correlation coefficient defined as Eq. (2.20) is used to measure the correlation coefficient between HFSs:

Example 2.4 (Liao et al. 2015b). The assessment of business failure risk, i.e., the assessment of firm performance and the prediction of failure events has drawn the attention of many researchers in recent years (Chen et al. 2013a). Suppose that there are 10 firms $A_i(i = 1, 2, \ldots, 10)$ to be evaluated by several risk evaluation organizations from different aspects. To get fair assessments for these firms, the risk evaluation organizations established five criteria: ζ_1 : managers work experience, ζ_2 : profitability, ζ_3 : operating capacity, ζ_4 : debt-paying ability, and ζ_5 : market competition, whose weighting vector is set as $w = (0.15, 0.3, 0.2, 0.25, 0.1)$. As the risk evaluation organizations have different backgrounds and knowledge, it is possible that they may get different evaluation values from their perspectives. To better reflect the opinions established by different organizations, the evaluation values given by them are represented by HFEs and displayed in Table 2.5.

In the following, we use Algorithm 2.1 and the weighted correlation coefficient to cluster the firms.

Step 1. By Eq. (2.20), we can calculate the weighted correlation coefficients between each pair of the alternatives $\rho(A_i, A_j)$, $i, j = 1, 2, \ldots, 10$:

$$
C_w = \begin{pmatrix}
1.0000 & -0.8347 & -0.6840 & -0.0619 & -0.7198 & 0.8272 & 0.4225 & -0.6728 & -0.2983 & -0.3817 \\
-0.8347 & 1.0000 & 0.9659 & 0.5143 & 0.6062 & -0.8432 & 0.0948 & 0.5874 & 0.7511 & -0.1724 \\
-0.6840 & 0.9659 & 1.0000 & 0.6586 & 0.5097 & -0.7364 & 0.3472 & 0.5766 & 0.8665 & -0.3965 \\
-0.0619 & 0.5143 & 0.6586 & 1.0000 & -0.3041 & -0.0295 & 0.7073 & -0.0649 & 0.9365 & -0.8463 \\
-0.7198 & 0.6062 & 0.5097 & -0.3041 & 1.0000 & -0.8852 & -0.2776 & 0.8068 & 0.0388 & 0.3949 \\
0.8272 & -0.8432 & -0.7364 & -0.0295 & -0.8852 & 1.0000 & 0.2176 & -0.6119 & -0.3648 & -0.1863 \\
0.4225 & 0.0948 & 0.3472 & 0.7073 & -0.2776 & 0.2176 & 1.0000 & -0.0034 & 0.6454 & -0.9438 \\
-0.6728 & 0.5874 & 0.5766 & -0.0649 & 0.8068 & -0.6119 & -0.0034 & 1.0000 & 0.1741 & 0.2108 \\
-0.2983 & 0.7511 & 0.8665 & 0.9365 & 0.0388 & -0.3648 & 0.6454 & 0.1741 & 1.0000 & -0.7634 \\
-0.3817 & -0.1724 & -0.3965 & -0.8463 & 0.3949 & -0.1863 & -0.9438 & 0.2108 & -0.7634 & 1.0000
\end{pmatrix}
$$

Table 2.5 The evaluation information for the 5 criteria of 10 firms

	ζ_1	ζ_2	ζ_3	ζ_4	ζ_5
A_1	{0.3,0.4,0.5}	{0.4,0.5}	{0.8}	{0.5}	{0.2,0.3}
A_2	{0.4,0.6}	{0.6,0.8}	{0.2,0.3}	{0.3,0.4}	{0.6,0.7,0.9}
A_3	{0.5,0.7}	{0.9}	{0.3,0.4}	{0.3}	{0.8,0.9}
A_4	{0.3,0.4,0.5}	{0.8,0.9}	{0.7,0.9}	{0.1,0.2}	{0.9,1.0}
A_5	{0.8,1.0}	{0.8,1.0}	{0.4,0.6}	{0.8}	{0.7,0.8}
A_6	{0.4,0.5,0.6}	{0.2,0.3}	{0.9,1.0}	{0.5}	{0.3,0.4,0.5}
A_7	{0.6}	{0.7,0.9}	{0.8}	{0.3,0.4}	{0.4,0.7}
A_8	{0.9,1.0}	{0.7,0.8}	{0.4,0.5}	{0.5,0.6}	{0.7}
A_9	{0.4,0.6}	{1.0}	{0.6,0.7}	{0.2,0.3}	{0.9,1.0}
A_{10}	{0.9}	{0.6,0.7}	{0.5,0.8}	{1.0}	{0.7,0.8,0.9}

Step 2. The equivalent correlation matrix is constructed as follows:

$$C_w^2 = \begin{pmatrix}
1.0000 & 0.0948 & 0.3472 & 0.4225 & -0.2776 & 0.8272 & 0.4225 & -0.0034 & 0.4225 & -0.1863 \\
0.0948 & 1.0000 & 0.9659 & 0.7511 & 0.6062 & 0.0948 & 0.6454 & 0.6062 & 0.8665 & 0.3949 \\
0.3472 & 0.9659 & 1.0000 & 0.8665 & 0.6062 & 0.2176 & 0.6586 & 0.5874 & 0.8665 & 0.3949 \\
0.4225 & 0.7511 & 0.8665 & 1.0000 & 0.5143 & 0.2176 & 0.7073 & 0.5766 & 0.9365 & -0.0649 \\
-0.2776 & 0.6062 & 0.6062 & 0.5143 & 1.0000 & -0.1863 & 0.3472 & 0.8068 & 0.6062 & 0.3949 \\
0.8272 & 0.0948 & 0.2176 & 0.2176 & -0.1863 & 1.0000 & 0.4225 & -0.0034 & 0.2176 & -0.1863 \\
0.4225 & 0.6454 & 0.6586 & 0.7073 & 0.3472 & 0.4225 & 1.0000 & 0.3472 & 0.7073 & -0.0034 \\
-0.0034 & 0.6062 & 0.5874 & 0.5766 & 0.8068 & -0.0034 & 0.3472 & 1.0000 & 0.5874 & 0.3949 \\
0.4225 & 0.8665 & 0.8665 & 0.9365 & 0.6062 & 0.2176 & 0.7073 & 0.5874 & 1.0000 & 0.1741 \\
-0.1863 & 0.3949 & 0.3949 & -0.0649 & 0.3949 & -0.1863 & -0.0034 & 0.3949 & 0.1741 & 1.0000
\end{pmatrix}$$

$$C_w^4 = \begin{pmatrix}
1.0000 & 0.4225 & 0.4225 & 0.4225 & 0.4225 & 0.8272 & 0.4225 & 0.4225 & 0.4225 & 0.3472 \\
0.4225 & 1.0000 & 0.9659 & 0.8665 & 0.6062 & 0.4225 & 0.7073 & 0.6062 & 0.8665 & 0.3949 \\
0.4225 & 0.9659 & 1.0000 & 0.8665 & 0.6062 & 0.4225 & 0.7073 & 0.6062 & 0.8665 & 0.3949 \\
0.4225 & 0.8665 & 0.8665 & 1.0000 & 0.6062 & 0.4225 & 0.7073 & 0.6062 & 0.9365 & 0.3949 \\
0.4225 & 0.6062 & 0.6062 & 0.6062 & 1.0000 & 0.3472 & 0.6062 & 0.8068 & 0.6062 & 0.3949 \\
0.8272 & 0.4225 & 0.4225 & 0.4225 & 0.3472 & 1.0000 & 0.4225 & 0.3472 & 0.4225 & 0.2176 \\
0.4225 & 0.7073 & 0.7073 & 0.7073 & 0.6062 & 0.4225 & 1.0000 & 0.6062 & 0.7073 & 0.3949 \\
0.4225 & 0.6062 & 0.6062 & 0.6062 & 0.8068 & 0.3472 & 0.6062 & 1.0000 & 0.6062 & 0.3949 \\
0.4225 & 0.8665 & 0.8665 & 0.9365 & 0.6062 & 0.4225 & 0.7073 & 0.6062 & 1.0000 & 0.3949 \\
0.3472 & 0.3949 & 0.3949 & -0.0649 & 0.3949 & 0.2176 & 0.3949 & 0.3949 & 0.3949 & 1.0000
\end{pmatrix}$$

$$C_w^8 = \begin{pmatrix}
1.0000 & 0.4225 & 0.4225 & 0.4225 & 0.4225 & 0.8272 & 0.4225 & 0.4225 & 0.4225 & 0.3949 \\
0.4225 & 1.0000 & 0.9659 & 0.8665 & 0.6062 & 0.4225 & 0.7073 & 0.6062 & 0.8665 & 0.3949 \\
0.4225 & 0.9659 & 1.0000 & 0.8665 & 0.6062 & 0.4225 & 0.7073 & 0.6062 & 0.8665 & 0.3949 \\
0.4225 & 0.8665 & 0.8665 & 1.0000 & 0.6062 & 0.4225 & 0.7073 & 0.6062 & 0.9365 & 0.3949 \\
0.4225 & 0.6062 & 0.6062 & 0.6062 & 1.0000 & 0.4225 & 0.6062 & 0.8068 & 0.6062 & 0.3949 \\
0.8272 & 0.4225 & 0.4225 & 0.4225 & 0.4225 & 1.0000 & 0.4225 & 0.4225 & 0.4225 & 0.3949 \\
0.4225 & 0.7073 & 0.7073 & 0.7073 & 0.6062 & 0.4225 & 1.0000 & 0.6062 & 0.7073 & 0.3949 \\
0.4225 & 0.6062 & 0.6062 & 0.6062 & 0.8068 & 0.4225 & 0.6062 & 1.0000 & 0.6062 & 0.3949 \\
0.4225 & 0.8665 & 0.8665 & 0.9365 & 0.6062 & 0.4225 & 0.7073 & 0.6062 & 1.0000 & 0.3949 \\
0.3949 & 0.3949 & 0.3949 & 0.3949 & 0.3949 & 0.3949 & 0.3949 & 0.3949 & 0.3949 & 1.0000
\end{pmatrix}$$

$$C_w^{16} = \begin{pmatrix}
1.0000 & 0.4225 & 0.4225 & 0.4225 & 0.4225 & 0.8272 & 0.4225 & 0.4225 & 0.4225 & 0.3949 \\
0.4225 & 1.0000 & 0.9659 & 0.8665 & 0.6062 & 0.4225 & 0.7073 & 0.6062 & 0.8665 & 0.3949 \\
0.4225 & 0.9659 & 1.0000 & 0.8665 & 0.6062 & 0.4225 & 0.7073 & 0.6062 & 0.8665 & 0.3949 \\
0.4225 & 0.8665 & 0.8665 & 1.0000 & 0.6062 & 0.4225 & 0.7073 & 0.6062 & 0.9365 & 0.3949 \\
0.4225 & 0.6062 & 0.6062 & 0.6062 & 1.0000 & 0.4225 & 0.6062 & 0.8068 & 0.6062 & 0.3949 \\
0.8272 & 0.4225 & 0.4225 & 0.4225 & 0.4225 & 1.0000 & 0.4225 & 0.4225 & 0.4225 & 0.3949 \\
0.4225 & 0.7073 & 0.7073 & 0.7073 & 0.6062 & 0.4225 & 1.0000 & 0.6062 & 0.7073 & 0.3949 \\
0.4225 & 0.6062 & 0.6062 & 0.6062 & 0.8068 & 0.4225 & 0.6062 & 1.0000 & 0.6062 & 0.3949 \\
0.4225 & 0.8665 & 0.8665 & 0.9365 & 0.6062 & 0.4225 & 0.7073 & 0.6062 & 1.0000 & 0.3949 \\
0.3949 & 0.3949 & 0.3949 & 0.3949 & 0.3949 & 0.3949 & 0.3949 & 0.3949 & 0.3949 & 1.0000
\end{pmatrix}$$

As $C^{16} = C^8$, then C^8 is an equivalent correlation matrix.

Step 3. For a confidence level λ, according to Eq. (2.24), we can construct a λ-cutting matrix $C_\lambda = \left(\rho_{ij}^\lambda\right)_{m \times m}$. Different λ produces different λ-cutting matrix $C_\lambda = \left(\rho_{ij}^\lambda\right)_{m \times m}$.

Step 4. Based on the derived λ-cutting matrix $C_\lambda = \left(\rho_{ij}^\lambda\right)_{m \times m}$, we can classify these 10 firms $A_j(j = 1, 2, \ldots, 10)$ into different clusters. The possible classifications of these firms are shown in Table 2.6.

Chen et al. (2013a) utilized the correlation coefficient formula in the form of Eq. (2.2) to conduct the cluster analysis and produced a correlation matrix and an equivalent correlation matrix as well, based on which, some clustering results were obtained (see Table 2.7).

Comparing our method with that of Chen et al. (2013a), the superiorities are significant. Firstly, in terms of the correlation matrix, our correlation matrix consists of different values varying from negative values to positive values; however, in Chen et al. (2013a)'s correlation matrix, only positive values can be used, which consequently cannot represent the negative correlation coefficients between the firms. Secondly, as to the equivalent correlation matrix, the value range in $C^{16\prime}$ is from 0.7984 to 1, which is quite narrow, and thus, it may be not quite convincing to distinguish different clusters. But if using our weighted correlation coefficient, the values in the produced equivalent correlation matrix vary from 0.3949 to 1, which is twice wider than that of $C^{16\prime}$, and thus can better reflect the differences between different clusters.

Table 2.6 Clustering results with respect to the correlation coefficient

Class	Confidence level	Clusters
10	$0.9659 < \lambda \leq 1$	$\{A_1\}, \{A_2\}, \{A_3\}, \{A_4\}, \{A_5\}, \{A_6\}, \{A_7\}, \{A_8\}, \{A_9\}, \{A_{10}\}$
9	$0.9365 < \lambda \leq 0.9659$	$\{A_1\}, \{A_2, A_3\}, \{A_4\}, \{A_5\}, \{A_6\}, \{A_7\}, \{A_8\}, \{A_9\}, \{A_{10}\}$
8	$0.8665 < \lambda \leq 0.9365$	$\{A_1\}, \{A_2, A_3\}, \{A_4, A_9\}, \{A_5\}, \{A_6\}, \{A_7\}, \{A_8\}, \{A_{10}\}$
7	$0.8272 < \lambda \leq 0.8665$	$\{A_1\}, \{A_2, A_3, A_4, A_9\}, \{A_5\}, \{A_6\}, \{A_7\}, \{A_8\}, \{A_{10}\}$
6	$0.8068 < \lambda \leq 0.8272$	$\{A_1, A_6\}, \{A_2, A_3, A_4, A_9\}, \{A_5\}, \{A_7\}, \{A_8\}, \{A_{10}\}$
5	$0.7073 < \lambda \leq 0.8068$	$\{A_1, A_6\}, \{A_2, A_3, A_4, A_9\}, \{A_5, A_8\}, \{A_7\}, \{A_{10}\}$
4	$0.6062 < \lambda \leq 0.7073$	$\{A_1, A_6\}, \{A_2, A_3, A_4, A_7, A_9\}, \{A_5, A_8\}, \{A_{10}\}$
3	$0.4225 < \lambda \leq 0.6062$	$\{A_1, A_6\}, \{A_2, A_3, A_4, A_5, A_7, A_8, A_9\}, \{A_{10}\}$
2	$0.3949 < \lambda \leq 0.4225$	$\{A_1, A_2, A_3, A_4, A_5, A_6 A_7, A_8, A_9\}, \{A_{10}\}$
1	$0 \leq \lambda \leq 0.3949$	$\{A_1, A_2, A_3, A_4, A_5, A_6, A_7, A_8, A_9, A_{10}\}$

Table 2.7 Clustering results with respect to Chen et al. (2013a)'s correlation coefficient

Class	Confidence level	Clusters
10	$0.9515 < \lambda \le 1$	$\{A_1\}, \{A_2\}, \{A_3\}, \{A_4\}, \{A_5\}, \{A_6\}, \{A_7\}, \{A_8\}, \{A_9\}, \{A_{10}\}$
9	$0.9306 < \lambda \le 0.9515$	$\{A_1\}, \{A_2\}, \{A_3\}, \{A_4\}, \{A_6\}, \{A_7\}, \{A_8\}, \{A_9\}, \{A_5, A_{10}\}$
8	$0.9238 < \lambda \le 0.9306$	$\{A_1\}, \{A_2\}, \{A_3\}, \{A_4, A_9\}, \{A_6\}, \{A_7\}, \{A_8\}, \{A_5, A_{10}\}$
7	$0.9104 < \lambda \le 0.9238$	$\{A_1\}, \{A_2\}, \{A_3\}, \{A_4, A_7, A_9\}, \{A_6\}, \{A_8\}, \{A_5, A_{10}\}$
6	$0.9025 < \lambda \le 0.9104$	$\{A_1, A_6\}, \{A_2\}, \{A_3\}, \{A_4, A_7, A_9\}, \{A_8\}, \{A_5, A_{10}\}$
5	$0.8997 < \lambda \le 0.9025$	$\{A_1, A_6\}, \{A_2\}, \{A_3\}, \{A_4, A_7, A_8, A_9\}, \{A_5, A_{10}\}$
4	$0.8520 < \lambda \le 0.8997$	$\{A_1, A_6\}, \{A_2\}, \{A_3, A_4, A_7, A_8, A_9\}, \{A_5, A_{10}\}$
3	$0.8200 < \lambda \le 0.8520$	$\{A_1, A_6\}, \{A_2\}, \{A_3, A_4, A_5, A_7, A_8, A_9, A_{10}\}$
2	$0.7984 < \lambda \le 0.8200$	$\{A_1, A_6\}, \{A_2, A_3, A_4, A_5, A_7, A_8, A_9, A_{10}\}$
1	$0 \le \lambda \le 0.7984$	$\{A_1, A_2, A_3, A_4, A_5, A_6, A_7, A_8, A_9, A_{10}\}$

2.2 Novel Entropy Measures of Hesitant Fuzzy Sets

2.2.1 The Existing Entropy Measures of Hesitant Fuzzy Sets

Motivated by the axiomatic definition of entropy for fuzzy sets, Xu and Xia (2012) proposed the principles of entropy measure for HFE in terms of the fuzziness of a HFE.

Definition 2.7 (*Xu and Xia* 2012). A real-valued function $E : H \to [0, 1]$ is called an entropy for the HFE α, if it satisfies:

(1) $E(\alpha) = 0$ if and only if $\alpha = \{0\}$ or $\alpha = \{1\}$.
(2) $E(\alpha) = 1$ if and only if $\alpha_{\sigma(i)} + \alpha_{\sigma(l-i+1)} = 1$, for $i = 1, 2, \ldots, l_\alpha$.
(3) $E(\alpha) \le E(\beta)$ if $\alpha_{\sigma(i)} \le \beta_{\sigma(i)}$ for $\beta_{\sigma(i)} + \beta_{\sigma(l-i+1)} \le 1$ or $\alpha_{\sigma(i)} \ge \beta_{\sigma(i)}$ for $\beta_{\sigma(i)} + \beta_{\sigma(l-i+1)} \ge 1$, $i = 1, 2, \ldots, l$.
(4) $E(\alpha) = E(\alpha^c)$.

Based on Definition 2.7, Xu and Xia (2012) introduced some entropy measures for a HFE α.

$$E_1(\alpha) = \frac{1}{l_\alpha(\sqrt{2} - 1)} \sum_{i=1}^{l_\alpha} \left(\sin \frac{\pi \left(\alpha_{\sigma(i)} + \alpha_{\sigma(l_\alpha - i + 1)} \right)}{4} + \sin \frac{\pi \left(2 - \alpha_{\sigma(i)} - \alpha_{\sigma(l_\alpha - i + 1)} \right)}{4} - 1 \right)$$

(2.25)

$$E_2(\alpha) = \frac{1}{l_\alpha(\sqrt{2} - 1)} \sum_{i=1}^{l_\alpha} \left(\cos \frac{\pi \left(\alpha_{\sigma(i)} + \alpha_{\sigma(l_\alpha - i + 1)} \right)}{4} + \cos \frac{\pi \left(2 - \alpha_{\sigma(i)} - \alpha_{\sigma(l_\alpha - i + 1)} \right)}{4} - 1 \right)$$

(2.26)

$$E_3(\alpha) = -\frac{1}{l_\alpha \ln 2} \sum_{i=1}^{l_\alpha} \left(\frac{\alpha_{\sigma(i)} + \alpha_{\sigma(l_\alpha-i+1)}}{2} \ln \frac{\alpha_{\sigma(i)} + \alpha_{\sigma(l_\alpha-i+1)}}{2} \right.$$
$$\left. + \frac{2 - \alpha_{\sigma(i)} - \alpha_{\sigma(l_\alpha-i+1)}}{2} \ln \frac{2 - \alpha_{\sigma(i)} - \alpha_{\sigma(l_\alpha-i+1)}}{2} \right) \tag{2.27}$$

$$E_4(\alpha) = -\frac{1}{l_\alpha(2^{(1-s)t} - 1)} \sum_{i=1}^{l_\alpha} \left[\left(\left(\frac{\alpha_{\sigma(i)} + \alpha_{\sigma(l_\alpha-i+1)}}{2} \right)^s + \left(\frac{2 - \alpha_{\sigma(i)} - \alpha_{\sigma(l_\alpha-i+1)}}{2} \right)^s \right)^t - 1 \right]$$

$$t \neq 0,\ s \neq 1,\ s > 0 \tag{2.28}$$

All the above entropy measures satisfy the conditions in Definition 2.7. However, if we apply them to the HFEs whose complements are equal to themselves, we can get the same entropy degree. This indicates that the entropy measures introduced by Xu and Xia (2012) cannot correctly discriminate different HFEs in some cases.

Example 2.5 (Zhao et al. 2015). Let $\alpha_1 = \{0.2, 0.5, 0.8\}$ and $\alpha_2 = \{0.4, 0.5, 0.6\}$ be two HFEs. Obviously, $\alpha_1 = \alpha_1^c$, $\alpha_2 = \alpha_2^c$ and the fuzziness of α_2 is greater than that of α_1. Applying the entropy measures E_i $(i = 1, 2, 3, 4)$ to the HFEs α_1 and α_2, we obtain $E_i(\alpha_1) = E_i(\alpha_2) = 1$, for $i = 1, 2, 3, 4$, which are not consistent with our intuition.

Based on the distance measure between HFEs (Xu and Xia 2011a), Farhadinia (2013) gave the following axiomatic definition of entropy to measure the fuzziness of a HFE.

Definition 2.8 (*Farhadinia* 2013). Let $d(\alpha, \{0.5\})$ be the distance between the HFE α and $\{0.5\}$. A real function $E_d : H \to [0, 1]$ is called a distance-based entropy for the HFE α, if it satisfies:

(1) $E_d(\alpha) = 0$ if and only if $\alpha = \{0\}$ or $\alpha = \{1\}$.
(2) $E_d(\alpha) = 1$ if and only if $\alpha = \{0.5\}$.
(3) $E_d(\alpha) \le E_d(\beta)$ if $d(\alpha, \{0.5\}) \ge d(\beta, \{0.5\})$.
(4) $E_d(\alpha) = E_d(\alpha^c)$.

Theorem 2.4 provided an approach to generate the distance-based entropy measures for HFEs.

Theorem 2.4 (Farhadinia 2013). *Let* $Z : [0, 1] \to [0, 1]$ *be a strictly monotone decreasing real function, and* $d(\alpha, \{0.5\})$ *be the distance between the HFE* α *and* $\{0.5\}$. *Then,*

$$E_d(\alpha) = \frac{Z(2d(\alpha, \{0.5\})) - Z(1)}{Z(0) - Z(1)} \tag{2.29}$$

is an entropy measure for the HFE α.

Xu and Xia (2011a) defined three kinds of distance measures which can be used to calculate the distance between the HFE α and $\{0.5\}$.

$$d_{1k}(\alpha, \{0.5\}) = \left[\frac{1}{l}\sum_{i=1}^{l}|\alpha_{\sigma(i)} - 0.5|^k\right]^{1/k}, k = 1, 2 \qquad (2.30)$$

$$d_{2k}(\alpha, \{0.5\}) = \max_i\left\{|\alpha_{\sigma(i)} - 0.5|^k\right\}, k = 1, 2 \qquad (2.31)$$

$$d_{3k}(\alpha, \{0.5\}) = \left\{\left[\frac{1}{l}\sum_{i=1}^{l}|\alpha_{\sigma(i)} - 0.5|^k\right]^{1/k} + \max_i\left\{|\alpha_{\sigma(i)} - 0.5|^k\right\}\right\}, k = 1, 2$$

$$(2.32)$$

Let α be three HFEs $\{0, 1\}$, $\{0\}$ and $\{1\}$, respectively. Then by Eqs. (2.30)–(2.32), we can calculate

$$d_{1k}(\{0, 1\}, \{0.5\}) = d_{1k}(\{0\}, \{0.5\}) = d_{1k}(\{1\}, \{0.5\}) = \frac{1}{2}, k = 1, 2$$

$$d_{2k}(\{0, 1\}, \{0.5\}) = d_{2k}(\{0\}, \{0.5\}) = d_{2k}(\{1\}, \{0.5\}) = \left(\frac{1}{2}\right)^k, k = 1, 2$$

$$d_{3k}(\{0, 1\}, \{0.5\}) = d_{3k}(\{0\}, \{0.5\}) = d_{3k}(\{1\}, \{0.5\}) = \frac{1}{2}\left[\frac{1}{2} + \left(\frac{1}{2}\right)^k\right], k = 1, 2$$

According to Theorem 2.4, we get

$$E_{d_{1k}}(\{0, 1\}) = E_{d_{1k}}(\{0\}) = E_{d_{1k}}(\{1\}) = 0, k = 1, 2$$

$$E_{d_{21}}(\{0, 1\}) = E_{d_{21}}(\{0\}) = E_{d_{21}}(\{1\}) = 0$$

$$E_{d_{22}}(\{0\}) = E_{d_{22}}(\{1\}) = \frac{Z(1/2) - Z(1)}{Z(0) - Z(1)} \neq 0$$

$$E_{d_{31}}(\{0, 1\}) = E_{d_{31}}(\{0\}) = E_{d_{31}}(\{1\}) = 0$$

$$E_{d_{32}}(\{0\}) = E_{d_{32}}(\{1\}) = \frac{Z(3/8) - Z(1)}{Z(0) - Z(1)} \neq 0$$

The above results reveal that no matter which distance measure we employ, the derived entropies for HFEs do not meet the first condition in Definition 2.8, which implies that the entropy measure in Eq. (2.29) is unreasonable. Moreover, for any two HFEs α and β, if $d(\alpha, \{0.5\}) = d(\beta, \{0.5\})$, then by Eq. (2.29), we have $E(\alpha) = E(\beta)$. That is to say, different HFEs that have the same distance from the HFE $\{0.5\}$ would yield the same entropy in case we use the entropy measure proposed in Theorem 2.4. This is definitely unreasonable.

Particularly, let $Z(t) = 1 - t$ and d be the hesitant normalized Hamming distance d_{11}, then the entropy measure in Eq. (2.29) turns out to be:

$$E_{d_{11}}(\alpha) = 1 - \frac{2}{l_\alpha} \sum_{i=1}^{l_\alpha} \left| \alpha_{\sigma(i)} - \frac{1}{2} \right| \qquad (2.33)$$

Example 2.6 (*Zhao et al.* 2015). In a multiple criteria decision making problem, two decision organizations consider the possible membership degrees of x to the set M. The experts in the first organization think that the membership degree should be 0.01, while in the second organization, some experts deem it as 0.01, and the others deem it as 0.99. Then, the membership degree provided by the first organization is 0.01, which is very small, and thus, we can easily deduce that the experts in the first organization are inclined to consider that x does not belong to the set M. Similarly, it can be easily deduced that some experts in the second organization tend to think that x does not belong to the set M, and the others tend to believe that x belongs to the set M, and the degrees that x belongs to and not to the set M are the same, which implies that according to the decision information provided by the second decision organization, we are not sure whether x belongs to the set M or not. Thus, we may say that the decision information offered by the first organization is more specific than that offered by the other one. We can use the HFEs $\alpha_1 = \{0.01\}$ and $\alpha_2 = \{0.01, 0.99\}$ to represent the possible membership degrees of x into M provided by these two organizations, respectively. Then, by Eq. (2.33), we get $E_{d_{11}}(\alpha_1) = E_{d_{11}}(\alpha_2) = 0.02$, which is unreasonable because these two HFEs are significantly different in terms of specificity based on the above analysis.

2.2.2 Novel Two-Tuple Entropy Measures of Hesitant Fuzzy Sets

As mentioned above, the entropy measures proposed by Xu and Xia (2012) and Farhadinia (2013) are incapable to effectively distinguish HFEs in many cases. In our opinion, for a HFE, except for the fuzziness, there exists another kind of uncertainty, i.e., non-specificity. The fuzziness of a HFE is related to the deviation between the HFE and its nearest crisp set, while the non-specificity is related to the imprecise knowledge contained in the HFE. Suppose that the membership degrees of the element x to the set A provided by a decision organization are presented by the HFE $h(x) = \{0, 1\}$. From the HFE $h(x) = \{0, 1\}$, we know that the membership degree of x to A may be 0 indicating that x absolutely does not belong to A, and may be 1 implying that x completely belongs to A. We are not sure whether the element x belongs to the set A or not. That is to say, this case involves non-specificity. Non-specificity is another kind of uncertainty associated with a HFE. In this section, we present a new axiomatic definition of the entropy for HFEs, which captures the two types of uncertainty associated with a HFE. Then, we introduce some methods to construct the entropy measures for HFE.

Definition 2.9 (*Zhao et al.* 2015). Let $E_F, E_{NS} : H \to [0, 1]$ be two real functions. The pair (E_F, E_{NS}) is called a two-tuple entropy measure for the HFE α if E_F satisfies the following axiomatic requirements:

(1) $E_F(\alpha) = 0$ if and only if α is crisp, that is, $\alpha = \{0\}$ or $\alpha = \{1\}$;
(2) $E_F(\alpha) = 1$ if and only if $\alpha = \{0.5\}$;
(3) $E_F(\alpha) = E_F(\alpha^c)$;
(4) For any $i = 1, 2, \ldots, l$, if $\alpha_{\sigma(i)} \leq \beta_{\sigma(i)}$ for $\beta_{\sigma(i)} \leq 0.5$ or if $\alpha_{\sigma(i)} \geq \beta_{\sigma(i)}$ for $\beta_{\sigma(i)} \geq 0.5$, then $E_F(\alpha) \leq E_F(\beta)$, and E_{NS} satisfies the following axiomatic requirements:
(5) $E_{NS}(\alpha) = 0$ if and only if there is only one value in α, that is, $\alpha = \{u\}$ with $0 \leq u \leq 1$;
(6) $E_{NS}(\alpha) = 1$ if and only if $\alpha = \{0, 1\}$;
(7) $E_{NS}(\alpha) = E_{NS}(\alpha^c)$;
(8) $E_{NS}(\alpha) \geq E_{NS}(\beta)$ if for any $i, j = 1, 2, \ldots, l$, $\left|\alpha_{\sigma(i)} - \alpha_{\sigma(j)}\right| \geq \left|\beta_{\sigma(i)} - \beta_{\sigma(j)}\right|$.

Definition 2.9 uses a pair (E_F, E_{NS}) to represent the two kinds of uncertainty linked to a HFE where E_F, called the fuzzy entropy, is considered as a measure of fuzziness to quantify how far the HFE is from its closest crisp set, and E_{NS}, called the non-specific entropy, is proposed to measure the non-specificity of a HFE. It is noticed that the proposed non-specificity measure differs from that linked to the fuzzy set or the IFS. The introduced two-tuple entropy measure (E_F, E_{NS}) not only maintains the traditional properties of entropy, i.e., measuring the fuzziness aspect of uncertainty, but also reflects another aspect of uncertainty, i.e., non-specificity.

(1) Fuzzy entropy E_F
In this part, we provide some methods to generate the measures to quantify the fuzziness of a HFE.

Theorem 2.5 (Zhao et al. 2015). *Let $R : [0, 1]^2 \to [0, 1]$ be a mapping and satisfy:*

(1) *$R(x, y) = 0$ if and only if $x = y = 0$ or $x = y = 1$.*
(2) *$R(x, y) = 1$ if and only if $x = y = 0.5$.*
(3) *$R(x, y) = R(1 - y, 1 - x)$ for all $x, y \in [0, 1]$.*
(4) *If $0 \leq x_1 \leq x_2 \leq 0.5, 0 \leq y_1 \leq y_2 \leq 0.5$, then $R(x_1, y_1) \leq R(x_2, y_2)$; if $0.5 \leq x_1 \leq x_2 \leq 1, 0.5 \leq y_1 \leq y_2 \leq 1$, then $R(x_1, y_1) \geq R(x_2, y_2)$.*

Then the mapping $E_F : H \to [0, 1]$ defined as

$$E_F(\alpha) = \frac{2}{l_\alpha(l_\alpha + 1)} \sum_{i=1}^{l_\alpha} \sum_{j \geq i} R\left(\alpha_{\sigma(i)}, \alpha_{\sigma(j)}\right) \tag{2.34}$$

fulfills the axioms (1)–(4) in Definition 2.9.

Proof

(1) If $\alpha = \{0\}$, then by Eq. (2.34), we get $E_F(\{0\}) = R(0,0) = 0$; if $\alpha = \{1\}$, then $E_F(\{1\}) = R(1,1) = 0$. Conversely, if $E_F(\alpha) = 0$, then $R(\alpha_{\sigma(i)}, \alpha_{\sigma(j)}) = 0$, for any $i,j = 1,2,\ldots,l_\alpha$, $j \geq i$. According to the property (1) in Theorem 2.5, we have $\alpha_{\sigma(i)} = 0$ or $\alpha_{\sigma(i)} = 1$, for any $i = 1,2,\ldots,l_\alpha$. Thus, the condition (1) in Definition 2.9 holds.

(2) If $\alpha = \{0.5\}$, then according to Eq. (2.34), we obtain $E_F(\{0.5\}) = R(0.5, 0.5) = 1$. On the contrary, if $E_F(\alpha) = 1$, then by Eq. (2.34), we have $R(\alpha_{\sigma(i)}, \alpha_{\sigma(j)}) = 1$, for any $i,j = 1,2,\ldots,l_\alpha$, $j \geq i$. According to the property (2) in Theorem 2.5, we have $\alpha_{\sigma(i)} = 0.5$, for any $i = 1,2,\ldots,l_\alpha$. Thus, the condition (2) in Definition 2.9 holds.

(3) By Eq. (2.34), we have $E_F(\alpha^c) = \frac{2}{l_\alpha(l_\alpha+1)} \sum_{i=1}^{l_\alpha} \sum_{j\geq i} R(\alpha_{\sigma(i)}^c, \alpha_{\sigma(j)}^c)$. Since $\alpha_{\sigma(i)}^c = 1 - \alpha_{\sigma(l_\alpha-i+1)}$, for $i = 1,2,\ldots,l_\alpha$, then, $E_F(\alpha^c) = \frac{2}{l_\alpha(l_\alpha+1)} \sum_{i=1}^{l_\alpha} \sum_{j\geq i} R(1 - \alpha_{\sigma(l_\alpha-i+1)}, 1 - \alpha_{\sigma(l_\alpha-j+1)})$. According to the property (3) in Theorem 2.5, we have $E_F(\alpha^c) = \frac{2}{l_\alpha(l_\alpha+1)} \sum_{i=1}^{l_\alpha} \sum_{j\geq i} R(\alpha_{\sigma(l_\alpha-j+1)}, \alpha_{\sigma(l_\alpha-i+1)}) = E_F(\alpha)$. Thus, the condition (3) in Definition 2.9 holds.

(4) For any $i = 1,2,\ldots,l$, $\alpha_{\sigma(i)} \leq \beta_{\sigma(i)} \leq 0.5$, according to the property (4) in Theorem 2.5, we get $R(\alpha_{\sigma(i)}, \alpha_{\sigma(j)}) \leq R(\beta_{\sigma(i)}, \beta_{\sigma(j)})$, $i,j = 1,2,\ldots,l$, $j \geq i$. Then according to Eq. (2.34), we gain $E_F(\alpha) \leq E_F(\beta)$. The other case can be illustrated in a similar way. Thus, the condition (4) in Definition 2.9 holds.

Remark It is observed that $E_F(\alpha)$ is a fuzzy entropy for the HFE α. By Eq. (2.34), we have $E_F(\{0,1\}) = \frac{1}{3}R(0,1) \neq 0$, which shows that the fuzzy entropy of $\{0,1\}$ is different from those of the HFEs $\{0\}$ and $\{1\}$.

Theorem 2.6 (Zhao et al. 2015). *Let $\bar{R} : [0,1]^2 \to [0,1]$ be a mapping and the mapping $E_F : H \to [0,1]$ defined as:*

$$E_F(\alpha) = \frac{2}{l_\alpha(l_\alpha+1)} \sum_{i=1}^{l_\alpha} \sum_{j\geq i} \bar{R}(\alpha_{\sigma(i)}, \alpha_{\sigma(j)}) \qquad (2.35)$$

satisfies the axioms (1)–(4) in Definition 2.9, then

(1) $\bar{R}(0,0) = 0$ *and* $\bar{R}(1,1) = 0$.
(2) $\bar{R}(0.5, 0.5) = 1$.
(3) *If for all $x,y \in [0,1]$, $\bar{R}(x,y) = \bar{R}(y,x)$, then $\bar{R}(x,y) = \bar{R}(1-y, 1-x)$ for all $x,y \in [0,1]$.*
(4) *If $0 \leq x_1 \leq x_2 \leq 0.5$, then $\bar{R}(x_1,x_1) \leq \bar{R}(x_2,x_2)$; if $0.5 \leq x_1 \leq x_2 \leq 1$, then $\bar{R}(x_1,x_1) \geq \bar{R}(x_2,x_2)$.*

Proof (1) and (2) are easy to check, thus, we here only give the proofs of (3) and (4).

(3) Suppose that there exist $x, y \in [0, 1]$ such that $\bar{R}(x, y) \neq \bar{R}(1 - y, 1 - x)$. Without loss of generality, assume that $x \leq y$ and $\bar{R}(x, y) > \bar{R}(1 - y, 1 - x)$. Given a HFE $\alpha = \{\alpha_{\sigma(1)}, \alpha_{\sigma(2)}\}$ where $\alpha_{\sigma(1)} = x$ and $\alpha_{\sigma(2)} = y$, Then by Eq. (2.35), we get

$$E_F(\alpha) = \frac{1}{3} \left[\bar{R}(\alpha_{\sigma(1)}, \alpha_{\sigma(1)}) + \bar{R}(\alpha_{\sigma(1)}, \alpha_{\sigma(2)}) + \bar{R}(\alpha_{\sigma(2)}, \alpha_{\sigma(2)}) \right]$$

and

$$E_F(\alpha^c) = \frac{1}{3} \left[\bar{R}(1 - \alpha_{\sigma(2)}, 1 - \alpha_{\sigma(2)}) + \bar{R}(1 - \alpha_{\sigma(2)}, 1 - \alpha_{\sigma(1)}) + \bar{R}(1 - \alpha_{\sigma(1)}, 1 - \alpha_{\sigma(1)}) \right]$$

According to the condition (3) in Definition 2.9, we have $E_F(\{\alpha_{\sigma(1)}\}) = E_F(\{1 - \alpha_{\sigma(1)}\})$ and $E_F(\{\alpha_{\sigma(2)}\}) = E_F(\{1 - \alpha_{\sigma(2)}\})$, that is, $\bar{R}(\alpha_{\sigma(1)}, \alpha_{\sigma(1)}) = \bar{R}(1 - \alpha_{\sigma(1)}, 1 - \alpha_{\sigma(1)})$ and $\bar{R}(\alpha_{\sigma(2)}, \alpha_{\sigma(2)}) = \bar{R}(1 - \alpha_{\sigma(2)}, 1 - \alpha_{\sigma(2)})$. Thus, $E_F(\alpha) > E_F(\alpha^c)$, which contradicts the condition (3) in Definition 2.9. In other words, the property (3) holds.

(4) Assume that there exist $x_1, x_2 \in [0, 0.5]$ with $x_1 \leq x_2$ such that $\bar{R}(x_1, x_1) > \bar{R}(x_2, x_2)$, then by Eq. (2.35), we get $E_F(\{x_1\}) = \bar{R}(x_1, x_1) > \bar{R}(x_2, x_2) = E_F(\{x_2\})$, which contradicts the condition (4) in Definition 2.9. Similarly, the other case can be proven. $\qed$

It is not easy to look for the bivariate function R in Theorem 2.5. In what follows, we try to reduce it to a univariate function.

Theorem 2.7 (Zhao et al. 2015). *Let $\varphi : [0, 1] \to [0, 1]$ be a mapping and satisfy:*

(1) $\varphi(x) = 0$ *if and only if $x = 0$.*
(2) $\varphi(x) = 1$ *if and only if $x = 0.75$.*
(3) φ *is monotone non-decreasing in $[0, 0.75)$ and monotone non-increasing in $(0.75, 1]$.*

Then, the mapping $E_F : H \to [0, 1]$ defined as

$$E_F(\alpha) = \frac{2}{l_\alpha(l_\alpha + 1)} \sum_{i=1}^{l_\alpha} \sum_{j \geq i} \varphi(1 - \alpha_{\sigma(i)}\alpha_{\sigma(j)}) \cdot \varphi(\alpha_{\sigma(i)} - \alpha_{\sigma(i)}\alpha_{\sigma(j)} + \alpha_{\sigma(j)}) \quad (2.36)$$

fulfills the axioms (1)–(4) in Definition 2.9.

Proof Suppose that φ is defined as the above statement and $R(x, y) = \varphi(1 - xy) \cdot \varphi(x - xy + y)$. Then we only need to prove that $R(x, y)$ possesses the properties (1)–(4) in Theorem 2.5.

(1) If $R(x, y) = 0$, that is, $\varphi(1 - xy) \cdot \varphi(x - xy + y) = 0$, then $\varphi(1 - xy) = 0$ or $\varphi(x - xy + y) = 0$. If $\varphi(1 - xy) = 0$, then by the condition (1) in this theorem, we get $xy = 1$. Thus, $x = y = 1$. If $\varphi(x - xy + y) = 0$, then $x - xy + y = 0$.

Therefore, we deduce that $x = y = 0$. The converse is easy to prove. Accordingly, the property (1) in Theorem 2.5 holds.

(2) If $R(x, y) = 1$, that is, $\varphi(1 - xy) \cdot \varphi(x - xy + y) = 1$, then we get $\varphi(1 - xy) = 1$ and $\varphi(x - xy + y) = 1$. By the condition (2) in this theorem, we deduce that $xy = 0.25$ and $x - xy + y = 0.75$, from which we get $x = y = 0.5$. It is easy to prove the converse. Then we finish the proof of property (2) in Theorem 2.5.

(3) $R(1 - y, 1 - x) = \varphi(1 - (1 - y)(1 - x)) \cdot \varphi(1 - y - (1 - y)(1 - x) + 1 - x)$
$$= \varphi(x - xy + y) \cdot \varphi(1 - xy) = R(x, y)$$

(4) Assume $0 \le x_1 \le x_2 \le 0.5$ and $0 \le y_1 \le y_2 \le 0.5$, then $0.75 \le 1 - x_2 y_2 \le 1 - x_1 y_1 \le 1$ and $0 \le x_1 - x_1 y_1 + y_1 \le x_2 - x_2 y_2 + y_2 \le 0.75$. By the condition (3) in this theorem, we obtain $\varphi(1 - x_1 y_1) \le \varphi(1 - x_2 y_2)$ and $\varphi(x_1 - x_1 y_1 + y_1) \le \varphi(x_2 - x_2 y_2 + y_2)$, from which we derive $R(x_1, y_1) \le R(x_2, y_2)$. Similarly, the other case can be illustrated. Thus, the property (4) in Theorem 2.5 holds.

Based on Theorem 2.7, we can set out two entropy measures for HFEs as illustrative examples.

(1) Let $\varphi(t) = 1 - \left(\frac{1}{3}|4t - 3|\right)^r$ with $r \ge 1$. Obviously, φ satisfies the conditions in Theorem 2.7. Then we get the following entropy measure for HFEs:

$$E_F^r(\alpha) = \frac{2}{l_\alpha(l_\alpha + 1)} \sum_{i=1}^{l_\alpha} \sum_{j \ge i} \left(1 - \left(\frac{1}{3}|4\alpha_{\sigma(i)}\alpha_{\sigma(j)} - 1|\right)^r\right) \cdot$$
$$\left(1 - \left(\frac{1}{3}|4\alpha_{\sigma(i)} - 4\alpha_{\sigma(i)}\alpha_{\sigma(j)} + 4\alpha_{\sigma(j)} - 3|\right)^r\right) \tag{2.37}$$

For the simplicity of calculation, we take $r = 1$. Then the entropy measure in Eq. (2.37) becomes

$$E_F^1(\alpha) = \frac{2}{l_\alpha(l_\alpha + 1)} \sum_{i=1}^{l_\alpha} \sum_{j \ge i} \frac{1}{9}\left(3 - |4\alpha_{\sigma(i)}\alpha_{\sigma(j)} - 1|\right) \cdot \left(3 - |4\alpha_{\sigma(i)} - 4\alpha_{\sigma(i)}\alpha_{\sigma(j)} + 4\alpha_{\sigma(j)} - 3|\right)$$
$$\tag{2.38}$$

(2) Let $\varphi(t) = \frac{2}{3}[\min(2t - 1, 2 - 2t) + 1]$. Then φ satisfies the conditions in Theorem 2.7, and the generated entropy measure for HFEs is

$$E_F(\alpha) = \frac{2}{l_\alpha(l_\alpha + 1)} \sum_{i=1}^{l_\alpha} \sum_{j \ge i} \frac{4}{9}\left(\min\left(1 - 2\alpha_{\sigma(i)}\alpha_{\sigma(j)}, 2\alpha_{\sigma(i)}\alpha_{\sigma(j)}\right) + 1\right) \cdot$$
$$\left(\min\left(1 - 2\left(1 - \alpha_{\sigma(i)}\right)\left(1 - \alpha_{\sigma(j)}\right), 2\left(1 - \alpha_{\sigma(i)}\right)\left(1 - \alpha_{\sigma(j)}\right)\right) + 1\right) \tag{2.39}$$

The following example illustrates that the entropy measures proposed in this chapter can produce better results than those introduced by Xu and Xia (2012) in distinguishing HFEs.

Table 2.8 The results obtained by different entropy measures

Results	$E_F^1(\alpha_i)$	$E_F^2(\alpha_i)$	$E_F^3(\alpha_i)$	$E_F(\alpha_i)$
$i = 1$	0.8696	0.9851	0.9981	0.8696
$i = 2$	0.5065	0.7385	0.8391	0.5065

Example 2.7 (Zhao et al. 2015). Consider two HFEs $\alpha_1 = \{0.4, 0.5, 0.6\}$ and $\alpha_2 = \{0.1, 0.5, 0.9\}$. Obviously, $\alpha_1 = \alpha_1^c$ and $\alpha_2 = \alpha_2^c$, and intuitively, the fuzziness of α_1 should be greater than that of α_2. Utilizing the entropy measures shown in Eqs. (2.25)–(2.28) to calculate the entropy of the HFE $\alpha_i(i = 1, 2)$, we get $E_j(\alpha_1) = E_j(\alpha_2) = 1$ $(j = 1, 2, 3, 4)$, which are counter-intuitive. On the contrary, if we use the entropy measures shown in Eqs. (2.37)–(2.39), we can get different results presented in Table 2.8.

From Table 2.8, we can find that no matter which entropy measures we use, the entropy of α_1 is always greater than that of α_2. This is consistent with our intuition. In other words, the proposed entropy measures are able to overcome the drawback of Xu and Xia (2012)'s entropy measures, that is, those measures cannot differentiate the different HFEs which are equal to their complements.

(2) **Non-specific entropy** E_{NS}

Now we pay attention to the other aspect of uncertainty associated with a HFE, i.e., the non-specificity, and introduce some measures to quantify the non-specificity of a HFE.

Let

$$\langle l_\alpha \rangle = \begin{cases} 2, & l_\alpha = 1 \\ l_\alpha(l_\alpha - 1), & l_\alpha \geq 2 \end{cases}$$

Firstly, we give the following general result:

Theorem 2.8 (Zhao et al. 2015). *Let* $F : [0, 1]^2 \to [0, 1]$ *be a mapping and satisfy:*

(1) $F(x, y) = 0$ *if and only if* $x = y$.
(2) $F(x, y) = 1$ *if and only if* $\{0, 1\} \cap \{x, y\} \neq \phi$.
(3) $F(x, y) = F(1 - y, 1 - x)$ *for all* $x, y \in [0, 1]$.
(4) *For* $x, y, z, w \in [0, 1]$, *if* $|x - y| \geq |z - w|$, *then* $F(x, y) \geq F(z, w)$.

Then the mapping $E_{NS} : H \to [0, 1]$ *defined as:*

$$E_{NS}(\alpha) = \frac{2}{\langle l_\alpha \rangle} \sum_{i=1}^{l_\alpha} \sum_{j \geq i} F\big(\alpha_{\sigma(i)}, \alpha_{\sigma(j)}\big) \tag{2.40}$$

satisfies the axioms (5)–(8) in Definition 2.9.

Proof

(1) If there is only one value in the HFE α, that is, $\alpha = \{u\}$, then by Eq. (2.40), we get $E_{NS}(\alpha) = F(u,u) = 0$.
    Conversely, if $E_{NS}(\alpha) = 0$, then $F\left(\alpha_{\sigma(i)}, \alpha_{\sigma(j)}\right) = 0$, for any $i,j = 1, 2, \ldots,$ $l_\alpha, j \geq i$. According to the property (1) in this theorem, we deduce that $\alpha_{\sigma(i)} = \alpha_{\sigma(j)}$, for any $i,j = 1, 2, \ldots, l_\alpha, j \geq i$. That is to say, the HFE α has only one value. Thus, the condition (5) in Definition 2.9 holds.

(2) If $\alpha = \{0, 1\}$, then according to Eq. (2.40) and the property (1) and property (2) in this theorem, we have $E_{NS}(\alpha) = F(0,0) + F(0,1) + F(1,1) = 1$.
    On the contrary, if $E_{NS}(\alpha) = 1$ with $\alpha = \left\{\alpha_{\sigma(1)}, \alpha_{\sigma(2)}, \ldots, \alpha_{\sigma(l_\alpha)}\right\}$, $l_\alpha \geq 2$, then by Eq. (2.40), we obtain $F\left(\alpha_{\sigma(i)}, \alpha_{\sigma(j)}\right) = 1$, for any $i,j = 1, 2, \ldots, l_\alpha, j > i$. If $l_\alpha = 2$, then according to the property (2) in this theorem, we obtain $\alpha_{\sigma(1)} = 0$ and $\alpha_{\sigma(2)} = 1$, that is, $\alpha = \{0, 1\}$. If $l_\alpha > 2$, for instance, let $l_\alpha = 3$, then we get $F\left(\alpha_{\sigma(1)}, \alpha_{\sigma(2)}\right) = 1$, $F\left(\alpha_{\sigma(1)}, \alpha_{\sigma(3)}\right) = 1$ and $F\left(\alpha_{\sigma(2)}, \alpha_{\sigma(3)}\right) = 1$. By the first two equations, we deduce that $\alpha_{\sigma(1)} = 0$, $\alpha_{\sigma(2)} = 1$ and $\alpha_{\sigma(3)} = 1$, and by the third equation, we deduce that $\alpha_{\sigma(2)} = 0$ and $\alpha_{\sigma(3)} = 1$, which is contradictory. In a similar way, we can illustrate that it is contradictory when l_α takes any value larger than 3. Thus, the condition (6) in Definition 2.9 holds.

(3) The proof of the condition (7) is similar to that of the condition (3) in Theorem 2.5.

(4) The proof of the condition (8) is straightforward according to the property (4) in this theorem.

Theorem 2.9 (Zhao et al. 2015). *Let $\bar{F} : [0, 1]^2 \to [0, 1]$ be a mapping and let the mapping $E_{NS} : H \to [0, 1]$ defined as:*

$$E_{NS}(\alpha) = \frac{2}{\langle l_\alpha \rangle} \sum_{i=1}^{l_\alpha} \sum_{j \geq i} \bar{F}\left(\alpha_{\sigma(i)}, \alpha_{\sigma(j)}\right) \tag{2.41}$$

satisfy the axioms (5)–(8) in Definition 2.9, then

(1) *$\bar{F}(x, y) = 0$ if and only if $x = y$.*
(2) *If $\bar{F}(x, y) = \bar{F}(y, x)$, for all $x, y \in [0, 1]$, then (i) $\bar{F}(x, y) = 1$ if and only if $\{0, 1\} \cap \{x, y\} \neq \phi$; (ii) $\bar{F}(x, y) = \bar{F}(1 - y, 1 - x)$ for all $x, y \in [0, 1]$; (iii) $\bar{F}(x, y) \geq \bar{F}(z, w)$ if $|x - y| \geq |z - w|$ for $x, y, z, w \in [0, 1]$.*

Proof

(1) Assume that there exist $x, y \in [0, 1]$ with $x \neq y$ such that $\bar{F}(x, y) = 0$. Without loss of generality, suppose $x < y$. Consider the HFEs $\beta = \left\{\alpha_{\sigma(1)}\right\}$ and $\gamma = \left\{\alpha_{\sigma(2)}\right\}$ assigned by $\alpha_{\sigma(1)} = x$ and $\alpha_{\sigma(2)} = y$, respectively. Then according to the requirement (5) in Definition 2.9, we have

$$E_{NS}(\beta) = \bar{F}\left(\alpha_{\sigma(1)}, \alpha_{\sigma(1)}\right) = 0,\ E_{NS}(\gamma) = \bar{F}\left(\alpha_{\sigma(2)}, \alpha_{\sigma(2)}\right) = 0$$

For the HFE $\alpha = \left\{\alpha_{\sigma(1)}, \alpha_{\sigma(2)}\right\}$, we have

$$E_{NS}(\alpha) = \bar{F}\left(\alpha_{\sigma(1)}, \alpha_{\sigma(1)}\right) + \bar{F}\left(\alpha_{\sigma(1)}, \alpha_{\sigma(2)}\right) + \bar{F}\left(\alpha_{\sigma(2)}, \alpha_{\sigma(2)}\right) = 0 \qquad (2.42)$$

By the requirement (5) in Definition 2.9, Eq. (2.42) holds if and only if $\alpha_{\sigma(1)} = \alpha_{\sigma(2)}$, that is, $x = y$, which is contradictory. Similarly, the converse can be proven.

(2) (i) Let $\alpha = \{0, 1\}$, then according to Eq. (2.41) and the requirement (6) in Definition 2.9, we get $E_{NS}(\alpha) = \bar{F}(0,0) + \bar{F}(0,1) + \bar{F}(1,1) = 1$. Since $\bar{F}(0,0) = \bar{F}(1,1) = 0$, then we obtain $\bar{F}(0,1) = 1$. Since $\bar{F}(x,y) = \bar{F}(y,x)$, for all $x, y \in [0, 1]$, then $\bar{F}(1,0) = 1$. Conversely, suppose that there exist $x, y \in [0, 1]$ with $\{0,1\} \cap \{x, y\} = \phi$ such that $\bar{F}(x,y) = 1$. Without loss of generality, assume that $x > y$. Let a HFE α be $\alpha = \left\{\alpha_{\sigma(1)}, \alpha_{\sigma(2)}\right\}$ defined by $\alpha_{\sigma(1)} = y$ and $\alpha_{\sigma(2)} = x$. Then by Eq. (2.41), we get

$$E_{NS}(\alpha) = \bar{F}\left(\alpha_{\sigma(1)}, \alpha_{\sigma(2)}\right) = \bar{F}(y,x) = 1 \qquad (2.43)$$

According to the requirement (6) in Definition 2.9, Eq. (2.43) holds if and only if $\alpha_{\sigma(1)} = 0$ and $\alpha_{\sigma(2)} = 1$, which is contradictory.

(ii) Suppose that there exist $x, y \in [0, 1]$ such that $\bar{F}(x,y) \neq \bar{F}(1 - y, 1 - x)$. Without loss of generality, assume that $x \leq y$ and $\bar{F}(x,y) > \bar{F}(1 - y, 1 - x)$. Given a HFE α as $\alpha = \left\{\alpha_{\sigma(1)}, \alpha_{\sigma(2)}\right\}$, where we assign $\alpha_{\sigma(1)} = x$ and $\alpha_{\sigma(2)} = y$, then we get

$$E_{NS}(\alpha) = \bar{F}\left(\alpha_{\sigma(1)}, \alpha_{\sigma(2)}\right) > \bar{F}\left(1 - \alpha_{\sigma(2)}, 1 - \alpha_{\sigma(1)}\right) = E_{NS}(\alpha^c)$$

which contradicts the axiomatic requirement (7) in Definition 2.9.

(iii) Suppose that there exist $x, y, z, w \in [0, 1]$ with $|x - y| \geq |z - w|$ such that $\bar{F}(x,y) < \bar{F}(z,w)$. Without loss of generality, assume that $x \leq y$ and $z \leq w$. Considering the HFE $\alpha = \left\{\alpha_{\sigma(1)}, \alpha_{\sigma(2)}\right\}$ defined as $\alpha_{\sigma(1)} = x$ and $\alpha_{\sigma(2)} = y$, and the HFE $\beta = \left\{\beta_{\sigma(1)}, \beta_{\sigma(2)}\right\}$ given by $\beta_{\sigma(1)} = z$ and $\beta_{\sigma(2)} = w$, we obtain

$$E_{NS}(\alpha) = \bar{F}\left(\alpha_{\sigma(1)}, \alpha_{\sigma(2)}\right) < \bar{F}\left(\beta_{\sigma(1)}, \beta_{\sigma(2)}\right) = E_{NS}(\beta)$$

which contradicts the requirement (8) in Definition 2.9. This completes the proof of Theorem 2.8.

Bustince et al. (2012) introduced the grouping function to measure to what extent an element belongs to at least one of two given classes.

Definition 2.10 (*Bustince et al.* 2012). A grouping function is a mapping $G:$ $[0, 1]^2 \rightarrow [0, 1]$ such that:

(1) $G(x, y) = G(y, x)$ for all $x, y \in [0, 1]$.
(2) $G(x, y) = 0$ if and only if $x = y = 0$.
(3) $G(x, y) = 1$ if and only if $x = 1$ or $y = 1$.
(4) G is monotonically increasing in both variables.

We can construct the non-specific entropy measure for HFE by means of the grouping function.

Theorem 2.10 (Zhao et al. 2015). *Let G be a grouping function. Then the mapping $E_{NSG} : H \rightarrow [0, 1]$ shown as:*

$$E_{FBG}(\alpha) = \frac{2}{\langle l_\alpha \rangle} \sum_{i=1}^{l_\alpha} \sum_{j \geq i} G\left(\left|\alpha_{\sigma(i)} - \alpha_{\sigma(j)}\right|, \left|\alpha_{\sigma(i)} - \alpha_{\sigma(j)}\right|\right)$$

defines a non-specific entropy measure for the HFE α satisfying the axioms (5)–(8) in Definition 2.9.

Proof It is observed that the mapping $F(x, y) = G(|x - y|, |x - y|)$ satisfies the properties (1)–(4) stated in Theorem 2.8. Thus, $E_{FBG}(\alpha)$ is a non-specific entropy measure for α.

If we define $G : [0, 1]^2 \rightarrow [0, 1]$ as $G(x, y) = x + y - xy$, then G satisfies the conditions in Definition 2.10. That is to say, G is a grouping function. Thus, based on Theorem 2.10, we get a non-specific entropy measure:

$$E_{NSG}^1(\alpha) = \frac{2}{\langle l_\alpha \rangle} \sum_{i=1}^{l_\alpha} \sum_{j \geq i} 2\left|\alpha_{\sigma(i)} - \alpha_{\sigma(j)}\right| - \left(\alpha_{\sigma(i)} - \alpha_{\sigma(j)}\right)^2 \qquad (2.44)$$

If we define $G : [0, 1]^2 \rightarrow [0, 1]$ as $G(x, y) = 1 - \dfrac{\sqrt{(1-x)(1-y)}}{\sqrt{(1-x)(1-y)} + 1 - (1-x)(1-y)}$, then G is a grouping function, and the corresponding non-specific entropy measure is

$$E_{NSG}^2(\alpha) = \frac{2}{\langle l_\alpha \rangle} \sum_{i=1}^{l_\alpha} \sum_{j \geq i} 1 - \frac{1 - \left|\alpha_{\sigma(i)} - \alpha_{\sigma(j)}\right|}{1 + \left|\alpha_{\sigma(i)} - \alpha_{\sigma(j)}\right| - \left(\alpha_{\sigma(i)} - \alpha_{\sigma(j)}\right)^2} \qquad (2.45)$$

Clearly, it is a bit difficult to look for such a bivariate function satisfying the conditions in Theorem 2.8. Below we attempt to reduce it to a univariate function.

Theorem 2.11 (Zhao et al. 2015). *Let $g : [0, 1] \rightarrow [0, 1]$ be a mapping and satisfy:*

(1) $g(x) = 0$ *if and only if $x = 0$.*
(2) $g(x) = 1$ *if and only if $x = 1$.*
(3) *g is monotone non-decreasing.*
   Then the mapping $E_{NS} : H \rightarrow [0, 1]$ defined as:

$$E_{NS}(\alpha) = \frac{2}{\langle l_\alpha \rangle} \sum_{i=1}^{l_\alpha} \sum_{j \geq i} g\left(\left|\alpha_{\sigma(i)} - \alpha_{\sigma(j)}\right|\right) \tag{2.46}$$

satisfies the axioms (5)–(8) in Definition 2.9.

Proof Let $F(x,y) = g(|x-y|)$, then the mapping $F(x,y)$ satisfies the properties in Theorem 2.8.

Below we give several specific examples to illustrate Theorem 2.11.

(1) Let $g : [0,1] \to [0,1]$ be defined as $g(t) = \frac{2t}{1+t}$. It satisfies the conditions in Theorem 2.11. Thus, the corresponding non-specific entropy measure is

$$E_{NS}^1(\alpha) = \frac{2}{\langle l_\alpha \rangle} \sum_{i=1}^{l_\alpha} \sum_{j \geq i} \frac{2\left|\alpha_{\sigma(i)} - \alpha_{\sigma(j)}\right|}{\left|\alpha_{\sigma(i)} - \alpha_{\sigma(j)}\right| + 1} \tag{2.47}$$

(2) Let $g : [0,1] \to [0,1]$ be $g(t) = \frac{\lg(1+t)}{\lg 2}$, then g satisfies the conditions in Theorem 2.11, and the corresponding non-specific entropy measure is

$$E_{NS}^2(\alpha) = \frac{2}{\langle l_\alpha \rangle} \sum_{i=1}^{l_\alpha} \sum_{j \geq i} \frac{\lg\left(1 + \left|\alpha_{\sigma(i)} - \alpha_{\sigma(j)}\right|\right)}{\lg 2} \tag{2.48}$$

(3) Let $g : [0,1] \to [0,1]$ be $g(t) = te^{t-1}$, then g satisfies the conditions in Theorem 2.11, and the corresponding non-specific entropy measure is

$$E_{NS}^3(\alpha) = \frac{2}{\langle l_\alpha \rangle} \sum_{i=1}^{l_\alpha} \sum_{j \geq i} \left|\alpha_{\sigma(i)} - \alpha_{\sigma(j)}\right| e^{\left|\alpha_{\sigma(i)} - \alpha_{\sigma(j)}\right| - 1} \tag{2.49}$$

It can be easily observed that the automorphisms of the unit interval satisfy the conditions in Theorem 2.11. In the following, we set out several non-specific entropy measures produced by them.

(4) Let $\phi : [0,1] \to [0,1]$ be defined as $\phi(t) = t^r$ with $r > 0$. Then ϕ is an automorphism of the unit interval, i.e., ϕ is continuous, strictly increasing and satisfies the conditions $\phi(0) = 0$, $\phi(1) = 1$ (Bustince et al. 2003). According to Theorem 2.11, we obtain the corresponding non-specific entropy measure as:

$$E_{NSA}^1(\alpha) = \frac{2}{\langle l_\alpha \rangle} \sum_{i=1}^{l_\alpha} \sum_{j \geq i} \left|\alpha_{\sigma(i)} - \alpha_{\sigma(j)}\right|^r \tag{2.50}$$

Especially, if we take $r = 1$, then the non-specific entropy measure becomes

Table 2.9 The results generated by different non-specific entropy formulas

Result	$E^1_{NSG}(\alpha_i)$	$E^2_{NSG}(\alpha_i)$	$E^1_{NS}(\alpha_i)$	$E^2_{NS}(\alpha_i)$	$E^3_{NS}(\alpha_i)$	$E^2_{NSA}(\alpha_i)$	$E^4_{NSA}(\alpha_i)$
$i = 1$	0.84	0.6774	0.75	0.6781	0.4022	0.6	0.84
$i = 2$	0.2467	0.2196	0.2323	0.1793	0.0571	0.1333	0.2467

$$E^2_{NSA}(\alpha) = \frac{2}{\langle l_\alpha \rangle} \sum_{i=1}^{l_\alpha} \sum_{j \geq i} \left| \alpha_{\sigma(i)} - \alpha_{\sigma(j)} \right| \tag{2.51}$$

(5) Let $\phi : [0, 1] \to [0, 1]$ be defined as $\phi(t) = 1 - (1 - t)^r$ with $r > 0$. Then ϕ is an automorphism of the unit interval. Based on Theorem 2.11, we get the generated non-specific entropy measure:

$$E^3_{NSA}(\alpha) = \frac{2}{\langle l_\alpha \rangle} \sum_{i=1}^{l_\alpha} \sum_{j \geq i} 1 - \left(1 - \left| \alpha_{\sigma(i)} - \alpha_{\sigma(j)} \right| \right)^r \tag{2.52}$$

In particular, when $r = 2$, the non-specific entropy measure becomes

$$E^4_{NSA}(\alpha) = \frac{2}{\langle l_\alpha \rangle} \sum_{i=1}^{l_\alpha} \sum_{j \geq i} 1 - \left(1 - \left| \alpha_{\sigma(i)} - \alpha_{\sigma(j)} \right| \right)^2 \tag{2.53}$$

It is noted that the entropy measures introduced by Farhadinia (2013) cannot discriminate the HFEs having the same distance from the HFE $\{0.5\}$. The following example shows that our entropy measures can overcome this drawback perfectly.

Example 2.8 (Zhao et al. 2015). Suppose two HFEs $\alpha_1 = \{0.2, 0.8\}$ and $\alpha_2 = \{0.1, 0.2, 0.3\}$. Clearly, the information expressed by α_2 is more specific than that of α_1. Nevertheless, by Eq. (2.33), we get $E_{d_{hnh}}(\alpha_1) = E_{d_{hnh}}(\alpha_2) = 0.4$, which is unreasonable. For α_1 and α_2, applying the proposed non-specific entropy measures (2.47)–(2.53), we can get different results, which are listed in Table 2.9.

From Table 2.9, it can be observed that no matter which measure is applied, we always get that the non-specificity of α_1 is greater than that of α_2, which is consistent with our intuition. From this example, we can see that our non-specific entropy measures can distinguish those HFEs that have the same distance from the HFE $\{0.5\}$, while the entropy measure in Eq. (2.33) cannot.

References

Burillo P, Bustince H (1996) Entropy on intuitionistic fuzzy sets and on interval-valued fuzzy sets. Fuzzy Sets Syst 78:305–316

Bustince H, Burillo P, Soria F (2003) Automorphisms, negations and implication operators. Fuzzy Sets Syst 134:209–229

Bustince H, Pagola M, Mesiar R, Hullermeier E, Herrera F (2012) Grouping, overlap, and generalized bientropic functions for fuzzy modeling of pairwise comparisons. IEEE Trans Fuzzy Syst 20:405–415

Chaudhuri BB, Bhattacharya A (2001) On correlation between two fuzzy sets. Fuzzy Sets Syst 118:447–456

Chen N, Xu ZS, Xia MM (2013) Correlation coefficients of hesitant fuzzy sets and their application to clustering analysis. Appl Math Model 37:2197–2211

Chiang DA, Lin NP (1999) Correlation of fuzzy sets. Fuzzy Sets Syst 102:221–226

De Luca A, Termini S (1972) A definition of a non probabilistic entropy in the setting of fuzzy sets theory. Inf Control 20:301–312

Farhadinia B (2013) Information measures for hesitant fuzzy sets and interval-valued hesitant fuzzy sets. Inf Sci 240:129–144

Friedman M, Last M, Makover Y, Kandel A (2007) Anomaly detection in web documents using crisp and fuzzy-based clustering methodology. Inf Sci 177:467–475

Garmendia L, Yager RR, Trillas E, Salvador A (2003) On t-norms based measures of specificity. Fuzzy Sets Syst 133:237–248

Gerstenkorn T, Manko J (1991) Correlation of intuitionistic fuzzy sets. Fuzzy Sets Syst 44:39–43

Hong DH (2006) Fuzzy measures for a correlation coefficient of fuzzy numbers under T_w (the weakest t-norm)-based fuzzy arithmetic operations. Inf Sci 176:150–160

Hong DH, Hwang SY (1995) Correlation of intuitionistic fuzzy sets in probability spaces. Fuzzy Sets Syst 75:77–81

Hung WL (2001) Using statistical viewpoint in developing correlation of intuitionistic fuzzy sets. Int J Uncertainty Fuzziness Knowl Based Syst 9:509–516

Hung WL, Wu JW (2002) Correlation of intuitionistic fuzzy sets by centroid method. Inf Sci 144:219–225

Kaufmann A, Swanson D (1975) Introduction to the theory of fuzzy subsets: fundamental theoretical elements. Academic Press, New York

Liao HC, Xu ZS, Zeng XJ (2015) Novel correlation coefficients between hesitant fuzzy sets and their application in decision making. Knowl-Based Syst 82:115–127

Liu ST, Gao C (2002) Fuzzy measures for correlation coefficient of fuzzy numbers. Fuzzy Sets Syst 128:267–275

Mitchell HB (2004) A correlation coefficient for intuitionistic fuzzy sets. Int J Intell Syst 19: 483–490

Murithy CA, Pal SK, Dutta-Majumder D (1985) Correlation between two fuzzy membership functions. Fuzzy Sets Syst 17:23–38

Pal NR, Bustince H, Pagola M, Mukherjee UK, Goswami DP, Beliakov G (2013) Uncertainties with Atanassov's intuitionistic fuzzy sets: Fuzziness and lack of knowledge. Inf Sci 228:61–74

Parkash O, Sharma P, Mahajan R (2008) New measures of weighted fuzzy entropy and their applications for the study of maximum weighted fuzzy entropy principle. Inf Sci 178: 2389–2395

Szmidt E, Kacprzyk J (2001) Entropy for intuitionistic fuzzy sets. Fuzzy Sets Syst 118:467–477

Szmidt E, Kacprzyk J (2010) Correlation of intuitionistic fuzzy sets. In: Hüllermeier E, Kruse R, Hoffmann F. (eds), IPMU 2010, LNAI 6178, pp. 169–177

Wei CP, Liang X, Zhang YZ (2012) A comparative analysis and improvement of entropy measures for intuitionistic fuzzy sets. J Syst Sci Math Sci 32:1437–1448

Xu ZS (2006) On correlation measures of intuitionistic fuzzy sets. Lect Notes Comput Sci 4224: 16–24

Xu ZS, Xia MM (2011a) On distance and correlation measures of hesitant fuzzy information. Int J Intell Syst 26:410–425

Xu ZS, Xia MM (2011b) Distance and similarity measures for hesitant fuzzy sets. Inf Sci 181:2128–2138

Xu ZS, Xia MM (2012) Hesitant fuzzy entropy and cross-entropy and their use in multiattribute decision-making. Int J Intell Syst 27(9):799–822

Yager RR (1979) On the measure of fuzziness and negation part I: membership in the unit interval. Int J Gen Syst 5:221–229

Yager RR (1992) Default knowledge and measures of specificity. Inf Sci 61:1–44

Yager RR (1998) Measures of specificity. In: Computational intelligence: soft computing and fuzzy-neuro integration with applications. Springer, pp 94–113

Yager RR (2008) Measures of specificity over continuous spaces under similarity relations. Fuzzy Sets Syst 159:2193–2210

Yu CH (1993) Correlation of fuzzy numbers. Fuzzy Sets Syst 55:303–307

Zadeh LA (1965) Fuzzy sets. Inf Control 8:338–353

Zhao N, Xu ZS, Liu FJ (2015) Uncertainty measures for hesitant fuzzy information. Int J Intell Syst 30(7):818–836

Chapter 3
Multiple Criteria Decision Making with Hesitant Fuzzy Hybrid Weighted Aggregation Operators

In the process of decision making with multiple experts, in order to select the optimal alternative(s) from a set of candidate alternatives with multiple attributes, the aggregation process is essential. In hesitant fuzzy decision making situation, the aggregation process is also the most significant step in searching the best alternative (s). Till now, many different kinds of hesitant fuzzy aggregation operators have been developed. Based on the relationship between the IFS and the HFS, Xia and Xu (2011a) developed a family of operators to fuse hesitant fuzzy information, such as the hesitant fuzzy weighted averaging (HFWA) operator, the hesitant fuzzy weighted geometric (HFWG) operator, the hesitant fuzzy ordered weighted averaging (HFOWA) operator, the hesitant fuzzy ordered weighted geometric (HFOWG) operator, the generalized hesitant fuzzy weighted averaging (GHFWA) operator, the generalized hesitant fuzzy weighted geometric (GHFWG) operator, the generalized hesitant fuzzy ordered weighted averaging (GHFOWA) operator, the generalized hesitant fuzzy ordered weighted geometric (GHFOWG) operator, the hesitant fuzzy hybrid averaging (HFHA) operator, the hesitant fuzzy hybrid geometric (HFHG) operator, the generalized hesitant fuzzy hybrid averaging (GHFHA) operator, and the generalized hesitant fuzzy hybrid geometric (GHFHG) operator. To aggregate the hesitant fuzzy information under confidence level, Xia et al. (2011) introduced a series of confidence induced hesitant fuzzy aggregation operators. Motivated by the quasi-arithmetic means (Hardy et al. 1934) and the induced idea (Yager and Filev 1999), Xia et al. (2013a) established a sort of induced aggregation operators for HFSs. Zhu and Xu (2013) and Zhu et al. (2012a) proposed the hesitant fuzzy Bonferroni means and hesitant fuzzy geometric Bonferroni means to aggregate hesitant fuzzy information. Motivated by the idea of prioritized aggregation operators (Yager 2008a), Wei (2012) introduced some prioritized aggregation operators, and then applied them to develop some models for hesitant fuzzy multiple criteria decision making problems in which the criteria are in different priority levels. Zhang (2013) proposed a wide range of hesitant fuzzy power aggregation operators and investigated their properties and relationships.

© Springer Nature Singapore Pte Ltd. 2017   73
H. Liao and Z. Xu, *Hesitant Fuzzy Decision Making Methodologies and Applications*, Uncertainty and Operations Research,
DOI 10.1007/978-981-10-3265-3_3

It is observed that the HFWA and HFWG operators can be used to weight the hesitant fuzzy arguments, but ignore the importance of the ordered position of the arguments, while the HFOWA and HFOWG operators only weight the ordered position of each given argument, but ignore the importance of the arguments. To solve this drawback, the hesitant fuzzy hybrid averaging (HFHA) operator and the hesitant fuzzy hybrid geometric (HFHG) operator were proposed to aggregate hesitant fuzzy arguments, which weight all the given arguments and their ordered positions simultaneously. Hence, these two operators have many advantages than the above mentioned operators in aggregating hesitant fuzzy information. However, these two operators do not satisfy the basic property named idempotency, which is desirable for aggregating a finite collection of HFSs. Therefore, in this chapter, we shall introduce some new hesitant fuzzy hybrid weighted aggregation operators which not only maintain the advantages of HFHA and HFHG but also keep some desirable properties, such as idempotency, boundedness, commutativity, etc. Inspired by the quasi hesitant fuzzy ordered weighted averaging (QHFOWA) operator proposed in Xia et al. (2013a), we extend our proposed operators to more general forms. In addition, inspired by the generalized ordered weighted averaging (GOWA) operator (Yager 2004), we also define a class of generalized hesitant fuzzy hybrid weighted aggregation operators and their induced forms. Considering the powerfulness of HFSs in multiple criteria decision making, we give some procedures with the operators for multiple criteria single person decision making and multiple criteria group decision making.

3.1 Hesitant Fuzzy Aggregation Operators

In order to export the operations on fuzzy sets to HFSs, Torra and Narukawa (2009) proposed an aggregation principle for HFEs.

Definition 3.1 (*Torra and Narukawa* 2009). Let $H = \{h_1, h_2, \ldots, h_n\}$ be a set of HFEs, Θ be a function on H, $\Theta : [0, 1]^n \to [0, 1]$, then

$$\Theta_H = \bigcup_{\gamma \in \{h_1 \times h_2 \times \ldots \times h_n\}} \{\Theta(\gamma)\} \tag{3.1}$$

Based on the above extension principle, Xia and Xu (2011a) developed a series of specific aggregation operators for HFEs:

Definition 3.2 (*Xia and Xu* 2011a). Let h_j $(j = 1, 2, \ldots, n)$ be a collection of HFEs. A HFWA operator is a mapping $H^n \to H$ such that

$$\mathrm{HFWA}(h_1, h_2, \ldots, h_n) = \overset{n}{\underset{j=1}{\oplus}} (\omega_j h_j) = \bigcup_{\gamma_1 \in h_1, \gamma_2 \in h_2, \ldots, \gamma_n \in h_n} \left\{ 1 - \prod_{j=1}^{n} (1 - \gamma_j)^{\omega_j} \right\}$$

$$\tag{3.2}$$

where $\omega = (\omega_1, \omega_2, \ldots, \omega_n)^T$ is the weight vector of h_j $(j = 1, 2, \ldots, n)$ with $\omega_j \in [0, 1]$ and $\sum_{j=1}^{n} \omega_j = 1$. Especially, if $\omega = (1/n, 1/n, \ldots, 1/n)^T$, then the HFWA operator reduces to the hesitant fuzzy averaging (HFA) operator:

$$\text{HFA}(h_1, h_2, \ldots, h_n) = \overset{n}{\underset{j=1}{\oplus}} \left(\frac{1}{n} h_j\right) = \bigcup_{\gamma_1 \in h_1, \gamma_2 \in h_2, \ldots, \gamma_n \in h_n} \left\{ 1 - \prod_{j=1}^{n} (1 - \gamma_j)^{1/n} \right\}$$

(3.3)

Definition 3.3 (*Xia and Xu* 2011a). Let h_j $(j = 1, 2, \ldots, n)$ be a collection of HFEs. A HFWG operator is a mapping $H^n \to H$ such that

$$\text{HFWG}(h_1, h_2, \ldots, h_n) = \overset{n}{\underset{j=1}{\otimes}} h_j^{\omega_j} = \bigcup_{\gamma_1 \in h_1, \gamma_2 \in h_2, \ldots, \gamma_n \in h_n} \left\{ \prod_{j=1}^{n} \gamma_j^{\omega_j} \right\} \quad (3.4)$$

where $\omega = (\omega_1, \omega_2, \ldots, \omega_n)^T$ is the weight vector of h_j $(j = 1, 2, \ldots, n)$, with $\omega_j \in [0, 1]$ and $\sum_{j=1}^{n} \omega_j = 1$. In the case where $\omega = (1/n, 1/n, \ldots, 1/n)^T$, the HFWA operator reduces to the hesitant fuzzy geometric (HFG) operator:

$$\text{HFG}(h_1, h_2, \ldots, h_n) = \overset{n}{\underset{j=1}{\otimes}} h_j^{1/n} = \bigcup_{\gamma_1 \in h_1, \gamma_2 \in h_2, \ldots, \gamma_n \in h_n} \left\{ \prod_{j=1}^{n} \gamma_j^{1/n} \right\} \quad (3.5)$$

According to the adjusted operation laws of HFEs given as Definition 1.9, Liao et al. (2014b) further introduced the following aggregation operators, which do not increase the dimensions of the HFEs:

Definition 3.4 (*Liao et al.* 2014b). Let $H = \{h_1, h_2, \ldots, h_n\}$ be a collection of HFEs. An adjusted hesitant fuzzy weighted averaging (AHFWA) operator is a mapping $H^n \to H$ such that

$$\text{AHFWA}(h_1, h_2, \ldots, h_n) = \overset{n}{\underset{j=1}{\oplus}} (\omega_j h_j) = \left\{ 1 - \prod_{j=1}^{n} (1 - h_j^{\sigma(t)})^{\omega_j} \middle| t = 1, 2, \ldots, l \right\}$$

(3.6)

where $h_j^{\sigma(t)}$ is the tth smallest value in h_j, and $\omega = (\omega_1, \omega_2, \ldots, \omega_n)^T$ is the weight vector of h_j $(j = 1, 2, \ldots, n)$ with $\omega_j \in [0, 1]$, $j = 1, 2, \ldots, n$, and $\sum_{j=1}^{n} \omega_j = 1$.

Especially, if $\omega = (1/n, 1/n, \ldots, 1/n)^T$, then the AHFWA operator reduces to an adjusted hesitant fuzzy averaging (AHFA) operator:

$$\text{AHFA}(h_1, h_2, \ldots, h_n) = \overset{n}{\underset{j=1}{\oplus}} \left(\frac{1}{n} h_j\right) = \left\{ 1 - \prod_{j=1}^{n} (1 - h_j^{\sigma(t)})^{1/n} \middle| t = 1, 2, \ldots, l \right\}$$

(3.7)

Definition 3.5 (*Liao et al.* 2014b). Let $H = \{h_1, h_2, \ldots, h_n\}$ be a collection of HFEs and let AHFWG: $H^n \to H$, if

$$\text{AHFWG}(h_1, h_2, \ldots, h_n) = \overset{n}{\underset{j=1}{\otimes}} (h_j)^{\omega_j} = \left\{ \prod_{j=1}^{n} (h_j^{\sigma(t)})^{\omega_j} \middle| t = 1, 2, \ldots, l \right\} \quad (3.8)$$

then AHFWG is called an adjusted hesitant fuzzy weighted geometric (AHFWG) operator, where $h_j^{\sigma(t)}$ is the tth smallest value in h_j, and $\omega = (\omega_1, \omega_2, \ldots, \omega_n)^T$ is the weight vector of h_j $(j = 1, 2, \ldots, n)$, with $\omega_j \in [0, 1]$, $j = 1, 2, \ldots, n$, and $\sum_{j=1}^{n} \omega_j = 1$.

In the case where $\omega = (1/n, 1/n, \ldots, 1/n)^T$, the AHFWA operator reduces to an adjusted hesitant fuzzy geometric (AHFG) operator:

$$\text{AHFG}(h_1, h_2, \ldots, h_n) = \overset{n}{\underset{j=1}{\otimes}} (h_j)^{1/n} = \left\{ \prod_{j=1}^{n} (h_j^{\sigma(t)})^{1/n} \middle| t = 1, 2, \ldots, l \right\} \quad (3.9)$$

Based on the idea of the ordered weighted averaging (OWA) operator (Yager 1988), the HFOWA and HFOWG operators were defined.

Definition 3.6 (*Xia and Xu* 2011a). Let h_j $(j = 1, 2, \ldots, n)$ be a collection of HFEs, $h_{\sigma(j)}$ be the jth largest of them, $\omega = (\omega_1, \omega_2, \ldots, \omega_n)^T$ be the aggregation-associated vector such that $\omega_j \in [0, 1]$ and $\sum_{j=1}^{n} \omega_j = 1$, then

(1) A hesitant fuzzy ordered weighted averaging (HFOWA) operator is a mapping HFOWA: $H^n \rightarrow H$, where

$$\text{HFOWA}(h_1, h_2, \ldots, h_n) = \overset{n}{\underset{j=1}{\oplus}} \left(\omega_j h_{\sigma(j)} \right)$$

$$= \bigcup_{\gamma_{\sigma(1)} \in h_{\sigma(1)}, \gamma_{\sigma(2)} \in h_{\sigma(2)}, \ldots, \gamma_{\sigma(n)} \in h_{\sigma(n)}} \left\{ 1 - \prod_{j=1}^{n} (1 - \gamma_{\sigma(j)})^{\omega_j} \right\}$$

$$(3.10)$$

(2) A hesitant fuzzy ordered weighted geometric (HFOWG) operator is a mapping HFOWG: $H^n \rightarrow H$, where

$$\text{HFOWG}(h_1, h_2, \ldots, h_n) = \overset{n}{\underset{j=1}{\otimes}} h_{\sigma(j)}^{\omega_j} = \bigcup_{\gamma_{\sigma(1)} \in h_{\sigma(1)}, \gamma_{\sigma(2)} \in h_{\sigma(2)}, \ldots, \gamma_{\sigma(n)} \in h_{\sigma(n)}} \left\{ \prod_{j=1}^{n} \gamma_{\sigma(j)}^{\omega_j} \right\}$$

$$(3.11)$$

In the case where $\omega = (1/n, 1/n, \ldots, 1/n)^T$, the HFOWA operator reduces to the HFA operator, and the HFOWG operator becomes the HFG operator.

It is noted that the HFWA and HFWG operators only weight the hesitant fuzzy arguments themselves, but ignore the importance of the ordered positions of the arguments, while the HFOWA and HFOWG operators only weight the ordered position of each given argument, but ignore the importance of the arguments. To solve this drawback, Xia and Xu (2011a) then introduced some hybrid aggregation operators for hesitant fuzzy arguments, which weight all the given arguments and their ordered positions:

Definition 3.7 (*Xia and Xu* 2011a). For a collection of HFEs h_j $(j = 1, 2, \ldots, n)$, $\lambda = (\lambda_1, \lambda_2, \ldots, \lambda_n)^T$ is the weight vector of them with $\lambda_j \in [0, 1]$ and $\sum_{j=1}^{n} \lambda_j = 1$, n is the balancing coefficient which plays a role of balance, then we define the following aggregation operators, which are all based on the mapping $H^n \to H$ with an aggregation-associated vector $\omega = (\omega_1, \omega_2, \ldots, \omega_n)^T$ such that $\omega_j \in [0, 1]$ and $\sum_{j=1}^{n} \omega_j = 1$:

(1) The hesitant fuzzy hybrid averaging (HFHA) operator:

$$\text{HFHA}(h_1, h_2, \ldots, h_n) = \overset{n}{\underset{j=1}{\oplus}} \left(\omega_j \dot{h}_{\sigma(j)} \right)$$

$$= \bigcup_{\dot{\gamma}_{\sigma(1)} \in \dot{h}_{\sigma(1)}, \dot{\gamma}_{\sigma(2)} \in \dot{h}_{\sigma(2)}, \ldots, \dot{\gamma}_{\sigma(n)} \in \dot{h}_{\sigma(n)}} \left\{ 1 - \prod_{j=1}^{n} (1 - \dot{\gamma}_{\sigma(j)})^{\omega_j} \right\}$$

(3.12)

where $\dot{h}_{\sigma(j)}$ is the jth largest of $\dot{h} = n\lambda_k h_k$ $(k = 1, 2, \ldots, n)$.

(2) The hesitant fuzzy hybrid geometric (HFHG) operator:

$$\text{HFHG}(h_1, h_2, \ldots, h_n) = \overset{n}{\underset{j=1}{\otimes}} \ddot{h}_{\sigma(j)}^{\omega_j} = \bigcup_{\ddot{\gamma}_{\sigma(1)} \in \ddot{h}_{\sigma(1)}, \ddot{\gamma}_{\sigma(2)} \in \ddot{h}_{\sigma(2)}, \ldots, \ddot{\gamma}_{\sigma(n)} \in \ddot{h}_{\sigma(n)}} \left\{ \prod_{j=1}^{n} \ddot{\gamma}_{\sigma(j)}^{\omega_j} \right\}$$

(3.13)

where $\ddot{h}_{\sigma(j)}$ is the jth largest of $\ddot{h}_k = h_k^{n\lambda_k}$ $(k = 1, 2, \ldots, n)$.

Especially, if $\omega = (1/n, 1/n, \ldots, 1/n)^T$, then the HFHA operator reduces to the HFOWA operator, the HFHG operator reduces to the HFOWG operator.

Although the HFHA (HFHG) operator generalizes both the HFWA (HFWG) and HFOWA (HFOWG) operators by weighting the given importance degrees and the ordered positions of the arguments, there is a flaw that the operator does not satisfy the desirable property, i.e., idempotency. An example can be used to illustrate this drawback.

Example 3.1 (Liao and Xu 2014a). Assume $h_1 = \{0.3, 0.3, 0.3\}$, $h_2 = \{0.3, 0.3, 0.3\}$ and $h_3 = \{0.3, 0.3, 0.3\}$ are three HFEs, whose weight vector is $\lambda = (1, 0, 0)^T$, and the aggregation-associated vector is also $\omega = (1, 0, 0)^T$. Then

$$\dot{h}_1 = 3 \times 1 \otimes h_1 = 3h_1 = \left(1 - (1 - 0.3)^3, 1 - (1 - 0.3)^3, 1 - (1 - 0.3)^3 \right) = (0.657, 0.657, 0.657)$$

$$\dot{h}_2 = 3 \times 0 \otimes h_2 = 0 \otimes h_2 = \left(1 - (1 - 0.3)^0, 1 - (1 - 0.3)^0, 1 - (1 - 0.3)^0 \right) = (0, 0, 0)$$

$$\dot{h}_3 = 3 \times 0 \otimes h_3 = 0 \otimes h_3 = \left(1 - (1 - 0.3)^0, 1 - (1 - 0.3)^0, 1 - (1 - 0.3)^0 \right) = (0, 0, 0)$$

Obviously, $s(\dot{h}_1) > s(\dot{h}_2) = s(\dot{h}_3)$. By using Eq. (3.12), we have

$$
\begin{aligned}
\mathrm{HFHA}(h_1, h_2, h_3) &= \mathop{\oplus}_{j=1}^{3} \left(\omega_j \dot{h}_{\sigma(j)}\right) \\
&= \bigcup_{\dot{\gamma}_{\sigma(1)} \in \dot{h}_{\sigma(1)}, \dot{\gamma}_{\sigma(2)} \in \dot{h}_{\sigma(2)}, \dot{\gamma}_{\sigma(3)} \in \dot{h}_{\sigma(3)}} \left\{1 - (1 - \dot{\gamma}_{\sigma(1)})^1 (1 - \dot{\gamma}_{\sigma(2)})^0 (1 - \dot{\gamma}_{\sigma(3)})^0\right\} \\
&= (0.657, 0.657, 0.657) \neq \{0.3, 0.3, 0.3\}
\end{aligned}
$$

Analogously,

$$
\begin{aligned}
\ddot{h}_1 &= h_1^{3 \times 1} = h_1^3 = (0.3^3, 0.3^3, 0.3^3) = (0.027, 0.027, 0.027) \\
\ddot{h}_2 &= h_2^{3 \times 0} = h_2^0 = (0.3^0, 0.3^0, 0.3^0) = (0, 0, 0) \\
\ddot{h}_3 &= h_3^{3 \times 0} = h_3^0 = (0.3^0, 0.3^0, 0.3^0) = (0, 0, 0)
\end{aligned}
$$

$$
\begin{aligned}
\mathrm{HFHG}(h_1, h_2, h_3) &= \mathop{\otimes}_{j=1}^{3} \ddot{h}_{\sigma(j)}^{\omega_j} = \bigcup_{\ddot{\gamma}_{\sigma(1)} \in \ddot{h}_{\sigma(1)}, \ddot{\gamma}_{\sigma(2)} \in \ddot{h}_{\sigma(2)}, \ddot{\gamma}_{\sigma(3)} \in \ddot{h}_{\sigma(3)}} \left\{\ddot{\gamma}_{\sigma(1)}^1 \ddot{\gamma}_{\sigma(2)}^0 \ddot{\gamma}_{\sigma(3)}^0\right\} \\
&= (0, 0, 0) \neq \{0.3, 0.3, 0.3\}
\end{aligned}
$$

Idempotency is the most important property for every aggregation operator (Lin and Jiang 2014; Liao and Xu 2014b), but the HFHA and HFWG operators do not meet this basic property, we need to develop some new hybrid aggregation operators which also weight the importance of each argument and its ordered position simultaneously.

3.2 Hesitant Fuzzy Hybrid Weighted Aggregation Operators

Considering the HFOWA operator given as Eq. (3.10), it is equivalent to the following form:

$$
\mathrm{HFOWA}(h_1, h_2, \ldots, h_n) = \mathop{\oplus}_{j=1}^{n} \left(\omega_{\varepsilon(j)} h_j\right) \tag{3.14}
$$

where h_j is the $\varepsilon(j)$ th largest element of h_j $(j = 1, 2, \ldots, n)$. Inspired by this, supposing the weight vector of the elements is $\lambda = (\lambda_1, \lambda_2, \ldots, \lambda_n)^T$, in order to weight the positions and the elements simultaneously, we can use such a form as $\mathop{\oplus}_{j=1}^{n} \lambda_j \omega_{\varepsilon(j)} h_j$, which weights both the elements and their positions. After normalization, a new hesitant fuzzy hybrid weighted averaging operator is generated.

Definition 3.8 (*Liao and Xu* 2014a). For a collection of HFEs h_j ($j = 1, 2, \ldots, n$), a hesitant fuzzy hybrid weighted averaging (HFHWA) operator is a mapping HFHWA: $H^n \rightarrow H$, defined by an associated weight vector $\omega = (\omega_1, \omega_2, \ldots, \omega_n)^T$ with $\omega_j \in [0, 1]$ and $\sum_{j=1}^{n} \omega_j = 1$, such that

$$\text{HFHWA}(h_1, h_2, \ldots, h_n) = \frac{\overset{n}{\underset{j=1}{\oplus}} \lambda_j \omega_{\varepsilon(j)} h_j}{\sum_{j=1}^{n} \lambda_j \omega_{\varepsilon(j)}} \tag{3.15}$$

where $\varepsilon : \{1, 2, \ldots, n\} \rightarrow \{1, 2, \ldots, n\}$ is the permutation such that h_j is the $\varepsilon(j)$th largest element of the collection of HFEs h_j ($j = 1, 2, \ldots, n$), and $\lambda = (\lambda_1, \lambda_2, \ldots, \lambda_n)^T$ is the weight vector of the HFEs h_j ($j = 1, 2, \ldots, n$), with $\lambda_j \in [0, 1]$ and $\sum_{j=1}^{n} \lambda_j = 1$.

Theorem 3.1 (Liao and Xu 2014a). *For a collection of HFEs h_j ($j = 1, 2, \ldots, n$), the aggregated value by using the HFHWA operator is also a HFE, and*

$$\text{HFHWA}(h_1, h_2, \ldots, h_n) = \bigcup_{\gamma_1 \in h_1, \gamma_2 \in h_2, \ldots, \gamma_n \in h_n} \left\{ 1 - \prod_{j=1}^{n} (1 - \gamma_j)^{\frac{\lambda_j \omega_{\varepsilon(j)}}{\sum_{j=1}^{n} \lambda_j \omega_{\varepsilon(j)}}} \right\} \tag{3.16}$$

where $\omega = (\omega_1, \omega_2, \ldots, \omega_n)^T$ is an associated weight vector with $\omega_j \in [0, 1]$ and $\sum_{j=1}^{n} \omega_j = 1$, $\varepsilon : \{1, 2, \ldots, n\} \rightarrow \{1, 2, \ldots, n\}$ is the permutation such that h_j is the $\varepsilon(j)$th largest element of the collection of HFEs h_j ($j = 1, 2, \ldots, n$), and $\lambda = (\lambda_1, \lambda_2, \ldots, \lambda_n)^T$ is the weight vector of the HFEs h_j ($j = 1, 2, \ldots, n$), with $\lambda_j \in [0, 1]$ and $\sum_{j=1}^{n} \lambda_j = 1$.

Proof. From the definition of HFS, it is obvious that the aggregated value by using the HFHWA operator is also a HFE.

By using the operational law (2) given in Definition 1.8, we have

$$\frac{\lambda_j \omega_{\varepsilon(j)}}{\sum_{j=1}^{n} \lambda_j \omega_{\varepsilon(j)}} h_j = \bigcup_{\gamma \in h_j} \left\{ 1 - (1 - \gamma)^{\frac{\lambda_j \omega_{\varepsilon(j)}}{\sum_{j=1}^{n} \lambda_j \omega_{\varepsilon(j)}}} \right\}, j = 1, 2, \ldots, n$$

Summing all these weighted HFEs $\frac{\lambda_j \omega_{\varepsilon(j)}}{\sum_{j=1}^{n} \lambda_j \omega_{\varepsilon(j)}} h_j$ ($j = 1, 2, \ldots, n$) by using the operational law (5) given in Definition 1.8, we can derive

$$\text{HFHWA}(h_1, h_2, \ldots, h_n) = \frac{\overset{n}{\underset{j=1}{\oplus}} \lambda_j \omega_{\varepsilon(j)} h_j}{\sum_{j=1}^{n} \lambda_j \omega_{\varepsilon(j)}} = \overset{n}{\underset{j=1}{\oplus}} \bigcup_{\gamma \in h_j} \left\{ 1 - (1-\gamma)^{\frac{\lambda_j \omega_{\varepsilon(j)}}{\sum_{j=1}^{n} \lambda_j \omega_{\varepsilon(j)}}} \right\}$$

$$= \bigcup_{\zeta_1 \in h_1', \zeta_2 \in h_2', \ldots, \zeta_n \in h_n'} \left\{ 1 - \prod_{j=1}^{n} (1-\zeta_j) \right\}$$

(3.17)

where

$$h_j' = \frac{\lambda_j \omega_{\varepsilon(j)}}{\sum_{j=1}^{n} \lambda_j \omega_{\varepsilon(j)}} h_j, \quad \zeta_j = 1 - (1-\gamma)^{\frac{\lambda_j \omega_{\varepsilon(j)}}{\sum_{j=1}^{n} \lambda_j \omega_{\varepsilon(j)}}}, \quad \gamma \in h_j, \, j = 1, 2, \ldots, n \quad (3.18)$$

Combining Eqs. (3.17) and (3.18), we obtain

$$\text{HFHWA}(h_1, h_2, \ldots, h_n) = \bigcup_{\gamma_1 \in h_1, \gamma_2 \in h_2, \ldots, \gamma_n \in h_n} \left\{ 1 - \prod_{j=1}^{n} \left[1 - \left(1 - (1-\gamma_j)^{\frac{\lambda_j \omega_{\varepsilon(j)}}{\sum_{j=1}^{n} \lambda_j \omega_{\varepsilon(j)}}} \right) \right] \right\}$$

$$= \bigcup_{\gamma_1 \in h_1, \gamma_2 \in h_2, \ldots, \gamma_n \in h_n} \left\{ 1 - \prod_{j=1}^{n} (1-\gamma_j)^{\frac{\lambda_j \omega_{\varepsilon(j)}}{\sum_{j=1}^{n} \lambda_j \omega_{\varepsilon(j)}}} \right\}$$

which completes the proof of Theorem 3.1.

Example 3.2 (Liao and Xu 2014a). Let $h_1 = \{0.2, 0.4, 0.5\}$, $h_2 = \{0.2, 0.6\}$ and $h_3 = \{0.1, 0.3, 0.4\}$ be three HFEs, whose weight vector is $\lambda = (0.15, 0.3, 0.55)^T$, and the aggregation-associated vector is $\omega = (0.3, 0.4, 0.3)^T$.

At first, comparing h_1, h_2 and h_3 by using the score function given as Eq. (1.17), we have

$$s(h_1) = \frac{0.2 + 0.4 + 0.5}{3} = 0.3667, \quad s(h_2) = \frac{0.2 + 0.6}{2} = 0.4$$

$$s(h_3) = \frac{0.1 + 0.3 + 0.4}{3} = 0.2667$$

Since $s(h_2) > s(h_1) > s(h_3)$, then we obtain $h_2 > h_1 > h_3$. Thus, $\varepsilon(1) = 2$, $\varepsilon(2) = 1$ and $\varepsilon(3) = 3$. Thus

$$\frac{\lambda_1 \omega_{\varepsilon(1)}}{\sum_{j=1}^{3} \lambda_j \omega_{\varepsilon(j)}} = \frac{0.15 \times 0.4}{0.15 \times 0.4 + 0.3 \times 0.3 + 0.55 \times 0.3} = 0.19$$

$$\frac{\lambda_2 \omega_{\varepsilon(2)}}{\sum_{j=1}^{3} \lambda_j \omega_{\varepsilon(j)}} = 0.286, \quad \frac{\lambda_3 \omega_{\varepsilon(3)}}{\sum_{j=1}^{3} \lambda_j \omega_{\varepsilon(j)}} = 0.524$$

By using Eq. (3.16), we can calculate that

$$\text{HFHWA}(h_1,h_2,h_3) = \frac{\overset{3}{\underset{j=1}{\oplus}}\lambda_j\omega_{\varepsilon(j)}h_j}{\sum_{j=1}^{n}\lambda_j\omega_{\varepsilon(j)}} = \bigcup_{\gamma_1\in h_1,\gamma_2\in h_2,\gamma_3\in h_3}\left\{1-\prod_{j=1}^{3}(1-\gamma_j)^{\frac{\lambda_j\omega_{\varepsilon(j)}}{\sum_{j=1}^{3}\lambda_j\omega_{\varepsilon(j)}}}\right\}$$

$$= \bigcup_{\gamma_1\in h_1,\gamma_2\in h_2,\gamma_3\in h_3}\left\{1-(1-\gamma_1)^{0.19}(1-\gamma_2)^{0.286}(1-\gamma_3)^{0.524}\right\}$$

$$= \{0.1490, 0.1943, 0.2217, 0.2541, 0.2938, 0.3020, 0.3119, 0.3178, 0.3392, 0.3485,$$
$$0.3617, 0.3707, 0.3882, 0.4207, 0.4356, 0.4405, 0.4656, 0.4838\}$$

Theorem 3.2 (Liao and Xu 2014a). (Idempotency) *If $h_j = h$ ($j = 1,2,\ldots,n$), then HFHWA $(h_1,h_2,\ldots,h_n) = h$.*

Proof According to Eq. (3.16), we have

$$\text{HFHWA}(h_1,h_2,\ldots,h_n) = \frac{\overset{n}{\underset{j=1}{\oplus}}\lambda_j\omega_{\varepsilon(j)}h_j}{\sum_{j=1}^{n}\lambda_j\omega_{\varepsilon(j)}} = \frac{\overset{n}{\underset{j=1}{\oplus}}\lambda_j\omega_{\varepsilon(j)}h}{\sum_{j=1}^{n}\lambda_j\omega_{\varepsilon(j)}} = h\frac{\sum_{j=1}^{n}\lambda_j\omega_{\varepsilon(j)}}{\sum_{j=1}^{n}\lambda_j\omega_{\varepsilon(j)}} = h$$

Thus, HFHWA $(h_1,h_2,\ldots,h_n) = h$, which completes the proof of Theorem 3.2.

Example 3.3 (Liao and Xu 2014a). Let us use the HFHWA operator to calculate Example 3.1, then

$$\text{HFHWA}(h_1,h_2,h_3) = \frac{\overset{3}{\underset{j=1}{\oplus}}\lambda_j\omega_{\varepsilon(j)}h_j}{\sum_{j=1}^{n}\lambda_j\omega_{\varepsilon(j)}} = \bigcup_{\gamma_1\in h_1,\gamma_2\in h_2,\gamma_3\in h_3}\left\{1-(1-\gamma_1)^1(1-\gamma_2)^0(1-\gamma_3)^0\right\}$$

$$= \{0.3, 0.3, 0.3\} = h_1 = h_2 = h_3$$

which satisfies the property of idempotency. This is also consistent with our intuition. From this example, we can see that the HFHWA operator is more reasonable than the HFHA operator developed by Xia and Xu (2011a).

Theorem 3.3 reveals that the HFHWA operator has the property of boundedness:

Theorem 3.3 (Liao and Xu 2015a). (Boundedness) *For a collection of HFEs $h_j = \left\{\gamma_{jt}|t = 1,2,\ldots,l_{h_j}\right\}$ ($j = 1,2,\ldots,n$), let $h^- = \left\{\underset{j=1}{\overset{n}{\min}}\ \underset{t=1}{\overset{l_{h_j}}{\min}}\ \gamma_{jt}\right\}$, and $h^+ = \left\{\underset{j=1}{\overset{n}{\max}}\ \underset{t=1}{\overset{l_{h_j}}{\max}}\ \gamma_{jt}\right\}$, then, $h^- \leq \text{HFHWA}\ (h_1,h_2,\ldots,h_n) \leq h^+$.*

Proof For the simplicity of presentation, let $\gamma^- \triangleq \min\limits_{j=1}^{n} \min\limits_{t=1}^{l_{h_j}} \gamma_{jt}$, $\gamma^+ \triangleq \max\limits_{j=1}^{n} \max\limits_{t=1}^{l_{h_j}} \gamma_{jt}$,

$\dot{h} \triangleq \text{HFHWA}(h_1, h_2, \ldots, h_n)$, and $\dot{\gamma} \triangleq 1 - \prod_{j=1}^{n} (1 - \gamma_{jt})^{\frac{\lambda_j \omega_{\varepsilon(j)}}{\sum_{j=1}^{n} \lambda_j \omega_{\varepsilon(j)}}}$, $(t = 1, 2, \ldots, l_{h_j})$.

Then, $h^- = \{\gamma^-\}$, $h^+ = \{\gamma^+\}$, $\text{HFHWA}(h_1, h_2, \ldots, h_n) = \dot{h} = \bigcup_{\dot{\gamma} \in \dot{h}} \{\dot{\gamma}\}$.

For any j, we have $\gamma^- = \min\limits_{j=1}^{n} \min\limits_{t=1}^{l_{h_j}} \gamma_{jt} \leq \gamma_{jt} \leq \max\limits_{j=1}^{n} \max\limits_{t=1}^{l_{h_j}} \gamma_{jt} = \gamma^+$. Since $y = x^a$

$(0 < a < 1)$ is a monotonic increasing function when $x > 0$, then we get

$$1 - \prod_{j=1}^{n} (1 - \gamma^-)^{\frac{\lambda_j \omega_{\varepsilon(j)}}{\sum_{j=1}^{n} \lambda_j \omega_{\varepsilon(j)}}} \leq 1 - \prod_{j=1}^{n} (1 - \gamma_{jt})^{\frac{\lambda_j \omega_{\varepsilon(j)}}{\sum_{j=1}^{n} \lambda_j \omega_{\varepsilon(j)}}}$$

$$\leq 1 - \prod_{j=1}^{n} (1 - \gamma^+)^{\frac{\lambda_j \omega_{\varepsilon(j)}}{\sum_{j=1}^{n} \lambda_j \omega_{\varepsilon(j)}}}$$

which is equivalent to

$$1 - (1 - \gamma^-)^{\frac{\sum_{j=1}^{n} \lambda_j \omega_{\varepsilon(j)}}{\sum_{j=1}^{n} \lambda_j \omega_{\varepsilon(j)}}} \leq \dot{\gamma} \leq 1 - (1 - \gamma^+)^{\frac{\sum_{j=1}^{n} \lambda_j \omega_{\varepsilon(j)}}{\sum_{j=1}^{n} \lambda_j \omega_{\varepsilon(j)}}}$$

i.e.,

$$\gamma^- \leq \dot{\gamma} \leq \gamma^+. \tag{3.19}$$

According to Eq. (1.17), it follows $s(\dot{h}) = \frac{1}{l_{\dot{h}}} \sum_{\dot{\gamma} \in \dot{h}} \dot{\gamma}$, $s(h^-) = \frac{1}{l_{h^-}} \sum_{\gamma \in h^-} \gamma = \gamma^-$,

$s(h^+) = \frac{1}{l_{h^+}} \sum_{\gamma \in h^+} \gamma = \gamma^+$. From Eq. (3.19), we have

$$\gamma^- = \frac{1}{l_{\dot{h}}} \sum \gamma^- \leq \frac{1}{l_{\dot{h}}} \sum_{\dot{\gamma} \in \dot{h}} \dot{\gamma} \leq \frac{1}{l_{\dot{h}}} \sum \gamma^+ = \gamma^+$$

Thus

$$s(h^-) \leq s(\dot{h}) \leq s(h^+)$$

which implies $h^- \leq \text{HFHWA}(h_1, h_2, \ldots, h_n) \leq h^+$. This completes the proof of Theorem 3.3.

In addition, it is easy to check that the HFHWA operator is also commutative.

Theorem 3.4 (Liao and Xu 2015a). (Commutativity) *For two collections of HFEs*

$h_j = \{\gamma_{jt} | t = 1, 2, \ldots, l_{h_j}\}$　　$(j = 1, 2, \ldots, n)$　　*and*　　$h'_j = \{\gamma'_{jt} | t = 1, 2, \ldots, l_{h'_j}\}$

$(j = 1, 2, \ldots, n)$, *it follows that*

$$\text{HFHWA}(h_1, h_2, \ldots, h_n) = \text{HFHWA}(h'_1, h'_2, \ldots, h'_n)$$

where $(h'_1, h'_2, \ldots, h'_n)$ is any permutation of $(h_1, h_2, \ldots, h_n)$.

A simple example can be given to illustrate this theorem and hereby the theoretical proof is omitted.

Example 3.4 (Liao and Xu 2015a). In Example 3.2, $h_1 = \{0.2, 0.4, 0.5\}$, $h_2 = \{0.2, 0.6\}$ and $h_3 = \{0.1, 0.3, 0.4\}$ are three HFEs, whose weight vector is $\lambda = (0.15, 0.3, 0.55)^T$, and the aggregation-associated vector is $\omega = (0.3, 0.4, 0.3)^T$. We have calculated that

$$\text{HFHWA}(h_1, h_2, h_3) = \{0.1490, 0.1943, 0.2217, 0.2541, 0.2938, 0.3020, 0.3119, 0.3178,$$
$$0.3392, 0.3485, 0.3617, 0.3707, 0.3882, 0.4207, 0.4356, 0.4405, 0.4656, 0.4838\}$$

Now we consider a permutation of the above three HFEs, for example, $h'_1 = \{0.2, 0.6\}$, $h'_2 = \{0.1, 0.3, 0.4\}$, and $h'_3 = \{0.2, 0.4, 0.5\}$. The weight vector is $\lambda = (0.3, 0.55, 0.15)^T$, and the aggregation-associated vector is also $\omega = (0.3, 0.4, 0.3)^T$. Then, it is easy to obtain that $\varepsilon(1) = 1, \varepsilon(2) = 3, \varepsilon(3) = 2$, thus, $\frac{\lambda_1 \omega_{\varepsilon(1)}}{\sum_{j=1}^{3} \lambda_j \omega_{\varepsilon(j)}} = 0.286$, $\frac{\lambda_2 \omega_{\varepsilon(2)}}{\sum_{j=1}^{3} \lambda_j \omega_{\varepsilon(j)}} = 0.524$, $\frac{\lambda_3 \omega_{\varepsilon(3)}}{\sum_{j=1}^{3} \lambda_j \omega_{\varepsilon(j)}} = 0.19$. It follows that

$$\text{HFHWA}(h'_1, h'_2, h'_3) = \frac{\overset{3}{\underset{j=1}{\oplus}} \lambda_j \omega_{\varepsilon(j)} h_j}{\sum_{j=1}^{n} \lambda_j \omega_{\varepsilon(j)}} = \bigcup_{\gamma_1 \in h_1, \gamma_2 \in h_2, \gamma_3 \in h_3} \left\{ 1 - \prod_{j=1}^{3} (1 - \gamma_j)^{\frac{\lambda_j \omega_{\varepsilon(j)}}{\sum_{j=1}^{3} \lambda_j \omega_{\varepsilon(j)}}} \right\}$$

$$= \bigcup_{\gamma_1 \in h_1, \gamma_2 \in h_2, \gamma_3 \in h_3} \left\{ 1 - (1 - \gamma_1)^{0.286} (1 - \gamma_2)^{0.524} (1 - \gamma_3)^{0.19} \right\}$$

$$= \{0.1490, 0.1943, 0.2217, 0.2541, 0.2938, 0.3020, 0.3119, 0.3178, 0.3392, 0.3485,$$
$$0.3617, 0.3707, 0.3882, 0.4207, 0.4356, 0.4405, 0.4656, 0.4838\}$$

$$= \text{HFHWA}(h_1, h_2, h_3)$$

By using the different manifestation of weight vector, the HFHWA operator can be reduced into some special cases. For example, if the associated weight vector $\omega = (1/n, 1/n, \ldots, 1/n)^T$, then the HFHWA operator reduces to the HFWA operator; if $\lambda = (1/n, 1/n, \ldots, 1/n)^T$, then the HFHWA operator reduces to the HFOWA operator. It must be pointed out that the weighting operation of the ordered positions can be synchronized with the weighting operation of the given importance by the HFHWA operator. This characteristic is different from the HFHA operator.

Analogously, we also can develop the HFHWG operator for HFEs:

Definition 3.9 (*Liao and Xu* 2014a). For a collection of HFEs h_j ($j = 1, 2, \ldots, n$), a hesitant fuzzy hybrid weighted geometric (HFHWG) operator is a mapping HFHWG: $H^n \to H$, defined by an associated weight vector $\omega = (\omega_1, \omega_2, \ldots, \omega_n)^T$ with $\omega_j \in [0, 1]$ and $\sum_{j=1}^{n} \omega_j = 1$, such that

$$\mathrm{HFHWG}(h_1, h_2, \ldots, h_n) = \overset{n}{\underset{j=1}{\otimes}} (h_j)^{\frac{\lambda_j \omega_{\varepsilon(j)}}{\sum_{j=1}^{n} \lambda_j \omega_{\varepsilon(j)}}} \tag{3.20}$$

where $\varepsilon : \{1, 2, \ldots, n\} \to \{1, 2, \ldots, n\}$ is the permutation such that h_j is the $\varepsilon(j)$th largest element of the collection of HFEs h_j $(j = 1, 2, \ldots, n)$, and $\lambda = (\lambda_1, \lambda_2, \ldots, \lambda_n)^T$ is the weight vector of the HFEs h_j $(j = 1, 2, \ldots, n)$, with $\lambda_j \in [0, 1]$ and $\sum_{j=1}^{n} \lambda_j = 1$.

Theorem 3.5 (Liao and Xu 2014a). *For a collection of HFEs h_j $(j = 1, 2, \ldots, n)$, the aggregated value by using the HFHWG operator is also a HFE, and*

$$\mathrm{HFHWG}(h_1, h_2, \ldots, h_n) = \bigcup_{\gamma_1 \in h_1, \gamma_2 \in h_2, \ldots, \gamma_n \in h_n} \left\{ \prod_{j=1}^{n} \gamma_j^{\frac{\lambda_j \omega_{\varepsilon(j)}}{\sum_{j=1}^{n} \lambda_j \omega_{\varepsilon(j)}}} \right\} \tag{3.21}$$

where $\omega = (\omega_1, \omega_2, \ldots, \omega_n)^T$ is an associated weight vector with $\omega_j \in [0, 1]$ and $\sum_{j=1}^{n} \omega_j = 1$, $\varepsilon : \{1, 2, \ldots, n\} \to \{1, 2, \ldots, n\}$ is the permutation such that h_j is the $\varepsilon(j)$ th largest element of the collection of HFEs $h_j (j = 1, 2, \ldots, n)$, and $\lambda = (\lambda_1, \lambda_2, \ldots, \lambda_n)^T$ is the weighting vector of the HFEs h_j $(j = 1, 2, \ldots, n)$, with $\lambda_j \in [0, 1]$ and $\sum_{j=1}^{n} \lambda_j = 1$.

Proof Similar to Theorem 3.1, the aggregated value by using the HFHWG operator is also a HFE.

By using the operational law (1) given in Definition 1.8, we have

$$(h_j)^{\frac{\lambda_j \omega_{\varepsilon(j)}}{\sum_{j=1}^{n} \lambda_j \omega_{\varepsilon(j)}}} = \bigcup_{\gamma \in h_j} \gamma^{\frac{\lambda_j \omega_{\varepsilon(j)}}{\sum_{j=1}^{n} \lambda_j \omega_{\varepsilon(j)}}}, j = 1, 2, \ldots, n.$$

According to the operational law (6) given in Definition 1.8, we can derive

$$\mathrm{HFHWG}(h_1, h_2, \ldots, h_n) = \overset{n}{\underset{j=1}{\otimes}} (h_j)^{\frac{\lambda_j \omega_{\varepsilon(j)}}{\sum_{j=1}^{n} \lambda_j \omega_{\varepsilon(j)}}} = \bigcup_{\xi_1 \in h_1'', \xi_2 \in h_2'', \ldots, \xi_n \in h_n''} \left\{ \prod_{j=1}^{n} \xi_j \right\} \tag{3.22}$$

where

$$h_j'' = (h_j)^{\frac{\lambda_j \omega_{\varepsilon(j)}}{\sum_{j=1}^{n} \lambda_j \omega_{\varepsilon(j)}}}, \ \xi_j = \gamma^{\frac{\lambda_j \omega_{\varepsilon(j)}}{\sum_{j=1}^{n} \lambda_j \omega_{\varepsilon(j)}}}, \ \gamma \in h_j, j = 1, 2, \ldots, n \tag{3.23}$$

Combining Eqs. (3.22) and (3.23), we can obtain

$$\text{HFHWG}(h_1, h_2, \ldots, h_n) = \bigcup_{\gamma_1 \in h_1, \gamma_2 \in h_2, \ldots, \gamma_n \in h_n} \left\{ \prod_{j=1}^{n} \gamma_j^{\frac{\lambda_j \omega_{\varepsilon(j)}}{\sum_{j=1}^{n} \lambda_j \omega_{\varepsilon(j)}}} \right\}$$

This completes the proof of the theorem.

Example 3.5 (Liao and Xu 2014a). Let us use the HFHWG operator to fuse the HFEs h_1, h_2 and h_3 in Example 3.3. According to Eq. (3.21), we have

$$\text{HFHWG}(h_1, h_2, h_3) = \bigcup_{\gamma_1 \in h_1, \gamma_2 \in h_2, \ldots, \gamma_n \in h_n} \left\{ \prod_{j=1}^{3} \gamma_j^{\frac{\lambda_j \omega_{\varepsilon(j)}}{\sum_{j=1}^{3} \lambda_j \omega_{\varepsilon(j)}}} \right\}$$

$$= \bigcup_{\gamma_1 \in h_1, \gamma_2 \in h_2, \gamma_3 \in h_3} \left\{ \gamma_1^{0.19} \gamma_2^{0.286} \gamma_3^{0.524} \right\}$$

$$= \{0.1391, 0.1587, 0.1655, 0.1904, 0.2172, 0.2266, 0.2473, 0.2822, 0.2876, 0.2944,$$
$$0.3281, 0.3387, 0.3423, 0.3863, 0.3938, 0.4031, 0.4492, 0.4686\}$$

Theorem 3.6 (Liao and Xu 2014a). (Idempotency) *If* $h_j = h$ $(j = 1, 2, \ldots, n)$, *then* HFHWG $(h_1, h_2, \ldots, h_n) = h$.

Proof. Since $h_j = h$, then $\gamma_j = \gamma$. Hence, according to Eq. (3.21), we have

$$\text{HFHWG}(h_1, h_2, \ldots, h_n) = \bigcup_{\gamma_1 \in h_1, \gamma_2 \in h_2, \ldots, \gamma_n \in h_n} \left\{ \prod_{j=1}^{n} \gamma_j^{\frac{\lambda_j \omega_{\varepsilon(j)}}{\sum_{j=1}^{n} \lambda_j \omega_{\varepsilon(j)}}} \right\}$$

$$= \bigcup_{\gamma_1 \in h_1, \gamma_2 \in h_2, \ldots, \gamma_n \in h_n} \left\{ \gamma^{\frac{\sum_{j=1}^{n} \lambda_j \omega_{\varepsilon(j)}}{\sum_{j=1}^{n} \lambda_j \omega_{\varepsilon(j)}}} \right\} = \bigcup_{\gamma \in h} \{\gamma\} = h$$

Thus, HFHWG $(h_1, h_2, \ldots, h_n) = h$, which completes the proof of the theorem.

Example 3.6 (Liao and Xu 2014a). Let us use the HFHWG operator to calculate Example 3.1, then we have

$$\text{HFHWG}(h_1, h_2, h_3) = \overset{n}{\underset{j=1}{\otimes}} (h_j)^{\frac{\lambda_j \omega_{\varepsilon(j)}}{\sum_{j=1}^{n} \lambda_j \omega_{\varepsilon(j)}}} = \bigcup_{\gamma_1 \in h_1, \gamma_2 \in h_2, \gamma_3 \in h_3} \{\gamma_1^1 \gamma_2^0 \gamma_3^0\}$$

$$= \{0.3, 0.3, 0.3\} = h_1 = h_2 = h_3$$

which means the HFHWG operator satisfies idempotency. In other words, the HFHWG operator is more reasonable than HFHG operator.

Theorem 3.7 (Liao and Xu 2015a). (Boundedness) *For a collection of HFEs*
$h_j = \{\gamma_{jt}|t = 1, 2, \ldots, l_{h_j}\}$ $(j = 1, 2, \ldots, n)$, let $h^- = \left\{\min\limits_{j=1}^{n} \min\limits_{t=1}^{l_{h_j}} \gamma_{jt}\right\}$, $h^+ = \left\{\max\limits_{j=1}^{n} \max\limits_{t=1}^{l_{h_j}} \gamma_{jt}\right\}$, *then,* $h^- \leq \text{HFHWG}(h_1, h_2, \ldots, h_n) \leq h^+$.

Proof For the simplicity of presentation, let $\gamma^- \triangleq \min\limits_{j=1}^{n} \min\limits_{t=1}^{l_{h_j}} \gamma_{jt}$, $\gamma^+ \triangleq \max\limits_{j=1}^{n} \max\limits_{t=1}^{l_{h_j}} \gamma_{jt}$, $\ddot{h} \triangleq \text{HFHWG}(h_1, h_2, \ldots, h_n)$, and $\ddot{\gamma} \triangleq \prod\limits_{j=1}^{n} \gamma_{jt}^{\frac{\lambda_j \omega_{\varepsilon(j)}}{\sum_{j=1}^{n} \lambda_j \omega_{\varepsilon(j)}}}$ $(t = 1, 2, \ldots, l_{h_j})$. Then, $h^- = \{\gamma^-\}$, $h^+ = \{\gamma^+\}$, $\text{HFHWG}(h_1, h_2, \ldots, h_n) = \ddot{h} = \bigcup_{\ddot{\gamma} \in \ddot{h}} \{\ddot{\gamma}\}$.

For any j, we have $\gamma^- \leq \gamma_{jt} \leq \gamma^+$. Since $y = x^a, (0 < a < 1)$ is a monotonic increasing function when $x > 0$, then we get

$$\prod_{j=1}^{n} \gamma^{-\frac{\lambda_j \omega_{\varepsilon(j)}}{\sum_{j=1}^{n} \lambda_j \omega_{\varepsilon(j)}}} \leq \prod_{j=1}^{n} \gamma_{jt}^{\frac{\lambda_j \omega_{\varepsilon(j)}}{\sum_{j=1}^{n} \lambda_j \omega_{\varepsilon(j)}}} \leq \prod_{j=1}^{n} \gamma^{+\frac{\lambda_j \omega_{\varepsilon(j)}}{\sum_{j=1}^{n} \lambda_j \omega_{\varepsilon(j)}}}$$

which is equivalent to

$$\gamma^- = \gamma^{-\frac{\sum_{j=1}^{n} \lambda_j \omega_{\varepsilon(j)}}{\sum_{j=1}^{n} \lambda_j \omega_{\varepsilon(j)}}} \leq \ddot{\gamma} \leq \gamma^{+\frac{\sum_{j=1}^{n} \lambda_j \omega_{\varepsilon(j)}}{\sum_{j=1}^{n} \lambda_j \omega_{\varepsilon(j)}}} = \gamma^+$$

Thus, we have

$$\gamma^- = \frac{1}{l_{\ddot{h}}} \sum \gamma^- \leq \frac{1}{l_{\ddot{h}}} \sum_{\ddot{\gamma} \in \ddot{h}} \ddot{\gamma} \leq \frac{1}{l_{\ddot{h}}} \sum \gamma^+ = \gamma^+$$

i.e.,

$$s(h^-) \leq s(\ddot{h}) \leq s(h^+)$$

which implies $h^- \leq \text{HFHWG}(h_1, h_2, \ldots, h_n) \leq h^+$. This completes the proof of the theorem.

Theorem 3.8 (Liao and Xu 2015a). (Commutativity) *For two collections of HFEs*
$h_j = \{\gamma_{jt}|t = 1, 2, \ldots, l_{h_j}\}$ $(j = 1, 2, \ldots, n)$ *and* $h_j' = \{\gamma_{jt}'|t = 1, 2, \ldots, l_{h_j'}\}$
$(j = 1, 2, \ldots, n)$, *it follows that*

$$\text{HFHWG}(h_1, h_2, \ldots, h_n) = \text{HFHWG}(h_1', h_2', \ldots, h_n')$$

where $(h_1', h_2', \ldots, h_n')$ *is any permutation of* $(h_1, h_2, \ldots, h_n)$.

Especially, if the associated weight vector $\omega = (1/n, 1/n, \ldots, 1/n)^T$, then the HFHWG operator reduces to the HFWG operator; if $\lambda = (1/n, 1/n, \ldots, 1/n)^T$, then the HFHWG operator reduces to the HFOWG operator. With the HFHWG operator, the weighting operation of the ordered position also can be synchronized with the weighting operation of the given importance, while the HFHG operator does not have this characteristic.

Lemma 3.1 (Xu 2000). *If $x_j > 0$, $\omega_j > 0$, $j = 1, 2, \ldots, n$, and $\sum_{j=1}^{n} \omega_j = 1$, then $\prod_{j=1}^{n} x_j^{\omega_j} \leq \sum_{j=1}^{n} \omega_j x_j$, with equality if and only if $x_1 = x_2 = \cdots = x_n$.*

Theorem 3.9 reveals the relationship between the HFHWA and HFHWG operators:

Theorem 3.9 (Liao and Xu 2015a). *For a collection of HFEs $h_j = \{\gamma_{jt} | t = 1, 2, \ldots, l_{h_j}\}$ $(j = 1, 2, \ldots, n)$, then*

$$\text{HFHWG}(h_1, h_2, \ldots, h_n) \leq \text{HFHWA}(h_1, h_2, \ldots, h_n)$$

Proof For any $\gamma_1 \in h_1, \gamma_2 \in h_2, \ldots, \gamma_n \in h_n$, according to Lemma 3.1, it follows

$$\prod_{j=1}^{n} \gamma_j^{\frac{\lambda_j \omega_{\varepsilon(j)}}{\sum_{j=1}^{n} \lambda_j \omega_{\varepsilon(j)}}} \leq \sum_{j=1}^{n} \frac{\lambda_j \omega_{\varepsilon(j)}}{\sum_{j=1}^{n} \lambda_j \omega_{\varepsilon(j)}} \gamma_j$$

$$= 1 - \sum_{j=1}^{n} \frac{\lambda_j \omega_{\varepsilon(j)}}{\sum_{j=1}^{n} \lambda_j \omega_{\varepsilon(j)}} (1 - \gamma_j) \leq 1 - \prod_{j=1}^{n} (1 - \gamma_j)^{\frac{\lambda_j \omega_{\varepsilon(j)}}{\sum_{j=1}^{n} \lambda_j \omega_{\varepsilon(j)}}} \tag{3.24}$$

Combining Eqs. (3.16), (3.21) and (3.24), we can derive HFHWG $(h_1, h_2, \ldots, h_n) \leq \text{HFHWA}(h_1, h_2, \ldots, h_n)$. This completes the proof.

3.3 Quasi Hesitant Fuzzy Hybrid Weighted Aggregation Operators

Combining the HFOWA operator with the quasi-arithmetical average (Fodor et al. 1995), Xia et al. (2013a) developed the QHFOWA operator:

Definition 3.10 (*Xia et al.* 2013a). Let h_j $(j = 1, 2, \ldots, n)$ be a collection of HFEs and $h_{\rho(j)}$ be the jth largest of them. Let QHFOWA: $H^n \to H$, if

$$(h_j | j = 1, 2, \ldots, n) = \bigcup_{\gamma_{\rho(j)} \in h_{\rho(j)}, j=1,2,\ldots,n} \left\{ g^{-1} \left(\sum_{j=1}^{n} \omega_j g(\gamma_{\rho(j)}) \right) \right\} \tag{3.25}$$

then QHFOWA is called a quasi hesitant fuzzy ordered weighted averaging (QHFOWA) operator, where $g(\gamma)$ is a strictly continuous monotonic function, $\omega = (\omega_1, \omega_2, \ldots, \omega_n)^T$ is the associated weight vector with $\sum_{j=1}^{n} \omega_j = 1$, and $\omega_j \geq 0$, $j = 1, 2, \ldots, n$.

Similarly, we can propose the QHFOWG operator as follows:

Definition 3.11 (*Liao and Xu* 2014a). Let h_j $(j = 1, 2, \ldots, n)$ be a collection of HFEs and $h_{\rho(j)}$ be the jth largest of them. Let QHFOWG: $H^n \rightarrow H$, if

$$\text{QHFOWG}\left(h_j \middle| j = 1, 2, \ldots, n\right) = \bigcup_{\gamma_{\rho(j)} \in h_{\rho(j)}, j=1,2,\ldots,n} \left\{ g^{-1}\left(\prod_{j=1}^{n} g^{\omega_j}\left(\gamma_{\rho(j)}\right)\right) \right\} \quad (3.26)$$

then QHFOWG is called a quasi hesitant fuzzy ordered weighted geometric (QHFOWG) operator, where $g(\gamma)$ is a strictly continuous monotonic function, $\omega = (\omega_1, \omega_2, \ldots, \omega_n)^T$ is the associated weight vector with $\sum_{j=1}^{n} \omega_j = 1$, and $\omega_j \geq 0$, $j = 1, 2, \ldots, n$.

Motivated by Definitions 3.10 and 3.11, if we replace the arithmetical average and the geometric average in Definitions 3.8 and 3.9 with the quasi arithmetical average, respectively, then the QHFHWA and QHFHWG operators will be obtained, which are in mathematical forms as follows:

Definition 3.12 (*Liao and Xu* 2014a). For a collection of HFEs h_j $(j = 1, 2, \ldots, n)$, $\lambda = (\lambda_1, \lambda_2, \ldots, \lambda_n)^T$ is the weight vector of them with $\lambda_j \in [0, 1]$ and $\sum_{j=1}^{n} \lambda_j = 1$, then we define the following aggregation operators, which are all based on the mapping $H^n \rightarrow H$ with an aggregation-associated vector $\omega = (\omega_1, \omega_2, \ldots, \omega_n)^T$ such that $\omega_j \in [0, 1]$ and $\sum_{j=1}^{n} \omega_j = 1$, and a continuous strictly monotonic function $g(\gamma)$:

(1) The quasi hesitant fuzzy hybrid weighted averaging (QHFHWA) operator:

$$\text{QHFHWA}(h_1, h_2, \ldots, h_n) = g^{-1}\left(\frac{\overset{n}{\underset{j=1}{\oplus}} \lambda_j \omega_{\varepsilon(j)} g(h_j)}{\sum_{j=1}^{n} \lambda_j \omega_{\varepsilon(j)}} \right)$$

$$= \bigcup_{\gamma_1 \in h_1, \gamma_2 \in h_2, \ldots, \gamma_n \in h_n} \left\{ g^{-1}\left(1 - \prod_{j=1}^{n} (1 - g(\gamma_j))^{\frac{\lambda_j \omega_{\varepsilon(j)}}{\sum_{j=1}^{n} \lambda_j \omega_{\varepsilon(j)}}} \right) \right\}$$

$$(3.27)$$

(2) The quasi hesitant fuzzy hybrid weighted geometric (QHFHWG) operator:

$$\text{QHFHWG}(h_1, h_2, \ldots, h_n) = g^{-1}\left(\overset{n}{\underset{j=1}{\otimes}} (g(h_j))^{\frac{\lambda_j \omega_{\varepsilon(j)}}{\sum_{j=1}^{n} \lambda_j \omega_{\varepsilon(j)}}}\right)$$

$$= \bigcup_{\gamma_1 \in h_1, \gamma_2 \in h_2, \ldots, \gamma_n \in h_n} \left\{ g^{-1}\left(\prod_{j=1}^{n} (g(\gamma_j))^{\frac{\lambda_j \omega_{\varepsilon(j)}}{\sum_{j=1}^{n} \lambda_j \omega_{\varepsilon(j)}}}\right) \right\}$$

$$(3.28)$$

where $\varepsilon : \{1, 2, \ldots, n\} \to \{1, 2, \ldots, n\}$ is the permutation such that h_j is the $\varepsilon(j)$th largest element of the collection of HFEs h_j $(j = 1, 2, \ldots, n)$.

Note that when assigning different weight vectors of ω or λ or choosing different types of functions of $g(\gamma)$, the QHFHWA and QHFHWG operators will reduce to many special cases, which can be set out as follows:

(1) If the associated weight vector $\omega = (1/n, 1/n, \ldots, 1/n)^T$, then the QHFWA operator reduces to the QHFWA operator shown as:

$$\text{QHFHWA}(h_1, h_2, \ldots, h_n) = g^{-1}\left(\overset{n}{\underset{j=1}{\oplus}} \lambda_j g(h_j)\right)$$

$$= \bigcup_{\gamma_1 \in h_1, \gamma_2 \in h_2, \ldots, \gamma_n \in h_n} \left\{ g^{-1}\left(1 - \prod_{j=1}^{n} (1 - g(\gamma_j))^{\lambda_j}\right) \right\}$$

while the QHFWG operator reduces to the QHFWG operator shown as:

$$\text{QHFHWG}(h_1, h_2, \ldots, h_n) = g^{-1}\left(\overset{n}{\underset{j=1}{\otimes}} (g(h_j))^{\lambda_j}\right)$$

$$= \bigcup_{\gamma_1 \in h_1, \gamma_2 \in h_2, \ldots, \gamma_n \in h_n} \left\{ g^{-1}\left(\prod_{j=1}^{n} (g(\gamma_j))^{\lambda_j}\right) \right\}$$

(2) If the arguments' weight vector $\lambda = (1/n, 1/n, \ldots, 1/n)^T$, then the QHFHWA operator reduces to the QHFOWA operator, while the QHFHWG operator reduces to the QHFOWG operator.
(3) If $g(\gamma) = \gamma$, then the QHFHWA operator reduces to the HFHWA operator, while the QHFHWG operator reduces to the HFHWG operator.
(4) If $g(\gamma) = \ln \gamma$, then the QHFHWA operator reduces to the HFHWG operator, while the QHFHWG operator reduces to the HFHWA operator. The derivation can be shown as below:

$$\text{QHFHWA}(h_1, h_2, \ldots, h_n) = e^{\frac{\overset{n}{\underset{j=1}{\oplus}} \lambda_j \omega_{\varepsilon(j)} \ln(h_j)}{\sum_{j=1}^{n} \lambda_j \omega_{\varepsilon(j)}}} = \left(e^{\overset{n}{\underset{j=1}{\oplus}} \lambda_j \omega_{\varepsilon(j)} \ln(h_j)} \right)^{1 / \sum_{j=1}^{n} \lambda_j \omega_{\varepsilon(j)}}$$

$$= \overset{n}{\underset{j=1}{\otimes}} \left(h_j \right)^{\frac{\lambda_j \omega_{\varepsilon(j)}}{\sum_{j=1}^{n} \lambda_j \omega_{\varepsilon(j)}}} = \text{HFHWG}(h_1, h_2, \ldots, h_n)$$

while,

$$\text{QHFHWG}(h_1, h_2, \ldots, h_n) = e^{\overset{n}{\underset{j=1}{\otimes}} \left(\ln(h_j) \right)^{\frac{\lambda_j \omega_{\varepsilon(j)}}{\sum_{j=1}^{n} \lambda_j \omega_{\varepsilon(j)}}}} = \frac{\overset{n}{\underset{j=1}{\otimes}} \left(\ln(h_j) \right)^{\lambda_j \omega_{\varepsilon(j)}}}{\sum_{j=1}^{n} \lambda_j \omega_{\varepsilon(j)}}$$

$$= \frac{\overset{n}{\underset{j=1}{\oplus}} \lambda_j \omega_{\varepsilon(j)} h_j}{\sum_{j=1}^{n} \lambda_j \omega_{\varepsilon(j)}} = \text{HFHWA}(h_1, h_2, \ldots, h_n)$$

Some other special cases can also be constructed by choosing different types of the functions of $g(\gamma)$ for the QHFHWA and QHFHWG operators, such as $g(\gamma) = \gamma^\lambda$, $g(\gamma) = 1 - (1 - \gamma)^\lambda$, $g(\gamma) = \sin((\pi/2)\gamma)$, $g(\gamma) = 1 - \sin((\pi/2)(1 - \gamma))$, $g(\gamma) = \cos((\pi/2)\gamma)$, $g(\gamma) = 1 - \cos((\pi/2)(1 - \gamma))$, $g(\gamma) = \tan((\pi/2)\gamma)$, $g(\gamma) = 1 - \tan((\pi/2)(1 - \gamma))$, $g(\gamma) = \lambda^\gamma$, $g(\gamma) = 1 - b^{1-\gamma}$, and so on.

Theorem 3.10 (Liao and Xu 2014a). (Idempotency) *If $h_j = h$ $(j = 1, 2, \ldots, n)$, then QHFHWA $(h_1, h_2, \ldots, h_n) = h$, QHFHWG $(h_1, h_2, \ldots, h_n) = h$.*

Proof According to Definition 3.12, we can obtain

$$\text{QHFHWA}(h_1, h_2, \ldots, h_n) = g^{-1}\left(\frac{\overset{n}{\underset{j=1}{\oplus}} \lambda_j \omega_{\varepsilon(j)} g\left(h_j\right)}{\sum_{j=1}^{n} \lambda_j \omega_{\varepsilon(j)}} \right) = g^{-1}\left(\frac{\overset{n}{\underset{j=1}{\oplus}} \lambda_j \omega_{\varepsilon(j)} g(h)}{\sum_{j=1}^{n} \lambda_j \omega_{\varepsilon(j)}} \right)$$

$$= g^{-1}\left(\frac{g(h) \sum_{j=1}^{n} \lambda_j \omega_{\varepsilon(j)}}{\sum_{j=1}^{n} \lambda_j \omega_{\varepsilon(j)}} \right) = g^{-1}(g(h)) = h$$

$$\text{QHFHWG}(h_1, h_2, \ldots, h_n) = g^{-1}\left(\overset{n}{\underset{j=1}{\otimes}} \left(g(h_j) \right)^{\frac{\lambda_j \omega_{\varepsilon(j)}}{\sum_{j=1}^{n} \lambda_j \omega_{\varepsilon(j)}}} \right) = g^{-1}\left(\overset{n}{\underset{j=1}{\otimes}} \left(g(h) \right)^{\frac{\lambda_j \omega_{\varepsilon(j)}}{\sum_{j=1}^{n} \lambda_j \omega_{\varepsilon(j)}}} \right)$$

$$= g^{-1}\left((g(h))^{\frac{\sum_{j=1}^{n} \lambda_j \omega_{\varepsilon(j)}}{\sum_{j=1}^{n} \lambda_j \omega_{\varepsilon(j)}}} \right) = g^{-1}(g(h)) = h$$

This completes the proof of the theorem.

Theorem 3.11 (Liao and Xu 2015a). (Boundedness) *For a collection of HFEs*

$h_j = \{\gamma_{jt} | t = 1, 2, \ldots, l_{h_j}\}$ $(j = 1, 2, \ldots, n)$, let $h^- = \left\{ \min_{j=1}^{n} \min_{t=1}^{l_{h_j}} \gamma_{jt} \right\}$, $h^+ =$

$\left\{ \max_{j=1}^{n} \max_{t=1}^{l_{h_j}} \gamma_{jt} \right\}$, *then,* $h^- \leq \text{QHFHWA}(h_1, h_2, \ldots, h_n) \leq h^+$, $h^- \leq \text{QHFHWG}$
$(h_1, h_2, \ldots, h_n) \leq h^+$.

Proof

(1) For the simplicity of presentation, let $\gamma^- \triangleq \min_{j=1}^{n} \min_{t=1}^{l_{h_j}} \gamma_{jt}$, $\gamma^+ \triangleq \max_{j=1}^{n} \max_{t=1}^{l_{h_j}} \gamma_{jt}$,

$\bar{h} \triangleq \text{QHFHWA}(h_1, h_2, \ldots, h_n)$, and $\bar{\gamma} \triangleq g^{-1} \left(1 - \prod_{j=1}^{n} (1 - g(\gamma_{jt}))^{\frac{\lambda_j \omega_{\varepsilon(j)}}{\sum_{j=1}^{n} \lambda_j \omega_{\varepsilon(j)}}} \right)$,

$(t = 1, 2, \ldots, l_{h_j})$. Then, $h^- = \{\gamma^-\}, h^+ = \{\gamma^+\}$, $\text{QHFHWA}(h_1, h_2, \ldots, h_n) =$
$\bar{h} = \bigcup_{\bar{\gamma}_t \in \bar{h}} \{\bar{\gamma}\}$.

As we have pointed out in the proof of Theorem 3.3 that for any j, $\gamma^- \leq \gamma_{jt} \leq \gamma^+$. Since $y = x^a (0 < a < 1)$ is a monotonic increasing function when $x > 0$, then we get

$$1 - \prod_{j=1}^{n} (1 - \gamma^-)^{\frac{\lambda_j \omega_{\varepsilon(j)}}{\sum_{j=1}^{n} \lambda_j \omega_{\varepsilon(j)}}} \leq 1 - \prod_{j=1}^{n} (1 - \gamma_{jt})^{\frac{\lambda_j \omega_{\varepsilon(j)}}{\sum_{j=1}^{n} \lambda_j \omega_{\varepsilon(j)}}} \leq 1 - \prod_{j=1}^{n} (1 - \gamma^+)^{\frac{\lambda_j \omega_{\varepsilon(j)}}{\sum_{j=1}^{n} \lambda_j \omega_{\varepsilon(j)}}}$$

$$(3.29)$$

Meanwhile, it is noted that $g(\gamma)$ is a continuous strictly monotonic function, which implies that $g^{-1}(\gamma)$ is also a continuous strictly monotonic function, and they are in the same trend of increasing or decreasing. Thus, combined by Eq. (3.29), we have

$$g^{-1} \left(1 - \prod_{j=1}^{n} (1 - g(\gamma^-))^{\frac{\lambda_j \omega_{\varepsilon(j)}}{\sum_{j=1}^{n} \lambda_j \omega_{\varepsilon(j)}}} \right) \leq g^{-1} \left(1 - \prod_{j=1}^{n} (1 - g(\gamma_{jt}))^{\frac{\lambda_j \omega_{\varepsilon(j)}}{\sum_{j=1}^{n} \lambda_j \omega_{\varepsilon(j)}}} \right)$$

$$\leq g^{-1} \left(1 - \prod_{j=1}^{n} (1 - g(\gamma^+))^{\frac{\lambda_j \omega_{\varepsilon(j)}}{\sum_{j=1}^{n} \lambda_j \omega_{\varepsilon(j)}}} \right)$$

which is equivalent to

$$\gamma^- = g^{-1}(1 - (1 - g(\gamma^-))) \leq \bar{\gamma} \leq g^{-1}(1 - (1 - g(\gamma^+))) = \gamma^+$$

i.e.,

$$\gamma^- \leq \bar{\gamma} \leq \gamma^+$$

Then according to Eq. (1.17) and similar to the proving process of Theorem 3.3, we have

$$s(h^-) \leq s(\bar{h}) \leq s(h^+)$$

which implies $h^- \leq \mathrm{QHFHWA}(h_1, h_2, \ldots, h_n) \leq h^+$.

(2) Let $\gamma^- \triangleq \min\limits_{j=1}^{n} \min\limits_{t=1}^{l_{h_j}} \gamma_{jt}$, $\gamma^+ \triangleq \max\limits_{j=1}^{n} \max\limits_{t=1}^{l_{h_j}} \gamma_{jt}$, $\tilde{h} \triangleq \mathrm{QHFHWG}\ (h_1, h_2, \ldots, h_n)$, and

$$\tilde{\gamma} \triangleq g^{-1}\left(\prod_{j=1}^{n} (g(\gamma_{jt}))^{\frac{\lambda_j \omega_{\varepsilon(j)}}{\sum_{j=1}^{n} \lambda_j \omega_{\varepsilon(j)}}} \right) \quad (t = 1, 2, \ldots, l_{h_j}).\ \ \text{Then,}\ \ h^- = \{\gamma^-\},\ \ h^+ =$$

$\{\gamma^+\}$, $\mathrm{QHFHWG}\ (h_1, h_2, \ldots, h_n) = \tilde{h} = \bigcup_{\tilde{\gamma} \in \tilde{h}} \{\tilde{\gamma}\}$.

For any j, we have $\gamma^- \leq \gamma_{jt} \leq \gamma^+$. According to Theorem 3.7, we get

$$\prod_{j=1}^{n} \gamma^{- \frac{\lambda_j \omega_{\varepsilon(j)}}{\sum_{j=1}^{n} \lambda_j \omega_{\varepsilon(j)}}} \leq \prod_{j=1}^{n} \gamma_{jt}^{\frac{\lambda_j \omega_{\varepsilon(j)}}{\sum_{j=1}^{n} \lambda_j \omega_{\varepsilon(j)}}} \leq \prod_{j=1}^{n} \gamma^{+ \frac{\lambda_j \omega_{\varepsilon(j)}}{\sum_{j=1}^{n} \lambda_j \omega_{\varepsilon(j)}}}$$

Since $g(\gamma)$ and $g^{-1}(\gamma)$ are both continuous strictly monotonic functions and they have the same trend of increasing or decreasing, then we can derive

$$g^{-1}\left(\prod_{j=1}^{n} (g(\gamma^-))^{\frac{\lambda_j \omega_{\varepsilon(j)}}{\sum_{j=1}^{n} \lambda_j \omega_{\varepsilon(j)}}} \right) \leq g^{-1}\left(\prod_{j=1}^{n} (g(\gamma_{jt}))^{\frac{\lambda_j \omega_{\varepsilon(j)}}{\sum_{j=1}^{n} \lambda_j \omega_{\varepsilon(j)}}} \right) \leq g^{-1}\left(\prod_{j=1}^{n} (g(\gamma^+))^{\frac{\lambda_j \omega_{\varepsilon(j)}}{\sum_{j=1}^{n} \lambda_j \omega_{\varepsilon(j)}}} \right)$$

i.e.,

$$\gamma^- = g^{-1}(g(\gamma^-)) \leq \tilde{\gamma} \leq g^{-1}(g(\gamma^+)) = \gamma^+$$

Thus, we have

$$\gamma^- = \frac{1}{l_{\tilde{h}}} \sum \gamma^- \leq \frac{1}{l_{\tilde{h}}} \sum_{\tilde{\gamma} \in \tilde{h}} \tilde{\gamma} \leq \frac{1}{l_{\tilde{h}}} \sum \gamma^+ = \gamma^+$$

i.e.,

$$s(h^-) \leq s(\tilde{h}) \leq s(h^+)$$

which implies $h^- \leq \text{QHFHWG}(h_1, h_2, \ldots, h_n) \leq h^+$. This completes the proof of the theorem.

Theorem 3.12 (Liao and Xu 2015a). *(Commutativity) For two collections of HFEs*
$h_j = \{\gamma_{jt} | t = 1, 2, \ldots, l_{h_j}\}$ $(j = 1, 2, \ldots, n)$ *and* $h'_j = \{\gamma'_{jt} | t = 1, 2, \ldots, l_{h'_j}\}$
$(j = 1, 2, \ldots, n)$, *it follows that*

$$\text{QHFHWA}(h_1, h_2, \ldots, h_n) = \text{QHFHWA}(h'_1, h'_2, \ldots, h'_n)$$
$$\text{QHFHWG}(h_1, h_2, \ldots, h_n) = \text{QHFHWG}(h'_1, h'_2, \ldots, h'_n)$$

where $(h'_1, h'_2, \ldots, h'_n)$ *is any permutation of* $(h_1, h_2, \ldots, h_n)$.

Theorem 3.13 (Liao and Xu 2015a). *For a collection of HFEs* $h_j = \{\gamma_{jt} | t = 1, 2, \ldots, l_{h_j}\}$ $(j = 1, 2, \ldots, n)$, *then*

$$\text{QHFHWG}(h_1, h_2, \ldots, h_n) \leq \text{QHFHWA}(h_1, h_2, \ldots, h_n)$$

3.4 Generalized Hesitant Fuzzy Hybrid Weighted Aggregation Operators

Combining the OWA operator with the generalized mean operator (Dyckhoff and Pedrycz 1984), Yager (2004) proposed the generalized OWA (GOWA) operator:

Definition 3.13 (*Yager* 2004). A GOWA operator of dimension n is a mapping GOWA: $I^n \rightarrow I$, which has the following form:

$$GOWA_\omega(a_1, a_2, \ldots, a_n) = \left(\sum_{j=1}^n \omega_j b_j^p\right)^{1/p} \tag{3.30}$$

where $p \neq 0$, $\omega = (\omega_1, \omega_2, \ldots, \omega_n)^T$ is the weight vector of $(a_1, a_2, \ldots, a_n)$, with $\omega_j \in [0, 1]$, $j = 1, 2, \ldots, n$, and $\sum_{j=1}^n \omega_j = 1$, b_j is the jth largest of a_i, $I = [0, 1]$.

When $p = 1$, the GOWA operator reduces to the OWA operator; when $\omega = (1/n, 1/n, \ldots, 1/n)^T$, the GOWA operator reduces to the generalized mean operator. Inspired by the GOWA operator, Xia and Xu (2011a) proposed a family of generalized hesitant fuzzy aggregation operators, such as the GHFWA, GHFWG, GHFOWA, GHFOWG, GHFHA and GHFHG operators. Similarly, as for the hesitant fuzzy hybrid weighted aggregation operators, Liao and Xu (2015a) extended them into the generalized forms.

Definition 3.14 (*Liao and Xu* 2015a). For a collection of the HFEs h_j ($j = 1, 2, \ldots, n$), a generalized hesitant fuzzy hybrid weighted averaging (GHFHWA) operator is a mapping GHFHWA: $H^n \longrightarrow H$, defined by an associated weight vector $\omega = (\omega_1, \omega_2, \ldots, \omega_n)^T$ with $\omega_j \in [0, 1]$ and $\sum_{j=1}^{n} \omega_j = 1$, such that

$$\text{GHFHWA}(h_1, h_2, \ldots, h_n) = \left(\frac{\overset{n}{\underset{j=1}{\oplus}} \lambda_j \omega_{\varepsilon(j)} h_j^p}{\sum_{j=1}^{n} \lambda_j \omega_{\varepsilon(j)}} \right)^{1/p} \tag{3.31}$$

where $\varepsilon : \{1, 2, \ldots, n\} \rightarrow \{1, 2, \ldots, n\}$ is the permutation such that h_j is the $\varepsilon(j)$th largest element of the collection of HFEs h_j ($j = 1, 2, \ldots, n$); $\lambda = (\lambda_1, \lambda_2, \ldots, \lambda_n)^T$ is the weight vector of the HFEs h_j ($j = 1, 2, \ldots, n$), with $\lambda_j \in [0, 1]$ and $\sum_{j=1}^{n} \lambda_j = 1$; p is a parameter such that $p \in (-\infty, +\infty)$. Especially, if $p = 1$, then the GHFHWA operator reduces to the HFHWA operator.

Theorem 3.14 (Liao and Xu 2015a). *For a collection of HFEs h_j ($j = 1, 2, \ldots, n$), the aggregated value by using the GHFHWA operator is also a HFE, and*

$$\text{GHFHWA}(h_1, h_2, \ldots, h_n) = \bigcup_{\gamma_1 \in h_1, \gamma_2 \in h_2, \ldots, \gamma_n \in h_n} \left\{ \left(1 - \prod_{j=1}^{n} (1 - \gamma_j^p)^{\frac{\lambda_j \omega_{\varepsilon(j)}}{\sum_{j=1}^{n} \lambda_j \omega_{\varepsilon(j)}}} \right)^{1/p} \right\} \tag{3.32}$$

Proof: From the definition of HFS, it is obvious that the aggregated value by using the GHFHWA operator is also a HFE.

According to the operational law of HFEs, we have

$$h_j^p = \bigcup_{\gamma_j \in h_j} \left\{ \gamma_j^p \right\} \tag{3.33}$$

Then, combining Eqs. (3.16), (3.32) and (3.33), it follows that

$$\text{GHFHWA}(h_1, h_2, \ldots, h_n) = \bigcup_{\gamma_1 \in h_1, \gamma_2 \in h_2, \ldots, \gamma_n \in h_n} \left\{ \left(1 - \prod_{j=1}^{n} (1 - \gamma_j^p)^{\frac{\lambda_j \omega_{\varepsilon(j)}}{\sum_{j=1}^{n} \lambda_j \omega_{\varepsilon(j)}}} \right)^{1/p} \right\}$$

This completes the proof.

Definition 3.15 (*Liao and Xu* 2015a). For a collection of the HFEs h_j $(j = 1, 2, \ldots, n)$, a generalized hesitant fuzzy hybrid weighted geometric (GHFHWG) operator is a mapping GHFHWG: $H^n \to H$, defined by an associated weight vector $\omega = (\omega_1, \omega_2, \ldots, \omega_n)^T$ with $\omega_j \in [0, 1]$ and $\sum_{j=1}^{n} \omega_j = 1$, such that

$$\text{GHFHWG}(h_1, h_2, \ldots, h_n) = \left(\bigotimes_{j=1}^{n} \left(h_j^p \right)^{\frac{\lambda_j \omega_{\varepsilon(j)}}{\sum_{j=1}^{n} \lambda_j \omega_{\varepsilon(j)}}} \right)^{1/p} \quad (3.34)$$

where $\varepsilon : \{1, 2, \ldots, n\} \to \{1, 2, \ldots, n\}$ is the permutation such that h_j is the $\varepsilon(j)$ th largest element of the collection of HFEs h_j $(j = 1, 2, \ldots, n)$; $\lambda = (\lambda_1, \lambda_2, \ldots, \lambda_n)^T$ is the weight vector of the HFEs h_j $(j = 1, 2, \ldots, n)$, with $\lambda_j \in [0, 1]$ and $\sum_{j=1}^{n} \lambda_j = 1$; p is a parameter such that $p \in (-\infty, +\infty)$. Especially, if $p = 1$, then the GHFHWG operator reduces to the HFHWG operator.

Theorem 3.15 (Liao and Xu 2015a). *For a collection of HFEs h_j $(j = 1, 2, \ldots, n)$, the aggregated value by using the GHFHWG operator is also a HFE, and*

$$\text{GHFHWG}(h_1, h_2, \ldots, h_n) = \bigcup_{\gamma_1 \in h_1, \gamma_2 \in h_2, \ldots, \gamma_n \in h_n} \left\{ \left(\prod_{j=1}^{n} \gamma_j^{\frac{p \lambda_j \omega_{\varepsilon(j)}}{\sum_{j=1}^{n} \lambda_j \omega_{\varepsilon(j)}}} \right)^{1/p} \right\} \quad (3.35)$$

Proof: The proof is similar to the proof of Theorem 3.14.

It is easy to check that the following theorems hold for both the GHFHWA operator and the GHFHWG operators.

Theorem 3.16 (Liao and Xu 2015a). *Both the GHFHWA operator and the GHFHWG operator satisfy the properties of idempotency, boundedness and commutativity.*

Theorem 3.17 (Liao and Xu 2015a). *For a collection of HFEs $h_j = \{\gamma_{jt} | t = 1, 2, \ldots, l_{h_j}\}$ $(j = 1, 2, \ldots, n)$, then*

$$\text{GHFHWG}(h_1, h_2, \ldots, h_n) \leq \text{GHFHWA}(h_1, h_2, \ldots, h_n) \quad (3.36)$$

Theorem 3.17 reveals the relationship between the GHFHWA operator and the GHFHWG operator. Similar to the QHFHWA and QHFHWG operators, the following operators can be developed immediately:

Definition 3.16 (*Liao and Xu* 2015a). For a collection of HFEs h_j $(j = 1, 2, \ldots, n)$, $\lambda = (\lambda_1, \lambda_2, \ldots, \lambda_n)^T$ is the weight vector of them with $\lambda_j \in [0, 1]$ and $\sum_{j=1}^{n} \lambda_j = 1$, then we define the following aggregation operators, which are all based on the mapping $H^n \to H$ with an aggregation-associated vector $\omega = (\omega_1, \omega_2, \ldots, \omega_n)^T$ such that $\omega_j \in [0, 1]$ and $\sum_{j=1}^{n} \omega_j = 1$, and a continuous strictly monotonic function $g(\gamma)$:

(1) The generalized quasi hesitant fuzzy hybrid weighted averaging (GQHFHWA) operator:

$$
\text{GQHFHWA}(h_1, h_2, \ldots, h_n) = \left(g^{-1} \left(\frac{\overset{n}{\underset{j=1}{\oplus}} \lambda_j \omega_{\varepsilon(j)} g\left(h_j^p\right)}{\sum_{j=1}^{n} \lambda_j \omega_{\varepsilon(j)}} \right) \right)^{1/p}
$$

$$
= \bigcup_{\gamma_1 \in h_1, \gamma_2 \in h_2, \ldots, \gamma_n \in h_n} \left\{ \left(g^{-1} \left(1 - \prod_{j=1}^{n} (1 - g(\gamma_j^p))^{\frac{\lambda_j \omega_{\varepsilon(j)}}{\sum_{j=1}^{n} \lambda_j \omega_{\varepsilon(j)}}} \right) \right)^{1/p} \right\}
$$

(3.37)

(2) The generalized quasi hesitant fuzzy hybrid weighted geometric (GQHFHWG) operator:

$$
\text{GQHFHWG}(h_1, h_2, \ldots, h_n) = \left(g^{-1} \left(\overset{n}{\underset{j=1}{\otimes}} \left(g(h_j^p) \right)^{\frac{\lambda_j \omega_{\varepsilon(j)}}{\sum_{j=1}^{n} \lambda_j \omega_{\varepsilon(j)}}} \right) \right)^{1/p}
$$

$$
= \bigcup_{\gamma_1 \in h_1, \gamma_2 \in h_2, \ldots, \gamma_n \in h_n} \left\{ \left(g^{-1} \left(\prod_{j=1}^{n} (g(\gamma_j^p))^{\frac{\lambda_j \omega_{\varepsilon(j)}}{\sum_{j=1}^{n} \lambda_j \omega_{\varepsilon(j)}}} \right) \right)^{1/p} \right\}
$$

(3.38)

where $\varepsilon : \{1, 2, \ldots, n\} \rightarrow \{1, 2, \ldots, n\}$ is the permutation such that h_j is the $\varepsilon(j)$ th largest element of the collection of HFEs h_j $(j = 1, 2, \ldots, n)$; p is a parameter such that $p \in (-\infty, +\infty)$.

Especially, if $p = 1$, then the GQHFHWA (GQHFHWG) operator reduces to the QHFHWA (QHFHWG) operator. If $g(\gamma) = \gamma$, then the GQHFHWA operator reduces to the GHFHWA operator, while the GQHFHWG operator reduces to the GHFHWG operator.

Theorem 3.18 (Liao and Xu 2015a). *The GQHFHWA operator and the GQHFHWG operator also satisfy the properties of idempotency, boundedness and commutativity.*

Theorem 3.19 (Liao and Xu 2015a). *For a collection of HFEs $h_j = \{\gamma_{jt} | t = 1, 2, \ldots, l_{h_j}\}$ $(j = 1, 2, \ldots, n)$, then*

$$
\text{GQHFHWG}(h_1, h_2, \ldots, h_n) \leq \text{GQHFHWA}(h_1, h_2, \ldots, h_n) \tag{3.39}
$$

3.5 Induced Hesitant Fuzzy Hybrid Weighted Aggregation Operators

As an extension of the OWA operator, the induced OWA (IOWA) operator was introduced by Yager and Filev (1999). The main difference between the OWA and IOWA operators is that the reordering step of the IOWA operator is carried out with order-inducing variables rather than depending on the values of the arguments (Merigó and Gil-Lafuente 2009). The IOWA operator can be used to aggregate tuples of the form (v_j, a_j), where v_j is called the order inducing value and a_j is called the argument value.

Definition 3.17 (*Yager and Filev* 1999). An IOWA operator of dimension n is a mapping IOWA: $I^n \rightarrow I$ defined by an aggregation-associated vector $\omega = (\omega_1, \omega_2, \ldots, \omega_n)^T$ such that $\omega_j \in [0, 1]$ and $\sum_{j=1}^{n} \omega_j = 1$, and a set of order inducing variables v_j, by a formula of the following form:

$$IOWA_\omega(<v_1, a_1>, <v_2, a_2>, \ldots, <v_n, a_n>) = \sum_{j=1}^{n} \omega_j b_j \qquad (3.40)$$

where b_j is simply $(a_1, a_2, \ldots, a_n)$ reordered in decreasing order of the values of v_j, v_j is the order inducing value, a_j is the argument value, and $I = [0, 1]$.

The IOWA operator is monotonic, bounded, idempotent and commutative (Yager and Filev 1999). Motivated by the IOWA operator, Liao and Xu (2015a) extended the hesitant fuzzy hybrid weighted aggregation operators to the induced forms with the difference that the reordering step of these operators is not defined by the values of the argument a_j, but by the order inducing value v_j. These induced hesitant fuzzy hybrid weighted aggregation operators are more general, and thus can be used to handle more complex cases.

Definition 3.18 (*Liao and Xu* 2015a). For a collection of HFEs h_j ($j = 1, 2, \ldots, n$), $\lambda = (\lambda_1, \lambda_2, \ldots, \lambda_n)^T$ is the weight vector of the HFEs h_j ($j = 1, 2, \ldots, n$), with $\lambda_j \in [0, 1]$ and $\sum_{j=1}^{n} \lambda_j = 1$, then the following aggregation operators are defined, which are all based on the mapping $H^n \rightarrow H$ with the aggregation-associated vector $\omega = (\omega_1, \omega_2, \ldots, \omega_n)^T$ such that $\omega_j \in [0, 1]$ and $\sum_{j=1}^{n} \omega_j = 1$:

(1) An induced hesitant fuzzy hybrid weighted averaging (IHFHWA) operator:

$$IHFHWA(<v_1, h_1>, <v_2, h_2>, \ldots, <v_n, h_n>) = \frac{\overset{n}{\underset{j=1}{\oplus}} \lambda_j \omega_{\varepsilon(j)} h_j}{\sum_{j=1}^{n} \lambda_j \omega_{\varepsilon(j)}}$$

$$= \bigcup_{\gamma_1 \in h_1, \gamma_2 \in h_2, \ldots, \gamma_n \in h_n} \left\{ 1 - \prod_{j=1}^{n} (1 - \gamma_j)^{\frac{\lambda_j \omega_{\varepsilon(j)}}{\sum_{j=1}^{n} \lambda_j \omega_{\varepsilon(j)}}} \right\} \qquad (3.41)$$

(2) An induced hesitant fuzzy hybrid weighted geometric (IHFHWG) operator:

$$
\text{IHFHWG}(<v_1, h_1 > , <v_2, h_2 > , \ldots, <v_n, h_n >) = \overset{n}{\underset{j=1}{\otimes}} \left(h_j \right)^{\frac{\lambda_j \omega_{\varepsilon(j)}}{\sum_{j=1}^{n} \lambda_j \omega_{\varepsilon(j)}}}
$$

$$
= \bigcup_{\gamma_1 \in h_1, \gamma_2 \in h_2, \ldots, \gamma_n \in h_n} \left\{ \prod_{j=1}^{n} \gamma_j^{\frac{\lambda_j \omega_{\varepsilon(j)}}{\sum_{j=1}^{n} \lambda_j \omega_{\varepsilon(j)}}} \right\}
$$

$$(3.42)$$

(3) An induced generalized hesitant fuzzy hybrid weighted averaging (IGHFHWA) operator:

$$
\text{IGHFHWA}(<v_1, h_1 > , <v_2, h_2 > , \ldots, <v_n, h_n >) = \left(\frac{\overset{n}{\underset{j=1}{\oplus}} \lambda_j \omega_{\varepsilon(j)} h_j^p}{\sum_{j=1}^{n} \lambda_j \omega_{\varepsilon(j)}} \right)^{1/p}
$$

$$
= \bigcup_{\gamma_1 \in h_1, \gamma_2 \in h_2, \ldots, \gamma_n \in h_n} \left\{ \left(1 - \prod_{j=1}^{n} (1 - \gamma_j^p)^{\frac{\lambda_j \omega_{\varepsilon(j)}}{\sum_{j=1}^{n} \lambda_j \omega_{\varepsilon(j)}}} \right)^{1/p} \right\}
$$

$$(3.43)$$

(4) An induced generalized hesitant fuzzy hybrid weighted geometric (IGHFHWG) operator:

$$
\text{IGHFHWG}(<v_1, h_1 > , <v_2, h_2 > , \ldots, <v_n, h_n >) = \left(\overset{n}{\underset{j=1}{\otimes}} \left(h_j^p \right)^{\frac{\lambda_j \omega_{\varepsilon(j)}}{\sum_{j=1}^{n} \lambda_j \omega_{\varepsilon(j)}}} \right)^{1/p}
$$

$$
= \bigcup_{\gamma_1 \in h_1, \gamma_2 \in h_2, \ldots, \gamma_n \in h_n} \left\{ \left(\prod_{j=1}^{n} \gamma_j^{\frac{p \lambda_j \omega_{\varepsilon(j)}}{\sum_{j=1}^{n} \lambda_j \omega_{\varepsilon(j)}}} \right)^{1/p} \right\}
$$

$$(3.44)$$

where $\varepsilon : \{1, 2, \ldots, n\} \to \{1, 2, \ldots, n\}$ is the permutation such that v_j is the $\varepsilon(j)$th largest element of the collection of v_j $(j = 1, 2, \ldots, n)$, v_j is the order inducing value and it can be represented in any different forms, and p is a parameter such that $p \in (-\infty, +\infty)$.

Example 3.7 (Liao and Xu 2015a). Given the following collection of tuples $<John, \{0.2, 0.4, 0.5\}>$, $<Smith, \{0.2, 0.4\}>$ and $<Frank, \{0.1, 0.2, 0.4\}>$, whose weight vector is $\lambda = (0.2, 0.3, 0.5)$, and the aggregation-associated vector is $\omega = (0.4, 0.3, 0.3)^T$. By using a lexicographic ordering of these tuples by the first argument, we obtain the ordered tuples:

$$<Frank, \{0.1, 0.2, 0.4\}>, \quad <John, \{0.2, 0.4, 0.5\}>, \quad <Smith, \{0.2, 0.4\}>$$

Thus, $\varepsilon(1) = 2$, $\varepsilon(2) = 3$, $\varepsilon(3) = 1$. Then

$$\frac{\lambda_1 \omega_{\varepsilon(1)}}{\sum_{j=1}^{3} \lambda_j \omega_{\varepsilon(j)}} = \frac{0.2 \times 0.3}{0.2 \times 0.3 + 0.3 \times 0.3 + 0.5 \times 0.4} = 0.17$$

$$\frac{\lambda_2 \omega_{\varepsilon(2)}}{\sum_{j=1}^{3} \lambda_j \omega_{\varepsilon(j)}} = 0.26, \quad \frac{\lambda_3 \omega_{\varepsilon(3)}}{\sum_{j=1}^{3} \lambda_j \omega_{\varepsilon(j)}} = 0.57$$

If we use the IHFHWA operator to fuse these tuples, then, by Eq. (3.41), we have

$$\text{IHFHWA}(<John, h_1>, <Smith, h_2>, <Frank, h_3>)$$
$$= \bigcup_{\gamma_1 \in h_1, \gamma_2 \in h_2, \gamma_3 \in h_3} \left\{ 1 - (1 - \gamma_1)^{0.17} (1 - \gamma_2)^{0.26} (1 - \gamma_3)^{0.57} \right\}$$
$$= \{0.1444, 0.1853, 0.2000, 0.2061, 0.2101, 0.2382, 0.2440, 0.2577, 0.2614, 0.2671, 0.2931,$$
$$0.3147, 0.3210, 0.3534, 0.3699, 0.3731, 0.4000, 0.4183\}.$$

If we use the IHFHWG operator to fuse these tuples, then, using Eq. (3.42), we have

$$\text{IHFHWG}(<ohn, h_1>, <Smith, h_2>, <Frank, h_3>)$$
$$= \bigcup_{\gamma_1 \in h_1, \gamma_2 \in h_2, \gamma_3 \in h_3} \left\{ \gamma_1^{0.17} \gamma_2^{0.26} \gamma_3^{0.57} \right\}$$
$$= \{0.1347, 0.1516, 0.1574, 0.1613, 0.1815, 0.1885, 0.2000, 0.2250, 0.2337, 0.2395, 0.2694,$$
$$0.2799, 0.2969, 0.3340, 0.3470, 0.3555, 0.4000, 0.4155\}.$$

According to Eq. (1.17), we can compute

$$s(\text{IHFHWA}(<John, h_1>, <Smith, h_2>, <Frank, h_3>)) = 0.2810$$
$$s(\text{IHFHWG}(<John, h_1>, <Smith, h_2>, <Frank, h_3>)) = 0.2540$$

and thus,

$$\text{IHFHWA}(<John, h_1>, <Smith, h_2>, <Frank, h_3>)$$
$$< \text{IHFHWG}(<John, h_1>, <Smith, h_2>, <Frank, h_3>)$$

Definition 3.19 (*Liao and Xu* 2015a). For a collection of HFEs h_j ($j = 1, 2, \ldots, n$), $\lambda = (\lambda_1, \lambda_2, \ldots, \lambda_n)^T$ is the weight vector of them with $\lambda_j \in [0, 1]$ and $\sum_{j=1}^n \lambda_j = 1$, then the following aggregation operators are defined, which are all based on the mapping $H^n \to H$ with an aggregation-associated vector $\omega = (\omega_1, \omega_2, \ldots, \omega_n)^T$ such that $\omega_j \in [0, 1]$ and $\sum_{j=1}^n \omega_j = 1$, and a continuous strictly monotonic function $g(\gamma)$:

(1) An induced quasi hesitant fuzzy hybrid weighted averaging (IQHFHWA) operator:

$$
\text{IQHFHWA}(<v_1, h_1>, <v_2, h_2>, \ldots, <v_n, h_n>) = g^{-1}\left(\frac{\overset{n}{\underset{j=1}{\oplus}} \lambda_j \omega_{\varepsilon(j)} g(h_j)}{\sum_{j=1}^n \lambda_j \omega_{\varepsilon(j)}}\right)
$$

$$
= \bigcup_{\gamma_1 \in h_1, \gamma_2 \in h_2, \ldots, \gamma_n \in h_n} \left\{ g^{-1}\left(1 - \prod_{j=1}^n (1 - g(\gamma_j))^{\frac{\lambda_j \omega_{\varepsilon(j)}}{\sum_{j=1}^n \lambda_j \omega_{\varepsilon(j)}}}\right) \right\}
$$

(3.45)

(2) An induced quasi hesitant fuzzy hybrid weighted geometric (IQHFHWG) operator:

$$
\text{IQHFHWG}(<v_1, h_1>, <v_2, h_2>, \ldots, <v_n, h_n>) = g^{-1}\left(\overset{n}{\underset{j=1}{\otimes}} (g(h_j))^{\frac{\lambda_j \omega_{\varepsilon(j)}}{\sum_{j=1}^n \lambda_j \omega_{\varepsilon(j)}}}\right)
$$

$$
= \bigcup_{\gamma_1 \in h_1, \gamma_2 \in h_2, \ldots, \gamma_n \in h_n} \left\{ g^{-1}\left(\prod_{j=1}^n (g(\gamma_j))^{\frac{\lambda_j \omega_{\varepsilon(j)}}{\sum_{j=1}^n \lambda_j \omega_{\varepsilon(j)}}}\right) \right\}
$$

(3.46)

(3) An induced generalized quasi hesitant fuzzy hybrid weighted averaging (IGQHFHWA) operator:

$$
\text{IGQHFHWA}(<v_1, h_1>, <v_2, h_2>, \ldots, <v_n, h_n>) = \left(g^{-1}\left(\frac{\overset{n}{\underset{j=1}{\oplus}} \lambda_j \omega_{\varepsilon(j)} g(h_j^p)}{\sum_{j=1}^n \lambda_j \omega_{\varepsilon(j)}}\right)\right)^{1/p}
$$

$$
= \bigcup_{\gamma_1 \in h_1, \gamma_2 \in h_2, \ldots, \gamma_n \in h_n} \left\{ \left(g^{-1}\left(1 - \prod_{j=1}^n (1 - g(\gamma_j^p))^{\frac{\lambda_j \omega_{\varepsilon(j)}}{\sum_{j=1}^n \lambda_j \omega_{\varepsilon(j)}}}\right)\right)^{1/p} \right\}
$$

(3.47)

(4) An induced generalized quasi hesitant fuzzy hybrid weighted geometric
(IGQHFHWG) operator:

$$
\mathrm{IGQHFHWG}(<v_1, h_1>, <v_2, h_2>, \ldots, <v_n, h_n>) = \left(g^{-1} \left(\mathop{\otimes}_{j=1}^{n} \left(g(h_j^p) \right)^{\frac{\lambda_j \omega_{\varepsilon(j)}}{\sum_{j=1}^{n} \lambda_j \omega_{\varepsilon(j)}}} \right) \right)^{1/p}
$$

$$
= \bigcup_{\gamma_1 \in h_1, \gamma_2 \in h_2, \ldots, \gamma_n \in h_n} \left\{ \left(g^{-1} \left(\prod_{j=1}^{n} (g(\gamma_j^p))^{\frac{\lambda_j \omega_{\varepsilon(j)}}{\sum_{j=1}^{n} \lambda_j \omega_{\varepsilon(j)}}} \right) \right)^{1/p} \right\}
$$

(3.48)

where $\varepsilon : \{1, 2, \ldots, n\} \to \{1, 2, \ldots, n\}$ is the permutation such that v_j is the
$\varepsilon(j)$-th largest element of the collection of v_j $(j = 1, 2, \ldots, n)$, v_j is the order
inducing value and it can be represented in any different forms, and p is a
parameter such that $p \in (-\infty, +\infty)$.

The following theorems hold for the induced hesitant fuzzy hybrid weighted
aggregation operators.

Theorem 3.20 (Liao and Xu 2015a). *All the induced hesitant fuzzy hybrid
weighted aggregation operators, including the IHFHWA, IHFHWG, IGHFHWA,
IGHFHWG, IQHFHWA, IQHFHWG, IGQHFHWA and IGQHFHWG operators
satisfy the properties of idempotency, boundedness and commutativity.*

Theorem 3.21 (Liao and Xu 2015a). *For a collection of HFEs $h_j = \{\gamma_{jt} | t = 1, 2, \ldots, l_{h_j}\}$ $(j = 1, 2, \ldots, n)$, then*

$$
\begin{aligned}
&\mathrm{IHFHWG}(<v_1, h_1>, <v_2, h_2>, \ldots, <v_n, h_n>) \\
&\leq \mathrm{IHFHWA}(<v_1, h_1>, <v_2, h_2>, \ldots, <v_n, h_n>)
\end{aligned}
$$
(3.49)

$$
\begin{aligned}
&\mathrm{IGHFHWG}(<v_1, h_1>, <v_2, h_2>, \ldots, <v_n, h_n>) \\
&\leq \mathrm{IGHFHWA}(<v_1, h_1>, <v_2, h_2>, \ldots, <v_n, h_n>)
\end{aligned}
$$
(3.50)

$$
\begin{aligned}
&\mathrm{IQHFHWG}(<v_1, h_1>, <v_2, h_2>, \ldots, <v_n, h_n>) \\
&\leq \mathrm{IQHFHWA}(<v_1, h_1>, <v_2, h_2>, \ldots, <v_n, h_n>)
\end{aligned}
$$
(3.51)

$$
\begin{aligned}
&\mathrm{IGQHFHWG}(<v_1, h_1>, <v_2, h_2>, \ldots, <v_n, h_n>) \\
&\leq \mathrm{IGQHFHWA}(<v_1, h_1>, <v_2, h_2>, \ldots, <v_n, h_n>)
\end{aligned}
$$
(3.52)

To be easily understood, the hesitant fuzzy hybrid weighted aggregation operators can be classified in Fig. 3.1.

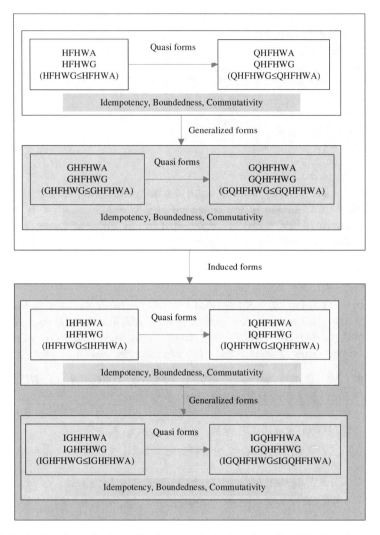

Fig. 3.1 Classification and relationships between the hesitant fuzzy hybrid weighted aggregation operators

3.6 Multiple Criteria Decision Making with Hesitant Fuzzy Hybrid Weighted Aggregation Operators

Since it was originally proposed by Torra (2010), the HFS has been investigated by many scholars from different points of view. It also has been applied to multiple criteria decision making problems and shows many advantages than other extended forms of fuzzy sets due to its particular powerfulness and efficiency in representing uncertainty and vagueness. The HFE is very close to the human's cognitive process when evaluating the candidate alternatives, because it can assign a set of possible values of membership degree of an element to a given set, while other extended forms of fuzzy set, like IFS, interval-valued IFS, linguistic fuzzy set, cannot be used to represent this situation in case people have several possible evaluation values at the same time. In addition, the HFE can be applied to maintain all the original assessments provided by the groups of decision makers, while all the other extended forms of fuzzy sets cannot be used to depict this situation exactly. Moreover, in practical group decision making problems, the anonymity is needed in order to protect the privacy of the decision makers or ensure non-interference opinions accumulated. Thus, it is natural to consider all the assessments so as to get more reasonable decision making results. This also can only be represented by HFEs.

Since the HFE can be used not only in multiple criteria single person decision making, but also in multiple criteria group decision making, in the following, we illustrate how to apply the hesitant fuzzy hybrid weighted aggregation operators to multiple criteria single person decision making and multiple criteria group decision making, respectively.

3.6.1   Hesitant Fuzzy Multiple Criteria Decision Making with Single Decision Maker

When a decision maker intends to evaluate a set of alternatives $A = \{A_1, A_2, \ldots, A_m\}$ with respect to a set of the predetermined criteria $C = \{C_1, C_2, \ldots, C_n\}$, he/she may find that it is hard to give a single value or a single interval for the membership degree of an element to a given set but a set of possible values due to the complexity of the problem and the incomplete information. For example, suppose that a person wants to buy a laptop, there are several brands for him/her to choose, such as ThinkPad, Apple, Acer, HP and so forth. The person finds that it is hard for him/her to decide which is the best even only over a single criterion named, for instance, appearance, because if he/she prefers the color of the alternative a to that of the alternative b with the membership degree 0.7, the design of the alternative b may be better than that of the alternative a with the membership degree 0.8. In this case, the decision maker cannot represent the judgments of the alternative a to the alternative b over the criterion "appearance", which consists of "color" and "design" as sub-criteria, in traditional fuzzy set but only in the HFE as $\{0.7, 0.8\}$. This case is common in our daily life. Hence, we need to develop some decision making models within the context of HFEs to aid the decision maker. Based on the hesitant fuzzy hybrid weighted aggregation operators, Liao and Xu (2014a) proposed a procedure for a decision maker to select the best choice with hesitant fuzzy information, which involves the following steps:

Algorithm 3.1

Step 1. Construct the hesitant fuzzy decision matrix. The decision maker determines the relevant criteria of the alternatives and gives the evaluation information of the alternatives with respect to the criteria in the form of HFEs. When the decision maker is asked to compare the alternatives over criteria, he/she may have several possible values regarding to the subcriteria. In this situation, it is natural to set out all the evaluation values given by the decision maker for an alternative under a criterion and represent them as HFE. The decision maker also determines the importance degrees $\lambda_j (j = 1, 2, \ldots, n)$ for the relevant criteria according to his/her preferences. Meanwhile, since different alternatives may have different focuses and advantages, to reflect this issue, the decision maker also gives the ordering weights $\omega_j (j = 1, 2, \ldots, n)$ for different criteria.

Step 2. Utilize the hesitant fuzzy hybrid weighted aggregation operators to obtain the HFEs h_i $(i = 1, 2, \ldots, m)$ for the alternatives A_i $(i = 1, 2, \ldots, m)$. Taking the HFHWA operator as an example, we have

$$h_i = \text{HFHWA}(h_{i1}, h_{i2}, \ldots, h_{in})$$

$$= \bigcup_{\gamma_{i1} \in h_{i1}, \gamma_{i2} \in h_{i2}, \ldots, \gamma_{in} \in h_{in}} \left\{ 1 - \prod_{j=1}^{n} (1 - \gamma_{ij})^{\frac{\lambda_j \omega_{\varepsilon(ij)}}{\sum_{j=1}^{n} \lambda_j \omega_{\varepsilon(ij)}}} \right\}, \quad i = 1, 2, \ldots, m$$

$$(3.53)$$

Step 3. Compute the score values $s(h_i)$ $(i = 1, 2, \ldots, m)$ of h_i $(i = 1, 2, \ldots, m)$ by Eq. (1.17) and the variance values $v(h_i)$ $(i = 1, 2, \ldots, m)$ of h_i $(i = 1, 2, \ldots, m)$ by Eq. (1.19).

Step 4. Get the rank of the alternatives A_i $(i = 1, 2, \ldots, m)$ according to $s(h_i)$ and $v(h_i)$ $(i = 1, 2, \ldots, m)$.

We now use a decision making problem to illustrate Algorithm 3.1.

Example 3.8 (Liao and Xu 2014a). Consider a customer who intends to buy a car. There are four alternatives A_1, A_2, A_3, A_4 under consideration and the customer takes three criteria into account to determine which car to buy:

C_1 : Quality of the car, which consists of three sub-criteria: S_1 (safety), S_2 (aerod. degree), and S_3 (remedy for quality problems).

C_2 : Overall cost of the product, which consists of four sub-criteria: S_4 (product price), S_5 (fuel economy), S_6 (tax), and S_7 (maintenance costs).

C_3 : Appearance of the car, which consists of three sub-criteria: S_8 (design), S_9 (color), and S_{10} (comfort).

As mentioned above, it is appropriate for the customer to represent his/her assessments in HFEs to maintain the original evaluation information adequately, which are shown in the hesitant fuzzy decision matrix $H = (h_{ij})_{4 \times 3}$ (see Table 3.1). Note that the criteria have two different types, i.e., benefit type and cost type. The customer should take this into account in the process of determining the preference values.

Table 3.1 Hesitant fuzzy decision matrix

	C_1	C_2	C_3
A_1	$\{0.6, 0.7, 0.9\}$	$\{0.6, 0.8\}$	$\{0.3, 0.6, 0.9\}$
A_2	$\{0.7, 0.9\}$	$\{0.4, 0.5, 0.8, 0.9\}$	$\{0.4, 0.8\}$
A_3	$\{0.6, 0.8\}$	$\{0.6, 0.7, 0.9\}$	$\{0.3, 0.5, 0.7\}$
A_4	$\{0.6, 0.8, 0.9\}$	$\{0.7, 0.9\}$	$\{0.2, 0.4, 0.7\}$

The weight information of these three criteria is also determined by the customer as $\lambda = (0.5, 0.3, 0.2)^T$. In addition, consider the fact that different cars may focus on different properties; for example, some cars are prominent in security with high price, while some cars are cheap but with low appearance. To reflect this concern, the customer gives another weight vector $\omega = (0.6, 0.2, 0.2)^T$ for the criteria, which denotes that the most prominent feature of the car assigns more weight while the remainders assign less weights.

To select the most desirable car, the hesitant fuzzy hybrid weighted aggregation operators can be used to aggregate the decision information. Suppose that we utilize the HFHWA operator to obtain the HFEs h_i $(i = 1, 2, 3, 4)$ for the cars A_1, A_2, A_3, A_4. Taking A_2 as an example, we have

$$h_2 = \text{HFHWA}(h_{21}, h_{22}, h_{23})$$
$$= \text{HFHWA}(\{0.7, 0.9\}, \{0.4, 0.5, 0.8, 0.9\}, \{0.4, 0.8\})$$

Since

$$s(h_{21}) = \frac{(0.7 + 0.9)}{2} = 0.8, \ s(h_{22}) = \frac{(0.4 + 0.5 + 0.8 + 0.9)}{4} = 0.65$$
$$s(h_{23}) = \frac{(0.4 + 0.8)}{2} = 0.6$$

then $h_{21} > h_{22} > h_{23}$, which implies $\varepsilon(21) = 1, \varepsilon(22) = 2, \varepsilon(23) = 3$. Thus,

$$\frac{\lambda_1 \omega_{\varepsilon(21)}}{\sum_{j=1}^3 \lambda_j \omega_{\varepsilon(2j)}} = \frac{0.5 \times 0.6}{0.5 \times 0.6 + 0.3 \times 0.2 + 0.2 \times 0.2} = 0.75$$

$$\frac{\lambda_2 \omega_{\varepsilon(22)}}{\sum_{j=1}^3 \lambda_j \omega_{\varepsilon(2j)}} = 0.15, \ \frac{\lambda_3 \omega_{\varepsilon(23)}}{\sum_{j=1}^3 \lambda_j \omega_{\varepsilon(2j)}} = 0.1$$

By using Eq. (3.16), we can calculate

$$h_2 = \text{HFHWA}(h_{21}, h_{22}, h_{23}) = \text{HFHWA}(\{0.7, 0.9\}, \{0.4, 0.5, 0.8, 0.9\}, \{0.4, 0.8\})$$
$$= \bigcup_{\gamma_{21} \in h_{21}, \gamma_{22} \in h_{22}, \gamma_{23} \in h_{23}} \left\{ 1 - (1 - \gamma_{21})^{0.75} (1 - \gamma_{22})^{0.15} (1 - \gamma_{23})^{0.1} \right\}$$
$$= \{0.6423, 0.6529, 0.6848, 0.6890, 0.6974, 0.7273, 0.7289, 0.7557, 0.8435,$$
$$0.8477, 0.8598, 0.8636 0.8673, 0.8804, 0.8811, 0.8928\}$$

Similarly, we can calculate different results for other alternatives A_1, A_3, and A_4 by using the HFHWA operator.

Finally, we compute the score values $s(h_i)$ $(i = 1, 2, 3, 4)$ and the variance values $v(h_i)$ $(i = 1, 2, 3, 4)$ of h_i $(i = 1, 2, 3, 4)$. Since $s(h_1) = 0.7329$, $s(h_2) = 0.6953$, $s(h_3) = 0.716$, and $s(h_4) = 0.7782$, then we get $s(h_4) > s(h_1) > s(h_3) > s(h_2)$, and thus, $h_4 \succ h_1 \succ h_3 \succ h_2$, i.e., the car A_4 is the most desirable choice for the customer.

If we use Xia and Xu (2011a)'s HFHA operator to solve this problem, then we have

$$h_1 = \text{HFHA}(h_{11}, h_{12}, h_{13}) = \text{HFHA}(\{0.6, 0.7, 0.9\}, \{0.6, 0.8\}, \{0.3, 0.6, 0.9\})$$
$$= \{0.6438, 0.6670, 0.7180, 0.6856, 0.7060, 0.7511, 0.7251, 0.7429, 0.7823, 0.7573,$$
$$0.7731, 0.8079, 0.8977, 0.9044, 0.9190, 0.9097, 0.9156, 0.9285\}$$

$$h_2 = \text{HFHA}(h_{21}, h_{22}, h_{23}) = \text{HFHA}(\{0.7, 0.9\}, \{0.4, 0.5, 0.8, 0.9\}, \{0.4, 0.8\})$$
$$= \{0.7097, 0.7456, 0.7191, 0.7538, 0.7618, 0.7912, 0.7897, 0.8157, 0.8920, 0.9053,$$
$$0.8955, 0.9084, 0.9114, 0.9223, 0.9218, 0.9314\}$$

$$h_3 = \text{HFHA}(h_{31}, h_{32}, h_{33}) = \text{HFHA}(\{0.6, 0.8\}, \{0.6, 0.7, 0.9\}, \{0.3, 0.5, 0.7\})$$
$$= \{0.6438, 0.6579, 0.6783, 0.6618, 0.6752, 0.6945, 0.7225, 0.7335, 0.7493, 0.8091,$$
$$0.8167, 0.8276, 0.8188, 0.8259, 0.8363, 0.8513, 0.8572, 0.8657\}$$

$$h_4 = \text{HFHA}(h_{41}, h_{42}, h_{43}) = \text{HFHA}(\{0.6, 0.8, 0.9\}, \{0.7, 0.9\}, \{0.2, 0.4, 0.7\})$$
$$= \{0.6564, 0.6680, 0.6945, 0.7180, 0.7276, 0.7493, 0.8158, 0.8221, 0.8363, 0.8489,$$
$$0.8540, 0.8657, 0.9013, 0.9047, 0.9123, 0.9190, 0.9218, 0.9280\}$$

Since $s(h_1) = 0.7908$, $s(h_2) = 0.8359$, $s(h_3) = 0.7625$, and $s(h_4) = 0.8191$, then we get $s(h_2) > s(h_4) > s(h_1) > s(h_3)$, and thus, $h_2 \succ h_4 \succ h_1 \succ h_3$. With the HFHA operator, the car A_2 turns out to be the most desirable choice for the customer, and all the other cars are in the same rank as that in the above result. Meanwhile, when using the HFHA operator, we need to calculate $\dot{h} = n\lambda_k h_k$ first and compare them, and then calculate $\omega_j \dot{h}_{\sigma(j)}$, after which, we shall compute the aggregation values with $\overset{n}{\underset{j=1}{\oplus}} (\omega_j \dot{h}_{\sigma(j)})$. Since the computation with HFEs is very complex, the results derived by the HFHA operator is hard to be obtained. As for the HFHWA operator, the weighting operation of the ordered positions is synchronized with the weighting operation of the given importance, which is in the

Table 3.2 Hesitant fuzzy decision matrix

	C_1	C_2	C_3	C_4
A_1	$\{0.6, 0.7, 0.9\}$	$\{0.6, 0.8\}$	$\{0.3, 0.6, 0.8\}$	$\{0.4, 0.6\}$
A_2	$\{0.7, 0.8\}$	$\{0.5, 0.7, 0.9\}$	$\{0.4, 0.8\}$	$\{0.6, 0.7\}$
A_3	$\{0.6, 0.8\}$	$\{0.6, 0.7, 0.9\}$	$\{0.4, 0.7\}$	$\{0.3, 0.4, 0.5\}$
A_4	$\{0.6, 0.7, 0.8\}$	$\{0.7, 0.9\}$	$\{0.2, 0.5, 0.7\}$	$\{0.2, 0.4\}$

mathematical form as $\lambda_j \omega_{\varepsilon(j)}$. Since both λ_j and $\omega_{\varepsilon(j)}$ are crisp numbers, we only need to calculate $\overset{n}{\underset{j=1}{\oplus}} \lambda_j \omega_{\varepsilon(j)} h_j \Big/ \sum_{j=1}^{n} \lambda_j \omega_{\varepsilon(j)}$, which makes the HFHWA operator easier to calculate than the HFHA operator.

Example 3.9 (Liao and Xu 2015a). Consider that a person is interested in investing his money to any one of the four portfolios: bank deposit (BD, A_1), debentures (DB, A_2), government bonds (GB, A_3), and shares (SH, A_4). Out of these portfolios he has to choose only one based on four criteria: return (C_1), risk (C_2), tax benefits (C_3), and liquidity (C_4). After some evaluations, the decision maker finds that it is hard for him/her to decide which portfolio should be invested due to his/her limited knowledge. Thus, he/she uses HFEs to represent his/her preference assessments to maintain the original evaluation information adequately, shown in the hesitant fuzzy decision matrix $H = (h_{ij})_{4 \times 4}$ (see Table 3.2). Note that the criteria have two different types such as benefit type and cost type. The customer should take this into account in the process of determining the preference values.

The weight information of these four criteria is determined by the decision maker as $\lambda = (0.5, 0.2, 0.2, 0.1)^T$. In addition, since different portfolios may focus on different points, the person gives another weight vector $\omega = (0.4, 0.2, 0.2, 0.2)^T$ for each criterion. To select the most desirable portfolio, the hesitant fuzzy hybrid weighted aggregation operators can be used to aggregate the decision information adequately. Hereby we utilize the GHFHWG operator as an example to obtain the HFEs h_i ($i = 1, 2, 3, 4$) for the portfolios A_1, A_2, A_3, A_4.

Take A_3 as an example and let $p = 1$. Since $s(h_{31}) = 0.7$, $s(h_{32}) = 0.73$, $s(h_{33}) = 0.55$, and $s(h_{34}) = 0.4$, then $h_{32} > h_{31} > h_{33} > h_{34}$. Thus, $\varepsilon(31) = 2$, $\varepsilon(32) = 1$, $\varepsilon(33) = 3$, $\varepsilon(34) = 4$. It follows

$$\frac{\lambda_1 \omega_{\varepsilon(31)}}{\sum_{j=1}^{4} \lambda_j \omega_{\varepsilon(3j)}} = 0.4167, \frac{\lambda_2 \omega_{\varepsilon(32)}}{\sum_{j=1}^{4} \lambda_j \omega_{\varepsilon(3j)}} = 0.3333$$

$$\frac{\lambda_3 \omega_{\varepsilon(33)}}{\sum_{j=1}^{4} \lambda_j \omega_{\varepsilon(3j)}} = 0.1667, \frac{\lambda_4 \omega_{\varepsilon(34)}}{\sum_{j=1}^{4} \lambda_j \omega_{\varepsilon(3j)}} = 0.0833$$

Thus, by Eq. (3.35), we can calculate

$$h_3 = \text{GHFHWG}(\{0.6, 0.8\}, \{0.6, 0.7, 0.9\}, \{0.4, 0.7\}, \{0.3, 0.4, 0.5\})$$

$$= \bigcup_{\gamma_{31} \in h_{31}, \gamma_{32} \in h_{32}, \gamma_{33} \in h_{33}, \gamma_{34} \in h_{34}} \{\gamma_{31}^{0.4167} \gamma_{32}^{0.3333} \gamma_{33}^{0.1667} \gamma_{34}^{0.0833}\}$$

$$= \{0.5293, 0.5422, 0.5523, 0.5811, 0.5952, 0.6063, 0.5572, 0.5707, 0.5815, 0.6117, 0.6266,$$
$$0.6383, 0.6059, 0.6206, 0.6323, 0.6652, 0.6813, 0.6941, 0.5967, 0.6112, 0.6227, 0.6551,$$
$$0.6710, 0.6836, 0.6282, 0.6434, 0.6555, 0.6896, 0.7063, 0.7196, 0.6831, 0.6997, 0.7128,$$
$$0.7499, 0.7681, 0.7825\}.$$

Similarly, the results for the alternatives A_1, A_2, and A_4 can be calculated by the GHFHWG operator:

$$h_1 = \text{GHFHWG}(h_{11}, h_{12}, h_{13}, h_{14})$$

$$= \{0.5324, 0.5470, 0.5840, 0.6000, 0.6068, 0.6235, 0.5533, 0.5684, 0.6068, 0.6235, 0.6305,$$
$$0.6478, 0.5901, 0.6063, 0.6472, 0.6649, 0.6725, 0.6909, 0.6131, 0.6300, 0.6725, 0.6909,$$
$$0.6988, 0.7179, 0.6977, 0.7168, 0.7653, 0.7862, 0.7952, 0.8170, 0.7250, 0.7449, 0.7952,$$
$$0.8170, 0.8263, 0.8489\}.$$

$$h_2 = \text{GHFHWG}(h_{21}, h_{22}, h_{23}, h_{24})$$

$$= \{0.6148, 0.6212, 0.6743, 0.6813, 0.6430, 0.6497, 0.7053, 0.7126, 0.6649, 0.6718, 0.7293,$$
$$0.7368, 0.6721, 0.6790, 0.7371, 0.7448, 0.7029, 0.7102, 0.7710, 0.7789, 0.7269, 0.7344,$$
$$0.7972, 0.8055\};$$

$$h_4 = \text{GHFHWG}(h_{41}, h_{42}, h_{43}, h_{44})$$

$$= \{0.4799, 0.5085, 0.5591, 0.5924, 0.5914, 0.6266, 0.5219, 0.5529, 0.6080, 0.6441, 0.6431,$$
$$0.6813, 0.5118, 0.5422, 0.5962, 0.6317, 0.6306, 0.6681, 0.5565, 0.5896, 0.6483, 0.6869,$$
$$0.6857, 0.7265, 0.5411, 0.5732, 0.6304, 0.6678, 0.6667, 0.7063, 0.5883, 0.6233,$$
$$0.6854, 0.7262, 0.7250, 0.7681\}.$$

Finally, we can compute the score values $s(h_i)$ $(i = 1, 2, 3, 4)$ and the variance values $v(h_i)$ $(i = 1, 2, 3, 4)$ of h_i $(i = 1, 2, 3, 4)$. Since $s(h_1) = 0.6590$, $s(h_2) = 0.7069$, $s(h_3) = 0.6436$, and $s(h_4) = 0.6218$, we get $s(h_2) > s(h_1) > s(h_3) > s(h_4)$. Thus, $h_2 \succ h_1 \succ h_3 \succ h_4$, i.e., the portfolio A_2 (debentures) is the most desirable choice for the decision maker.

In Example 3.9, the decision maker's judgments are represented in HFEs, which is suitable to express the vague and uncertain information of the decision maker. Especially at the beginning of evaluation, the decision maker may hesitate among the different values of membership degree. Using HFEs to represent his/her preference assessments can maintain his original evaluation information adequately. In addition, in Example 3.9, we use two weight vectors to show the importance of different criteria from different aspects. The first weight vector λ is used to weight the hesitant judgments, while the latter one ω is used to weight the positions of these judgments. Based on the proposed hesitant fuzzy hybrid weighted aggregation operators, we can fuse the hesitant fuzzy evaluating information more comprehensively, taking both of the two weight vectors into account. Thus, this approach is very useful in tackling some complicated decision making problems. It should be stated that although here we only implement the approach in evaluating candidate portfolios, our approach and the proposed hesitant fuzzy hybrid weighted aggregation operators can be applied into quite a wide range of practical complicated multiple criteria decision making problems.

3.6.2   Hesitant Fuzzy Multiple Criteria Decision Making with Multiple Decision Makers

For a group decision making problem under uncertainty, let $A = \{A_1, A_2, \ldots, A_m\}$ be the set of alternatives, $C = \{C_1, C_2, \ldots, C_n\}$ be the set of criteria and $E = \{e_1, e_2, \ldots, e_p\}$ be the set of decision makers. Suppose that the decision maker e_k provides all the possible evaluated values under the criterion C_j for the alternative A_i denoted by a HFE $h_{ij}^{(k)}$ and constructs the decision matrix $H_k = (h_{ij}^{(k)})_{m \times n}$. He/she also determines the importance degrees $\lambda_j^{(k)} (j = 1, 2, \ldots, n)$ for the relevant criteria according to his/her preferences. Meanwhile, since different alternatives may have different focuses and advantages, to reflect this issue, the decision maker also gives the ordering weights $\omega_j^{(k)} (j = 1, 2, \ldots, n)$ for different criteria. Suppose that the weight vector of the decision makers is $\sigma = (\sigma_1, \sigma_2, \ldots, \sigma_p)^T$. Then, based on the hesitant fuzzy hybrid weighted aggregation operators, Liao and Xu (2014a) developed a method for group decision making with hesitant fuzzy information, which involves the following steps:

Algorithm 3.2

Step 1.   Utilize the HFWA (or HFWG) operator to aggregate all the individual hesitant fuzzy decision matrix $H_k = (h_{ij}^{(k)})_{m \times n}$ $(k = 1, 2, \ldots, p)$ into the collective hesitant fuzzy decision matrix $H = (h_{ij})_{n \times n}$, where

$$h_{ij} = \bigcup_{\gamma_{ij}^{(k)} \in h_{ij}^{(k)}, k=1,2,\ldots,p} \left\{1 - \prod_{k=1}^{p} (1 - \gamma_{ij}^{(k)})^{\sigma_k}\right\}$$
$$i = 1, 2, \ldots, m, j = 1, 2, \ldots, n$$

or

$$h_{ij} = \bigcup_{\gamma_{ij}^{(k)} \in h_{ij}^{(k)}, k=1,2,\ldots,p} \left\{\prod_{k=1}^{p} (\gamma_{ij}^{(k)})^{\sigma_k}\right\}, i = 1, 2, \ldots, m, j = 1, 2, \ldots, n$$

$$(3.55)$$

Step 2. Utilize the hesitant fuzzy hybrid weighted aggregation operators, such as the QHFHWA (or QHFHWG) operator to obtain the HFEs h_i ($i = 1, 2, \ldots, m$) for the alternatives A_i ($i = 1, 2, \ldots, m$), where

$$h_i = \text{QHFHWA}(h_{i1}, h_{i2}, \ldots, h_{im})$$

$$= \bigcup_{\gamma_{ij} \in h_{ij}, j=1,2,\ldots,n} \left\{ g^{-1} \left(1 - \prod_{j=1}^{n} (1 - g(\gamma_j))^{\frac{\lambda_j \omega_{\varepsilon(j)}}{\sum_{j=1}^{n} \lambda_j \omega_{\varepsilon(j)}}} \right) \right\}, \ i = 1, 2, \ldots, m$$

(3.56)

or

$$h_i = \text{QHFHWG}(h_1, h_2, \ldots, h_n)$$

$$= \bigcup_{\gamma_{ij} \in h_{ij}, j=1,2,\ldots,n} \left\{ g^{-1} \left(\prod_{j=1}^{n} (g(\gamma_j))^{\frac{\lambda_j \omega_{\varepsilon(j)}}{\sum_{j=1}^{n} \lambda_j \omega_{\varepsilon(j)}}} \right) \right\}, \ i = 1, 2, \ldots, m$$

(3.57)

Step 3. Compute the score values $s(h_i)$ ($i = 1, 2, \ldots, m$) of h_i ($i = 1, 2, \ldots, m$) by Eq. (1.17) and the variance values $v(h_i)$ ($i = 1, 2, \ldots, m$) of h_i ($i = 1, 2, \ldots, m$) by Eq. (1.19).

Step 4. Get the rank of the alternatives A_i ($i = 1, 2, \ldots, m$) according to $s(h_i)$ and $v(h_i)$ ($i = 1, 2, \ldots, m$).

We now consider a multiple criteria group decision making problem that concerns evaluating and ranking the safety of work systems to illustrate the method:

Example 3.10 (Liao and Xu 2014a). Maintaining the safety of work systems in workplace is one of the most important components of safety management within an effective manufacturing organization. There are many factors which affect the safety system simultaneously. According to the statistical analysis of the past work accidents in a manufacturing company in Ankara, Turkey, Dağdeviren and Yüksel (2008) found that there are four sorts of factors which affect the safety system:

C_1: Organizational factors, which involve job rotation, working time, job completion pressure, and insufficient control.

C_2: Personal factors, which consist of insufficient preparation, insufficient responsibility, tendency of risky behavior, and lack of adaptation.

C_3: Job related factors, which can be divided into job related fatigue, reduced operation times due to dangerous behaviors, and variety and dimension of job related information.

In addition, it is not possible to assume that the effects of all factors of work safety are the same in all cases. Hence, by using the fuzzy analytic hierarchy process method, Dağdeviren and Yüksel (2008) constructed a hierarchical structure to depict the factors and sub-factors, and then determined the weight vector of these three factors, which is $\lambda = (0.388, 0.3, 0.312)^T$.

Three experts e_1, e_2, e_3 from different departments, whose weight vector is $\sigma = (0.4, 0.3, 0.3)^T$, are gathered together to evaluate three candidate work systems A_1, A_2, A_3 according to the above predetermined factors $C_j(j = 1, 2, 3)$. However, since these factors have non-physical structures, it is hard for the experts to represent their preferences by crisp fuzzy numbers. HFEs are appropriate for them to use in expressing these preferences since they may have a set of possible values when evaluating these behavioral and qualitative factors. Thus, the hesitant fuzzy judgment matrices $H_k = (h_{ij}^{(k)})_{3 \times 3}$ $(k = 1, 2, 3)$ were constructed by the experts, shown as Tables 3.3, 3.4 and 3.5. Furthermore, we consider the fact that different experts are familiar with different research fields, and meanwhile, different work systems may focus on different partitions, the experts may want to give more weights to the criterion which is more prominent. Hence, another weight vectors are determined by the experts according to their preferences, which are $\omega^{(1)} = (0.4, 0.3, 0.3)^T$, $\omega^{(2)} = (0.5, 0.3, 0.2)^T$ and $\omega^{(3)} = (0.4, 0.4, 0.2)^T$.
To get the optimal work system, the following steps can be given:

Step 1. Utilize the aggregation operator to fuse all the individual hesitant fuzzy decision matrices $H_k = (h_{ij}^{(k)})_{3 \times 3}$ $(k = 1, 2, 3)$ into the collective hesitant fuzzy decision matrix $H = (h_{ij})_{3 \times 3}$. Here we use the HFWA operator to fuse the individual hesitant fuzzy decision matrix. Thus, we have

$$h_{ij} = \bigcup_{\gamma_{ij}^{(k)} \in h_{ij}^{(k)}, k=1,2,3} \left\{ 1 - \prod_{k=1}^{3} (1 - \gamma_{ij}^{(k)})^{\sigma_k} \right\}, i = 1, 2, 3, j = 1, 2, 3$$

Table 3.3 The hesitant fuzzy decision matrix $H^{(1)}$

	C_1	C_2	C_3
A_1	{0.6}	{0.7}	{0.4,0.5}
A_2	{0.6,0.8}	{0.5,0.9}	{0.7}
A_3	{0.4,0.5}	{0.3}	{0.6}

Table 3.4 The hesitant fuzzy decision matrix $H^{(2)}$

	C_1	C_2	C_3
A_1	{0.2,0.4}	{0.3,0.5}	{0.4}
A_2	{0.8}	{0.7}	{0.6,0.7}
A_3	{0.4}	{0.3,0.6}	{0.5,0.7}

Table 3.5 The hesitant fuzzy decision matrix $H^{(3)}$

	C_1	C_2	C_3
A_1	{0.5}	{0.3,0.4}	{0.6}
A_2	{0.7,0.9}	{0.8}	{0.5,0.6}
A_3	{0.3,0.4}	{0.4,0.5}	{0.8}

Here we take h_{23} as an example:

$$h_{23} = \bigcup_{\gamma_{23}^{(k)} \in h_{23}^{(k)}, k=1,2,3} \left\{ 1 - \prod_{k=1}^{3} \left(1 - \gamma_{23}^{(k)}\right)^{\sigma_k} \right\}$$
$$= \{0.619, 0.644, 0.65, 0.673\}$$

Similarly, other fused values can be obtained, and then the collective hesitant fuzzy matrix can be derived as follows:

$$H = \begin{pmatrix} \{0.473, 0.517\} & \{0.501, 0.524, 0.549, 0.57\} & \{0.469, 0.506\} \\ \{0.702, 0.774, 0.786, 0.838\} & \{0.674, 0.829\} & \{0.619, 0.644, 0.65, 0.673\} \\ \{0.372, 0.4, 0.442, 0.416\} & \{0.332, 0.367, 0.435, 0.465\} & \{0.653, 0.702\} \end{pmatrix}$$

Step 2. Utilize the aggregation operator (such as the QHFHWA or QHFHWG operator) to obtain the HFEs h_i $(i = 1, 2, 3)$ for the alternatives A_i $(i = 1, 2, 3)$. Here we use the QHFHWA operator to fuse the collective HFEs and let $g(\gamma) = \gamma$, and then we get

$h_1 = \{0.4825, 0.4913, 0.493, 0.4983, 0.5012, 0.5017, 0.5068, 0.5084, 0.5098, 0.5113,$
$\qquad 0.5164, 0.5168, 0.5197, 0.5247, 0.5262, 0.5343\}$

$h_2 = \{0.6811, 0.685, 0.686, 0.6898, 0.7269, 0.7302, 0.7303, 0.731, 0.7336, 0.7343, 0.7344,$
$\qquad 0.7351, 0.7383, 0.7391, 0.7377, 0.7423, 0.769, 0.7718, 0.7725, 0.7733, 0.7753, 0.776,$
$\qquad 0.7761, 0.7768, 0.7787, 0.7794, 0.7795, 0.7821, 0.8083, 0.8106, 0.8112, 0.8135\}$

$h_3 = \{0.4894, 0.4942, 0.4999, 0.5043, 0.5046, 0.506, 0.509, 0.5107, 0.5145, 0.5162, 0.5171,$
$\qquad 0.5191, 0.5204, 0.5208, 0.5217, 0.525, 0.5271, 0.5303, 0.5312, 0.5316, 0.5329, 0.5348,$
$\qquad 0.5357, 0.5373, 0.5409, 0.5425, 0.5453, 0.5465, 0.5468, 0.5508, 0.5558, 0.5601\}$

The computational process of h_2 can be illustrated as an example. Since $s(h_{21}) = 0.775$, $s(h_{22}) = 0.7515$, $s(h_{23}) = 0.6465$, then, $h_{21} > h_{22} > h_{23}$. Thus, $\varepsilon(21) = 1$, $\varepsilon(22) = 2$, $\varepsilon(23) = 3$, and

$$\frac{\lambda_1 \omega_{\varepsilon(21)}}{\sum_{j=1}^{3} \lambda_j \omega_{\varepsilon(2j)}} = 0.56, \quad \frac{\lambda_2 \omega_{\varepsilon(22)}}{\sum_{j=1}^{3} \lambda_j \omega_{\varepsilon(2j)}} = 0.2598, \quad \frac{\lambda_3 \omega_{\varepsilon(23)}}{\sum_{j=1}^{3} \lambda_j \omega_{\varepsilon(2j)}} = 0.1801$$

Therefore, by using Eq. (3.27), we can calculate

$$h_2 = \text{QHFHWA}(h_{21}, h_{22}, h_{23}) = \bigcup_{\gamma_{21} \in h_{21}, \gamma_{22} \in h_{22}, \gamma_{23} \in h_{23}} \left\{ 1 - (1 - \gamma_{21})^{0.56}(1 - \gamma_{22})^{0.2598}(1 - \gamma_{23})^{0.1801} \right\}$$
$$= \{0.6811, 0.685, 0.686, 0.6898, 0.7269, 0.7302, 0.7303, 0.731, 0.7336, 0.7343, 0.7344,$$
$$0.7351, 0.7383, 0.7391, 0.7377, 0.7423, 0.769, 0.7718, 0.7725, 0.7733, 0.7753, 0.776,$$
$$0.7761, 0.7768, 0.7787, 0.7794, 0.7795, 0.7821, 0.8083, 0.8106, 0.8112, 0.8135\}$$

Step 3.   Compute the score values $s(h_i)$ $(i = 1, 2, \ldots, n)$ of h_i $(i = 1, 2, \ldots, n)$ by
Eq. (1.17), and then we have $s(h_1) = 0.5089$, $s(h_2) = 0.7534$, and
$s(h_3) = 0.5257$.

Step 4.   Since $s(h_2) > s(h_3) > s(h_1)$, then we get $h_2 \succ h_3 \succ h_1$, which means that
C_2 is the most desirable work system.

References

Dağdeviren M, Yüksel İ (2008) Developing a fuzzy analytic hierarchy process (AHP) model for
behavior-based safety management. Inf Sci 178:1717–1733

Dyckhoff H, Pedrycz W (1984) Generalized means as model of compensative connectives. Fuzzy
Sets Syst 14:143–154

Fodor J, Marichal JL, Roubens M (1995) Characterization of the ordered weighted averaging
operators. IEEE Trans Fuzzy Syst 3:236–240

Hardy GH, Littlewood JE, Pólya G (1934) Inequalities. Cambridge University Press, Cambridge

Liao HC, Xu ZS (2014a) Some new hybrid weighted aggregation operators under hesitant fuzzy
multi-criteria decision making environment. J Intell Fuzzy Syst 26(4):1601–1617

Liao HC, Xu ZS (2014b) Intuitionistic fuzzy hybrid weighted aggregation operators. Int J Intell
Syst 29(11):971–993

Liao HC, Xu ZS (2015) Extended hesitant fuzzy hybrid weighted aggregation operators and their
application in decision making. Soft Comput 19(9):2551–2564

Liao HC, Xu ZS, Xia MM (2014) Multiplicative consistency of hesitant fuzzy preference relation
and its application in group decision making. Int J Inf Tech Decis Making 13(1):47–76

Lin J, Jiang Y (2014) Some hybrid weighted averaging operators and their application to decision
making. Inf Fusion 16:18–28

Merigó JM, Gil-Lafuente AM (2009) The induced generalized OWA operator. Inf Sci
179:729–741

Torra V (2010) Hesitant fuzzy sets. Int J Intell Syst 25:529–539

Torra V, Narukawa Y (2009) On hesitant fuzzy sets and decision. The 18th IEEE international
conference on fuzzy systems (FS'09). Jeju Island, Korea, pp 1378–1382

Wei GW (2012) Hesitant fuzzy prioritized operators and their application to multiple attribute
decision making. Knowl-Based Syst 31:176–182

Xia MM, Xu ZS (2011) Hesitant fuzzy information aggregation in decision making. Int J
Approximate Reasoning 52(3):395–407

Xia MM, Xu ZS, Chen N (2011) Induced aggregation under confidence levels. Int J Uncertainty
Fuzziness Knowl-Based Syst 19:201–227

Xia MM, Xu ZS, Chen N (2013) Some hesitant fuzzy aggregation operators with their application
in group decision making. Group Decis Negot 22(2):259–279

Xu ZS (2000) On consistency of the weighted geometric mean complex judgment matrix in
AHP. Eur J Oper Res 126:683–687

Yager RR (1988) On ordered weighted averaging aggregation operators in multi-criteria decision
making. IEEE Trans Sys Man Cybern 18(1):183–190

Yager RR (2004) Generalized OWA aggregation operators. Fuzzy Optim Decis Making 3:93–107

Yager RR (2008) Prioritized aggregation operators. Int J Approximate Reasoning 48:263–274

Yager RR, Filev DP (1999) Induced ordered weighted averaging operators. IEEE Trans Syst Man
Cybern B Cybern 29(2):141–150

Zhang ZM (2013) Hesitant fuzzy power aggregation operators and their application to multiple attribute group decision making. Inf Sci 234:150–181

Zhu B, Xu ZS (2013) Hesitant fuzzy Bonferroni means for multi-criteria decision making. J Oper Res Soc 64:1831–1840

Zhu B, Xu ZS, Xia MM (2012) Dual hesitant fuzzy set. J Appl Math, Article ID 879629, 13 pages. doi:10.1155/2012/879629

Chapter 4
Hesitant Fuzzy Multiple Criteria Decision Making Methods with Complete Weight Information

Multiple criteria decision making takes place in many areas of operations research and management sciences. Generally, a multiple criteria decision making problem involves several alternatives, criteria and the ratings of these alternatives with respect to different criteria. Sometimes the weights of criteria are completely given while sometime are not. Different methods have been proposed regarding to different scenarios to obtain the final optimal solutions for the multiple criteria decision making problems. In this chapter, we introduce two methods for multiple criteria decision making problems in which the ratings of alternatives over the criteria are given as HFEs and the weights of criteria are completely known.

The VIKOR (vlsekriterijumska optimizacija i kompromisno resenje in serbian, meaning multicriteria optimization and compromise solution) method is an efficient tool to find a compromise solution from a set of conflicting criteria. It has been applied widely in many fields. Recently, Park et al. (2011) extended the VIKOR method to develop an approach to solve the multiple criteria decision making problems with interval-valued intuitionistic fuzzy numbers (Xu and Yager 2009). For the situations where the criteria values take the form of intuitionistic trapezoidal fuzzy numbers, Du and Liu (2011) developed three extensions of the VIKOR method based on the expected values of the intuitionistic trapezoidal fuzzy numbers, the distances between the intuitionistic trapezoidal fuzzy numbers and the distances between the interval numbers, respectively. As HFS is suitable and powerful in describing uncertainty and vagueness, in this chapter, we address the VIKOR method for multiple criteria decision making under hesitant fuzzy environments. The hesitant fuzzy group utility measure, the hesitant fuzzy individual regret measure, and the hesitant fuzzy compromise measure are introduced. Based on these new measures, the hesitant fuzzy VIKOR (HF-VIKOR) method with complete weight information is provided. A practical example is provided to show that the HF-VIKOR method is very effective in solving multiple criteria decision making problems with hesitant fuzzy information.

The ELECTRE (ELimination Et Choix Traduisant la REalité) method is another popular method for multiple criteria decision making. In this chapter, we introduce

© Springer Nature Singapore Pte Ltd. 2017
H. Liao and Z. Xu, *Hesitant Fuzzy Decision Making Methodologies and Applications*, Uncertainty and Operations Research,
DOI 10.1007/978-981-10-3265-3_4

two hesitant fuzzy ELECTRE (HF-ELECTRE) approaches which combine the HFS with the ELECTRE methods to efficiently handle different opinions of group members that are frequently encountered when handling the multiple criteria decision making problems. We define the concepts of hesitant fuzzy concordance and discordance sets and construct the strong and weak outranking relations which are employed to decide the ranking for a set of alternatives. Numerical examples are presented to exhibit the applications of the proposed methods. Furthermore, some comparison analyses between the HF-ELECTRE methods and the aggregation operator based method as well as the fuzzy group ELECTRE approach are provided. After that, some Algorithms based on the HF-ELECTRE methods are constructed to aid decision making, and the prominent characteristics of the HF-ELECTRE methods and future research challenges are also discussed.

4.1 Hesitant Fuzzy VIKOR Method for Multiple Criteria Decision Making

4.1.1 Multiple Criteria Decision Making and the VIKOR Method

A multiple criteria decision making problem can be interpreted simply as selecting the best alternative(s) from a set of alternatives to attain a goal (or goals). Each alternative has several criteria. The multiple criteria decision making procedure consists of five steps: ① defining a goal (or goals), ② generating alternatives, ③ establishing criteria and weighs regarding to the criteria, ④ evaluating alternatives with respect to different criteria, ⑤ ranking the alternatives and determining the acceptable choice. Duckstein and Opricovic (1980) divided the multiple criteria decision making process into two levels: managerial level and engineering level. The managerial level defines the goal to attain and chooses the final optimal solution, while the alternatives are generated and the consequences of implementing any one of them are also pointed out with respect to various criteria in engineering level. The weights of criteria are obtained and the ranking of alternatives is also performed in this level. In other words, the managerial level is in charge of the first and final steps of the multiple criteria decision making process, while the engineering level takes responsibility for the middle three of them.

At the beginning of multiple criteria decision making, the goal (or goals) should be established according to the preference structure provided by the managerial level. If there are a set of conflicting goals which cannot be achieved simultaneously, the handling methodology should be in another way, which is called multi-objective decision making (Ribeiro 1996). This book only considers the issues where there is only a single goal to attain. After the goal is established, the engineering level then sets out to formulate the alternatives, which are usually physical elements, actions, objects or likely scenarios, such as places for facility to locate or a specific policy to implement. Generating alternatives is a complex process as all the constraints must be satisfied. They sometimes only depend on the experts' experiences and the experiments about the existing system or other similar systems because no general procedure or mathematical model could replace human creativity in generating alternatives. The ratings of alternatives are based on the relative merits of the criteria. The criteria in different problems are quite different. For example, if we are going to buy a car, the criteria can be formulated as price, maximum velocity and comfort; while if we are planning to choose a school for children, the criteria would be distance, price, teaching quality and fame of the school. So the criteria should be established according to the reality of the problem itself. A large quantity of research has been done on the evaluation of alternatives and many different tools have been proposed, such as the non-fuzzy methods, the fuzzy hierarchical aggregation methods, the conjunction implication methods, the weighted average aggregation methods, and so on. Riberio (1996) reviewed and compared all these methods in details. After all these work was done, it is needed to rank the alternatives and choose the one with the highest degree of satisfaction for the multiple criteria decision making problem.

However, in many cases, the criteria involved in the multiple criteria decision making problem conflict with each other, and thus, there perhaps is no solution satisfying all criteria simultaneously, which makes the ranking process very intractable. One common example is the relationship between the development possibility and the protection of environment. The desirable solution is the one where both objectives are maximized; however, this alternative is impossible or infeasible in most cases. Pareto optimal solution is proposed to illustrate this situation, which has the property that if one criterion is to be improved, at least one other criterion has to be made worse (Pareto 1986). Since the Pareto optimal solution is a set of non-inferior alternatives, it cannot satisfy our initial purpose which is to select one alternative from a set of alternatives to implement. Although in more cases, we can aid the decision maker by making comprehensive analysis and listing the important characterization of the Pareto optimal solutions for interactive decision making, it is time consuming and not practicable. We prefer to find a unique solution which maximally achieves the overall criteria. By defining the concept of compromise solution, Yu (1973) overcame this problem and gave the method to find the compromise solution by using compromise programming.

The classical compromise programming is based on the distance measure which determines the closeness of a particular solution to the ideal/infeasible solution. Based on the compromise programming proposed by Yu (1973), many multiple criteria decision making methods have been investigated, such as the TOPSIS (technique for order preference by similarity to an ideal solution) method (Hwang and Yoon 1981), the VIKOR method (Opricovic 1998), the PROMETHEE (preference ranking organization method for enrichment evaluations) method (Liao and Xu 2014c), the ELECTRE (elimination et choice translating reality) method (Chen et al. 2015), and so on. After comparing the TOPSIS method and the VIKOR method in terms of aggregation functions and normalization effects, Opricovic and Tzeng (2004) pointed out that although both the TOPSIS method and the VIKOR method are on the basis of the distance of an alternative to the ideal solution, the VIKOR method determines a compromise solution which is established by mutual concessions, while the TOPSIS method determines a solution with the shortest distance from the ideal solution and the farthest distance from the negative-ideal solution, but it does not consider the relative importance degrees of these distances. Furthermore, Opricovic and Tzeng (2007) extended the VIKOR method with a stability analysis determining the weight stability intervals and with trade-offs analysis, and also compared it deeply with the TOPSIS method, the PROMETHEE method and the ELECTRE method. According to the comparisons, it is obvious that the VIKOR method has many advantages in handling the multiple criteria decision making problems especially when there are conflicting and noncommensurable criteria. Based on the analytic hierarchy process (AHP) (Saaty 1980), Kaya and Kahraman (2011) proposed an integrated VIKOR-AHP method, in which the weights of the criteria are derived by fuzzy pairwise comparison matrices of AHP. Up to now, the VIKOR method has been applied widely in many fields, such as the mountain destination choosing problem (Opricovic and Tzeng 2004), the alternative hydropower system evaluation problem (Opricovic and Tzeng 2004), the forestation and forest preservation problem (Kaya and Kahraman 2011), the post-earthquake sustainable reconstruction problem (Opricovic 2002), and so forth.

The VIKOR method is based on the particular measure of "closeness" to the "ideal" solution. It was firstly introduced by Opricovic (1998). The VIKOR method is an efficient tool to find a compromise solution from a set of conflicting criteria. The basic measure for compromise ranking is developed from the L_p—metric used as an aggregation function in the compromise programming (Yu 1973). Suppose that the ratings of the alternatives $A_i(i = 1, 2, \ldots, m)$ regarding to the criteria $C_j(j = 1, 2, \ldots, n)$ are given as $f_{ij}(i = 1, 2, \ldots, m; j = 1, 2, \ldots, n)$. Mathematically, the discrete form of L_p—metric distance measure over the alternatives $A_i(i = 1, 2, \ldots, m)$ in compromise programming can be given as:

$$L_{p,i} = \left(\sum_{j=1}^{n} \left(\omega_j \frac{f_j^* - f_{ij}}{f_j^* - f_j^-} \right)^p \right)^{1/p}, \quad 1 \le p \le \infty; \quad i = 1, 2, \ldots, m \qquad (4.1)$$

where $\omega_j(j = 1, 2, \ldots, n)$ are the corresponding weights of criteria, $f_j^* = \max_i f_{ij}$ and $f_j^- = \min_i f_{ij}$ are the best and worst values of A_i over the benefit-type criterion C_j, respectively. The parameter p plays an important role and has different meanings with respect to different values. Varying the parameter p from 1 to infinity is to move from minimizing the sum of individual regrets to minimizing the maximum regret. If $p = 1$, then all deviations are weighted equally; if $p = 2$, then the deviations are weighted according to their magnitudes; if $p = \infty$, then the deviations can be interpreted as the maximum individual regret. This is just the critical idea of the VIKOR method, which uses $L_{1,i}$ and $L_{\infty,i}$ to formulate group utility and individual regret of the opponent.

The compromise ranking procedure of the VIKOR method can be set out as follows:

(1) Find f_j^* and f_j^-.
(2) Compute the values of group utility and individual regret over the alternatives $A_i(i = 1, 2, \ldots, m)$ by the equations:

$$S_i = L_{1,i} = \sum_{j=1}^{n} \omega_j \frac{f_j^* - f_{ij}}{f_j^* - f_j^-} \tag{4.2}$$

$$R_i = L_{\infty,i} = \max_j \left(\omega_j \frac{f_j^* - f_{ij}}{f_j^* - f_j^-} \right) \tag{4.3}$$

(3) Calculate the values of $Q_i(i = 1, 2, \ldots, m)$ by the relation:

$$Q_i = \upsilon \frac{S_i - S^*}{S^- - S^*} + (1 - \upsilon) \frac{R_i - R^*}{R^- - R^*} \tag{4.4}$$

where $S^* = \min_i S_i$, $S^- = \max_i S_i$, $R^* = \min_i R_i$, $R^- = \max_i R_i$, and υ is the weight of the strategy of the majority of criteria or the maximum group utility. Without loss of generality, it takes the value 0.5.
(4) Rank the alternatives $A_i(i = 1, 2, \ldots, m)$ according to the values of S_i, R_i and Q_i. The results are three ranking lists.
(5) Determinate the best solution or a compromise solution.

It is clear that the smaller the value of Q_i is, the better the solution should be. To ensure the uniqueness of the final alternative, the following two qualifications must be satisfied simultaneously:

C1: $Q(a^{(2)}) - Q(a^{(1)}) \geq \frac{1}{m-1}$, where $a^{(1)}$ and $a^{(2)}$ are the alternatives with the first and second positions in the ranking list, respectively;
C2: $a^{(1)}$ should also be the best ranked by S_i and R_i.

Unfortunately, these two conditions often cannot be attained simultaneously. Thus, a set of compromise solutions are derived, which is of the most critical significance for the VIKOR method.

If the condition **C1** is not satisfied, then we shall explore the maximum value of M according to the equation:

$$Q\left(a^{(M)}\right) - Q\left(a^{(1)}\right) < \frac{1}{m-1} \tag{4.5}$$

All the alternatives $a^{(i)}(i = 1, 2, \ldots, M)$ are the compromise solutions.
If the condition **C2** is not satisfied, then the alternatives $a^{(1)}$ and $a^{(2)}$ are the compromise solutions.

4.1.2   Hesitant Fuzzy VIKOR Method for Multiple Criteria Decision Making

In many situations, the decision makers may be unwilling to give their preference information in just a single value, but a set of values, especially at the beginning of evaluation. Additionally, in the situation where a decision making problem solved by many people from different areas, it usually cannot achieve a consentaneous preference value over the considered alternative with respect to a criterion. Thus, it is more suitable to represent the rating information by HFE. However, handling hesitant fuzzy information is a tough work because it contains many evaluation values, and as pointed by Theorem 1.1, with the calculation going on, the dimension of the HFE will be larger and larger, which will increase the computational complexity as well. Although some methods have been proposed to fuse hesitant fuzzy information (Liao et al. 2014b; Liao and Xu 2015c; Xia and Xu 2011a), how to deal with the hesitant fuzzy preferences without loss any original information is still an open question. In what follows, we shall apply the VIKOR method to solve the multiple criteria decision making problems with hesitant fuzzy information.

Consider a multiple criteria decision making problem with a discrete set of m alternatives, $A = \{A_1, A_2, \ldots, A_m\}$, and let $C = \{C_1, C_2, \ldots, C_n\}$ be the set of all criteria. The evaluation value of the ith alternative on the criterion C_j is represented as the HFE h_{ij} with $h_{ij} = \{\gamma | 0 \leq \gamma \leq 1\}$, $i = 1, 2, \ldots, m, j = 1, 2, \ldots, n$. h_{ij} indicates the possible membership degrees of the ith alternative A_i under the j th criterion C_j. The hesitant fuzzy decision matrix H can be written as:

$$
H = \begin{bmatrix}
h_{11} & h_{12} & \cdots & h_{1n} \\
h_{21} & h_{22} & \cdots & h_{2n} \\
\vdots & \vdots & \ddots & \vdots \\
h_{m1} & h_{m2} & \cdots & h_{mn}
\end{bmatrix}
\tag{4.6}
$$

The weights $\omega_j (j = 1, 2, \ldots, n)$ of criteria represent the relative importance degrees of the criteria, where $0 \leq \omega_j \leq 1, j = 1, 2, \ldots, n$, and $\sum_{j=1}^{n} \omega_j = 1$.

Since the traditional VIKOR method is based on the particular distance measure of closeness to the ideal solution, in order to derive the HF-VIKOR method, firstly, we need to find the so called ideal solution. For the benefit-type criterion, the ideal solution is the maximum value in each column of the hesitant decision matrix, while for the cost-type criterion, it is the minimum one. We need to find both of them for the next step of computation. To do so, we only need to calculate the score function and the variance function of each HFE in the hesitant fuzzy decision matrix H, and then apply the comparison law of HFEs, which is given in Sect. 1.1.3. The outputs of this step are $h_j^* = \max_i h_{ij}$ and $h_j^- = \min_i h_{ij}$, which are the best and worst values of A_i over the benefit-type criterion C_j, respectively. The best and worst values of A_i over the cost-type criterion can be derived similarly.

Xu and Xia (2011a) defined several distance measures for HFEs. Here we use the Manhattan distance of two HFEs h_M and h_N as a representation, which is in the mathematical form as follows:

$$d(h_M, h_N) = \frac{1}{l} \sum_{k=1}^{l} \left| h_M^{\sigma(k)} - h_N^{\sigma(k)} \right| \qquad (4.7)$$

where $h_M^{\sigma(k)}$ and $h_N^{\sigma(k)}$ are the kth largest values in h_M and h_N, respectively.

Example 4.1 (Liao and Xu 2013). Let $h_1 = (0.7, 0.8, 0.9)$ and $h_2 = (0.5, 0.6, 0.8)$ be two HFEs. Then, $l = 3$. The Manhattan distance of h_1 and h_2 is

$$\begin{aligned} d(h_M, h_N) &= \frac{1}{l} \sum_{k=1}^{l} \left| h_M^{\sigma(k)} - h_N^{\sigma(k)} \right| \\ &= \frac{1}{3} \left(|0.7 - 0.5| + |0.8 - 0.6| + |0.9 - 0.8| \right) = 0.1667 \end{aligned}$$

From Example 4.1, we can draw a very interesting conclusion. Although h_1 and h_2 are HFEs, the values of their Manhattan distance are crisp numbers. This is a good way to convert two HFEs into a crisp number, which is very important in developing the hesitant fuzzy VIKOR method.

The ideal solution is always infeasible because the conflicting criteria cannot be satisfied simultaneously. Thus, we need to find the compromise solution, i.e., the closest solution to the ideal solution. Compromise solution is established by mutual concessions, and in the VIKOR method, it integrates with the maximum group utility and the minimum individual regret of the opponent. The group utility and the individual regret of the opponent are formulated by $L_{1,i}$ and $L_{\infty,i}$. Due to the fact that the evaluation values are HFEs, the computations of $L_{1,i}$ and $L_{\infty,i}$ will be very difficult. Motivated by the adjusted operation laws of HFEs given in Definition 1.9 and the hesitant distance given as Eq. (4.7), the traditional discrete form of L_p—metric distance measure over the alternatives $A_i (i = 1, 2, \ldots, m)$ in compromise programming given as Eq. (4.1) can be generalized into hesitant fuzzy environments.

Definition 4.1 (*Liao and Xu* 2013). The Manhattan L_p—metric of HFEs over the benefit-type criterion is in terms of the following form:

$$\widetilde{L}_{p,i} = \left(\sum_{j=1}^{n} \left(\omega_j \frac{d(h_j^*, h_{ij})}{d(h_j^*, h_j^-)} \right)^p \right)^{1/p}, 1 \le p \le \infty; \quad i = 1, 2, \ldots, m \qquad (4.8)$$

where $\omega_j (j = 1, 2, \ldots, n)$ are the corresponding weights of criteria, and satisfy $0 \le \omega_j \le 1, j = 1, 2, \ldots, n$, and $\sum_{j=1}^{n} \omega_j = 1$. $d\left(h_j^*, h_{ij} \right)$ is the Manhattan distance between h_j^* and h_{ij}, which is in the following mathematical form:

$$d(h_j^*, h_{ij}) = \frac{1}{l_j} \sum_{k=1}^{l_j} \left| h_j^{*\sigma(k)} - h_{ij}^{\sigma(k)} \right| \tag{4.9}$$

where $h_j^{*\sigma(k)}$ and $h_{ij}^{\sigma(k)}$ are the kth largest values in h_j^* and h_{ij}, respectively, and $l = \max\left\{ l_{h_j^*}, l_{h_{ij}} \right\}$. $d\left(h_j^*, h_j^- \right)$ can also be defined similarly.

It is similar to derive the Manhattan L_p—metric of HFSs over the cost-type criterion. There is one thing we need to point out. In the conventional VIKOR method, the normalization is needed because there are many noncommensurable (different units) criteria values. Under hesitant fuzzy circumstances, all criteria values are in [0, 1]; however, the normalization is also needed. The reason for this is that for different criteria, the cognitions and preferences of people over the same alternative may be different. For example, if we plan to buy a house, generally, we may be more sensitive to the price but less sensitive to the comfortableness or location (different people still may have different preferences because of their particular demands). Thus, the HFEs over price would be varied largely while the preference values of the criterion comfortableness or location may be closer. Hence, in order to eliminate the influence over different criteria, we still need to normalize the distance.

Based on Definitions 4.1, the hesitant fuzzy group utility and the individual regret of the opponent are formulated easily. For simplicity, in the following, we take the Manhattan L_p—metric of HFEs as an example to derive the HF-VIKOR method. Considering the original explanations of $L_{1,i}$ and $L_{\infty,i}$ in the classical VIKOR method under crisp environments, the hesitant fuzzy group utility and the individual regret of the opponent can be measured obviously.

Definition 4.2 (*Liao and Xu* 2013). The hesitant fuzzy group utility measurement over the benefit-type criterion is based on the formula:

$$\widetilde{S}_i = \widetilde{L}_{1,i} = \sum_{j=1}^{n} \omega_j \frac{d\left(h_j^*, h_{ij} \right)}{d\left(h_j^*, h_j^- \right)} \tag{4.10}$$

where $\omega_j(j = 1, 2, \ldots, n)$ are the corresponding weights of criteria satisfying $0 \leq \omega_j \leq 1$, $j = 1, 2, \ldots, n$, $\sum_{j=1}^{n} \omega_j = 1$, $d\left(h_j^*, h_{ij} \right)$ and $d\left(h_j^*, h_j^- \right)$ can be determined through Eq. (4.9).

Definition 4.3 (*Liao and Xu* 2013). The hesitant fuzzy individual regret measure over the benefit-type criterion is based on the relationship:

$$\widetilde{R}_i = \widetilde{L}_{\infty,i} = \max_j \left(\omega_j \frac{d\left(h_j^*, h_{ij} \right)}{d\left(h_j^*, h_j^- \right)} \right) \tag{4.11}$$

where $\omega_j(j = 1, 2, \ldots, n)$ are the corresponding weights of criteria satisfying $0 \le \omega_j \le 1$, $j = 1, 2, \ldots, n$, $\sum_{j=1}^{n} \omega_j = 1, d\left(h_j^*, h_{ij}\right)$ and $d\left(h_j^*, h_j^-\right)$ can be determined through Eq. (4.9).

Definition 4.4 (*Liao and Xu* 2013). The hesitant fuzzy compromise measure is based on the relationship:

$$\tilde{Q}_i = \upsilon \frac{\tilde{S}_i - \tilde{S}^*}{\tilde{S}^- - \tilde{S}^*} + (1 - \upsilon) \frac{\tilde{R}_i - \tilde{R}^*}{\tilde{R}^- - \tilde{R}^*} \tag{4.12}$$

where $\tilde{S}^* = \min_i \tilde{S}_i$, $\tilde{S}^- = \max_i \tilde{S}_i$, $\tilde{R}^* = \min_i \tilde{R}_i$, $\tilde{R}^- = \max_i \tilde{R}_i$, and υ is the weight of the strategy of the majority of criteria or the maximum overall utility. The larger the value of υ, the preferences of the decision maker over different criteria will be more average. Without loss of generality, it also takes the value 0.5.

From Eq. (4.12), we can see that the hesitant fuzzy compromise measure integrates two parts: the former is the distance in terms of group utility; the latter is the distance in terms of individual regret. The smaller the value of hesitant fuzzy compromise measure, the better the alternative will be. So we need to pick out the smallest one among $\tilde{Q}_i(i = 1, 2, \ldots, m)$.

For the convenience of application, the procedure of the HF-VIKOR method can be described as follows:

Algorithm 4.1

Step 1. Construct the hesitant fuzzy decision matrix. The decision makers determine the relevant criteria of the potential alternatives and give the evaluation information in the form of HFEs of the alternatives with respect to the criteria. When the decision makers are asked to compare the alternatives over criteria, different people may have different preferences. Meanwhile, anonymity is required in order to protect the decision makers' privacy or avoid influencing each other. Thus, it is natural to set out all the possible evaluations given by the decision makers for an alternative under certain criteria, which are represented as HFEs. They also determine the importance degrees $\omega_j(j = 1, 2, \ldots, n)$ for the relevant criteria according to his/her preferences.

Step 2. Calculate the score function and the variance function for the evaluation values of the considered alternatives over different criteria in the form of HFEs according to Eqs. (1.17) and (1.18) or Eq. (1.19). Then we choose the ideal values $h_j^* = \max_i h_{ij}$ and $h_j^- = \min_i h_{ij}$, which are the best and worst values of A_i over the benefit criterion C_j, respectively. The best and worst values of A_i over the cost criterion are $h_j^* = \min_i h_{ij}$ and $h_j^- = \max_i h_{ij}$. Go to Step 3.

Step 3. Compute the values of $\tilde{S}_i$, $\tilde{R}_i$ and $\tilde{Q}_i (i = 1, 2, \ldots, m)$ for different alternatives according to Eqs. (4.10), (4.11) and (4.12), and then go to the next step. $\tilde{S}_i$ indicates the hesitant fuzzy group utility value as it fuses the normalized distances between the ideal and each actual evaluation value determined by the decision makers with respect to each criterion. $\tilde{R}_i$ represents the hesitant fuzzy individual regret value by using the weighted normalized Manhattan distances between the ideal and each actual evaluation value determined by the decision makers. The hesitant fuzzy compromise measure $\tilde{Q}_i$ is the summation of the normalized hesitant fuzzy group utility value and the normalized hesitant fuzzy individual regret value.

Step 4. Rank the alternatives according to the values of $\tilde{S}_i$, $\tilde{R}_i$ and $\tilde{Q}_i (i = 1, 2, \ldots, m)$. The desirable solution must satisfy the following two conditions:

C1: $\tilde{Q}(A^{(2)}) - \tilde{Q}(A^{(1)}) \geq \frac{1}{m-1}$, where $A^{(1)}$ and $A^{(2)}$ are the alternatives with the first and second positions in the ranking list, respectively;

C2: $A^{(1)}$ should also be the best ranked by $\tilde{S}_i$ and $\tilde{R}_i$.

The condition **C1** is called as an acceptable advantage, while the condition **C2** is called as an acceptable stability. If these two conditions are not satisfied simultaneously, then go to the next step.

Step 5. Derive the compromise solutions:
If the condition **C1** is not satisfied, then we explore the maximum value of M according to the equation:

$$\tilde{Q}\left(A^{(M)}\right) - \tilde{Q}\left(A^{(1)}\right) < \frac{1}{m-1} \qquad (4.13)$$

All the alternatives $A^{(i)} (i = 1, 2, \ldots, M)$ are the compromise solutions.

If the condition **C2** is not satisfied, then the alternatives $A^{(1)}$ and $A^{(2)}$ are the compromise solutions.

4.1.3 Application of the Hesitant Fuzzy VIKOR Method in Airline Service Quality Evaluation

We now consider a decision making problem that concerns the evaluation of the service quality among domestic airlines to illustrate our method:

Example 4.2 (Liao and Xu 2013). Due to the development of high-speed railroad, the domestic airline market has faced a stronger challenger in Taiwan. Especially after 2008 the global economic downturn, more and more airlines have attempted to attract customers by reducing price. Unfortunately, they soon found that it was a no-win situation and only service quality is the critical and fundamental element to survive in this highly competitive domestic market. In order to improve the service quality of domestic airline, the civil aviation administration of Taiwan (CAAT) wants to know which airline is the best in Taiwan and then calls for the others to learn from it. So the CAAT constructs a committee to investigate the four major domestic airlines, which are UNI Air, Transasia, Mandarin, and Daily Air and four major criteria are given based on the research of Liou et al. (2011) to evaluate these four domestic airlines. These four main criteria are:

C_1: Booking and ticketing service, which involves convenience of booking or buying ticket, promptness of booking or buying ticket, courtesy of booking or buying ticket.

C_2: Check-in and boarding process, which consists of convenience check-in, efficient check-in, courtesy of employee, clarity of announcement and so on.

C_3: Cabin service, which can be divided into cabin safety demonstration, variety of newspapers and magazines, courtesy of flight attendants, flight attendant willing to help, clean and comfortable interior, in-flight facilities, and captain's announcement.

C_4: Responsiveness, which consists of fair waiting-list call, handing of delayed flight, complaint handing, and missing baggage handling.

After the survey about passengers' importance and perception for service criteria done by Liou et al. (2011), they found that cabin service is considered the most important factor of service quality, which can be interpreted easily because cabin service occupies more of a passenger's travelling time than other aspects. Meanwhile, booking and ticketing service is less important due to the fact that these works are mainly done by a computer. Therefore, the weight vector of the criteria is $\omega = (0.1, 0.2, 0.4, 0.3)^T$, which is consistent with the result of the survey done by Liou et al. (2011). Suppose that the committee gives the rating values by HFEs, and then the hesitant fuzzy decision matrix is presented in Table 4.1.

Table 4.1 Hesitant fuzzy decision matrix for Example 4.1

	C_1	C_2	C_3	C_4
UNI Air	$\{0.6, 0.7, 0.9\}$	$\{0.6, 0.8\}$	$\{0.3, 0.6, 0.9\}$	$\{0.4, 0.5, 0.9\}$
Transasia	$\{0.7, 0.8, 0.9\}$	$\{0.5, 0.8, 0.9\}$	$\{0.4, 0.8\}$	$\{0.5, 0.6, 0.7\}$
Mandarin	$\{0.5, 0.6, 0.8\}$	$\{0.6, 0.7, 0.9\}$	$\{0.3, 0.5, 0.7\}$	$\{0.5, 0.7\}$
Daily Air	$\{0.6, 0.9\}$	$\{0.7, 0.9\}$	$\{0.2, 0.4, 0.7\}$	$\{0.4, 0.5\}$

Table 4.2 Score values obtained by the score function

	C_1	C_2	C_3	C_4
UNI air	0.7333	0.7000	0.6000	0.6000
Transasia	0.8000	0.7333	0.6000	0.6000
Mandarin	0.6333	0.7333	0.5000	0.6000
Daily air	0.7500	0.8000	0.4333	0.4500

Table 4.3 Variance values obtained by the variance function

	C_1	C_2	C_3	C_4
UNI air	–	–	0.7348	0.6481
Transasia	–	–	0.4000	0.2449
Mandarin	–	–	–	0.2000
Daily air	–	–	–	–

Table 4.4 Weighted distance ratios

	C_1	C_2	C_3	C_4
UNI air	0.004	0.2	0.3	0.2
Transasia	0	0.0667	0	0.0667
Mandarin	0.1	0.3	0.2	0
Daily air	0.078	0	0.4	0.3

In the following, we use the HF-VIKOR method to solve this problem.

Step 1. Construct the hesitant fuzzy decision matrix, see Table 4.1.

Step 2. The values of the score function and the variance function for the preference values of the considered alternatives over different criteria in the form of HFEs can be calculated easily according to Eqs. (1.17) and (1.18), which are set out in Tables 4.2 and 4.3, respectively. Since the purpose of computing variance values is to find the best and worst values of domestic airlines over the criterion C_j, to simplify the computation, we only calculate the variance values of those HFEs which may be the best or worst value of C_j and have equal score value in each column.

Since all the criteria are benefit, the best and worst values of domestic airlines over the criterion C_j can be found easily by the operators $h_j^* = \max_i h_{ij}$ and $h_j^- = \min_i h_{ij}$. Now we take the first column as an example:

$$s(h_{11}) = \frac{0.6+0.7+0.9}{3} = 0.7333, \quad s(h_{21}) = \frac{0.7+0.8+0.9}{3} = 0.8000$$

$$s(h_{31}) = \frac{0.5+0.6+0.8}{3} = 0.6333, \quad s(h_{41}) = \frac{0.6+0.9}{2} = 0.7500$$

Since $s(h_{21}) > s(h_{41}) > s(h_{11}) > s(h_{31})$, according to the scheme in Sect. 1.1.3, there is no need to calculate the variance values and we can derive that $h_{21} \succ h_{41} \succ h_{11} \succ h_{31}$. Since C_1 is a benefit-type criterion, then we obtain $h_1^* = h_{21} = \{0.7, 0.8, 0.9\}$ and $h_1^- = h_{31} = \{0.5, 0.6, 0.8\}$.

If we take the last column as an example, then

$$s(h_{14}) = \frac{0.4 + 0.5 + 0.9}{3} = 0.6000, \quad s(h_{24}) = \frac{0.5 + 0.6 + 0.7}{3} = 0.6000$$

$$s(h_{34}) = \frac{0.5 + 00.7}{2} = 0.6000, \quad s(h_{44}) = \frac{0.4 + 0.5}{2} = 0.4500$$

Since $s(h_{14}) = s(h_{24}) = s(h_{34}) > s(h_{44})$, then according to the scheme in Sect. 1.1.3, we need to calculate the variance values of h_{14}, h_{24} and h_{34}:

$$v(h_{14}) = \frac{\sqrt{0.1^2 + 0.4^2 + 0.5^2}}{3} = 0.6481$$

$$v(h_{24}) = \frac{\sqrt{0.1^2 + 0.1^2 + 0.2^2}}{3} = 0.2449$$

$$v(h_{34}) = \frac{\sqrt{0.2^2}}{2} = 0.2000$$

Because $v(h_{14}) > v(h_{24}) > v(h_{34})$, then $h_{34} \succ h_{24} \succ h_{14} \succ h_{44}$. Since C_4 is a benefit-type criterion, then we obtain $h_4^* = h_{34} = \{0.5, 0.7\}$ and $h_4^- = h_{44} = \{0.4, 0.5\}$.

Step 3. Compute the values of $\tilde{S}_i$, $\tilde{R}_i$ and $\tilde{Q}_i(i = 1, 2, 3, 4)$ for different alternatives according to Eqs. (4.10), (4.11) and (4.12), and the weighted distance ratios with the equation:

$$r_{ij} = \omega_j \frac{d\left(h_j^*, h_{ij}\right)}{d\left(h_j^*, h_j^-\right)} \tag{4.14}$$

are shown in Table 4.4. Below we take the fourth alternative as an example. According to Example 4.1, we have

$$d\left(h_1^*, h_1^-\right) = \frac{1}{3}(|0.7 - 0.5| + |0.8 - 0.6| + |0.9 - 0.8|) = 0.1667$$

Similarly, we can obtain

$$d\left(h_2^*, h_2^-\right) = 0.1, d\left(h_3^*, h_3^-\right) = 0.1333, d\left(h_4^*, h_4^-\right) = 0.15, d\left(h_1^*, h_{41}\right) = 0.1333,$$
$$d\left(h_2^*, h_{42}\right) = 0, d\left(h_3^*, h_{43}\right) = 0.1333, d\left(h_4^*, h_{44}\right) = 0.15.$$

Hence,

$$
\begin{aligned}
\widetilde{S}_4 &= \sum_{j=1}^{4} \omega_j \frac{d\left(h_j^*, h_{4j}\right)}{d\left(h_j^*, h_j^-\right)} \\
&= 0.1 \times \frac{0.1333}{0.1667} + 0.1 \times \frac{0}{0.1} + 0.1 \times \frac{0.1333}{0.1333} + 0.1 \times \frac{0.15}{0.15} = 0.778
\end{aligned}
$$

$$
\widetilde{R}_4 = \max_j \left(\omega_j \frac{d\left(h_j^*, h_{4j}\right)}{d\left(h_j^*, h_j^-\right)} \right) = 0.4
$$

Similarly, we can get

$$
\widetilde{S}_1 = 0.704, \widetilde{S}_2 = 0.1334, \widetilde{S}_3 = 0.6, \widetilde{R}_1 = 0.3, \widetilde{R}_2 = 0.0667 \text{ and } \widetilde{R}_3 = 0.3
$$

So

$$
\widetilde{S}^* = \min_i \widetilde{S}_i = 0.1334, \quad \widetilde{S}^- = \max_i \widetilde{S}_i = 0.778
$$

$$
\widetilde{R}^* = \min_i \widetilde{R}_i = 0.0667, \quad \widetilde{R}^- = \max_i \widetilde{R}_i = 0.4
$$

Then,

$$
\begin{aligned}
\widetilde{Q}_4 &= \upsilon \frac{\widetilde{S}_4 - \widetilde{S}^*}{\widetilde{S}^- - \widetilde{S}^*} + (1 - \upsilon) \frac{\widetilde{R}_4 - \widetilde{R}^*}{\widetilde{R}^- - \widetilde{R}^*} \\
&= 0.5 \times \frac{0.778 - 0.1334}{0.778 - 0.1334} + 0.5 \times \frac{0.4 - 0.0667}{0.4 - 0.0667} = 0.5
\end{aligned}
$$

Similarly,

$$
\widetilde{Q}_1 = 0.7926, \widetilde{Q}_2 = 0, \widetilde{Q}_3 = 0.7119
$$

Therefore, $\widetilde{Q}_2 < \widetilde{Q}_4 < \widetilde{Q}_3 < \widetilde{Q}_1$, $\widetilde{S}_2 < \widetilde{S}_3 < \widetilde{S}_1 < \widetilde{S}_4$, and $\widetilde{R}_2 < \widetilde{R}_3 = \widetilde{R}_1 < \widetilde{R}_4$. In order to be clear at a glance, the overall computation process can be displayed in Table 4.5.

Step 4. Rank the alternatives according to the values of $\widetilde{S}_i$, $\widetilde{R}_i$ and $\widetilde{Q}_i (i = 1, 2, 3, 4)$.

Step 5. Derive the compromise solutions. Due to the fact that $\widetilde{Q}_2 < \widetilde{Q}_4 < \widetilde{Q}_3 < \widetilde{Q}_1$, $\widetilde{S}_2 < \widetilde{S}_3 < \widetilde{S}_1 < \widetilde{S}_4$, and $\widetilde{R}_2 < \widetilde{R}_3 = \widetilde{R}_1 < \widetilde{R}_4$, meanwhile, $\widetilde{Q}_2 - \widetilde{Q}_4 =$

Table 4.5 The overall computation process

	$\omega_1 = 0.1$	$\omega_2 = 0.2$	$\omega_3 = 0.4$	$\omega_4 = 0.3$	$\tilde{S}_i$	$\tilde{R}_i$	$\tilde{Q}_i$
	C_1	C_2	C_3	C_4			
UNI Air	{0.6, 0.7, 0.9}	{0.6, 0.8}	{0.3, 0.6, 0.9}	{0.4, 0.5, 0.9}	0.704	0.3	0.7926
Transasia	{0.7, 0.8, 0.9}	{0.5, 0.8, 0.9}	{0.4, 0.8}	{0.5, 0.6, 0.7}	0.1334	0.0667	0
Mandarin	{0.5, 0.6, 0.8}	{0.6, 0.7, 0.9}	{0.3, 0.5, 0.7}	{0.5, 0.7}	0.6	0.3	0.7119
Daily Air	{0.6, 0.9}	{0.7, 0.9}	{0.2, 0.4, 0.7}	{0.4, 0.5}	0.778	0.4	0.5
h_j^*	{0.7, 0.8, 0.9}	{0.7, 0.9}	{0.4, 0.8}	{0.5, 0.7}			
h_j^-	{0.5, 0.6, 0.8}	{0.6, 0.8}	{0.2, 0.4, 0.7}	{0.4, 0.5}			
$d_{hnh}\left(h_j^*, h_j^-\right)$	0.1667	0.1	0.1333	0.15			

$0.5 > \frac{1}{4-1} = 0.3333$, we can see that it satisfies the two conditions given in Algorithm 4.1. That is to say, Transasia is the best alternative, which means its service quality is with the highest satisfaction degree.

4.2 Hesitant Fuzzy ELECTRE II Methods for Multiple Criteria Decision Making

4.2.1 Main Characteristics of the ELECTRE Methods

As a major category of multiple criteria decision making, outranking can be used to select which alternative is preferable, incomparable or indifferent by pairwise comparisons between alternatives under each criterion. The advantages of the outranking method lies in that it is able to take into account purely ordinal scales, and in that indifference and preference thresholds can be considered when modeling the imperfect knowledge of data. Among the outranking methods, the ELECTRE method is the most popular one, whose main characteristic is the utilization of outranking relations (Wang and Triantaphyllou 2008). Since the first version of the ELECTRTE method, referred as the ELECTRE I (Roy 1968), was introduced, the ELECTRE approach has been generalized into a number of variants, including the ELECTRE II, III and IV as well as the ELECTRE-A and the ELECTRE TRI methods, which constitute a family of ELECTRE methods (Figueira et al. 2005). The ELECTRE method has been further developed to treat groups with imprecise information on parameter values (Dias and Clímaco 2005), to solve inconsistencies among constraints on the parameters (Mousseau and Dias 2004), to assist a group of decision makers with different value systems (Leyva and Fernandez 2003), and to incorporate the ideas of concordance and discordance for group ranking problems (Fernandez and Olmedo 2005), etc. In addition, the ELECTRE method has been applied to project selection (Buchanan and Vanderpooten 2007), transportation (Roy et al. 1986) and environment management (Salminen et al. 1998).

Intensive efforts have been made to deal with various types of fuzzy multiple criteria decision making problems with different kinds of ELECTRE methods. For example, Hatami-Marbini and Tavana (2011) proposed the extended ELECTRE I method to take account of the uncertain linguistic assessments. Hatami-Marbini et al. (2013) further applied an integrated fuzzy group ELECTRE method to safety and health assessments in hazardous waste recycling facilities. Wu and Chen (2011) adopted a similar approach to solve the multiple criteria decision making problems under intuitionistic fuzzy environments. Vahdani et al. (2010, 2013) performed an extension of the ELECTRE I for multiple criteria group decision making problems with IFSs and interval-valued fuzzy sets.

The ELECTRE I method is suitable to construct a partial prioritization and to choose a set of promising alternatives. Different from the ELECTRE I, the ELECTRE II method considers several concordance and discordance levels, which can be used to construct two embedded outranking relations (i.e., strong and weak outranking relations). With these relations, the strong and weak graphs can be depicted and the ranking of alternatives is finally derived.

The prominent features of the ELECTRE method include the four binary relations, the preference modeling through outranking relations, and the concepts of concordance and discordance.

Definition 4.5 (*Figueira et al.* 2010). For a pair $(a, b) \in A \times A$, to compare the two actions a and b, four binary relations are defined on $A \times A$:

- P denotes the strict preference relation, and aPb means that "a is strictly preferred to b";
- I denotes the indifference relation, and aIb means that "a is indifferent to b";
- Q denotes the weak preference relation, and aQb means that "a is weakly preferred to b, which expresses hesitation between the indifference (I) and the preference (P);
- R denotes the incomparability relation, and aRb means that "a is not comparable to b".

Definition 4.6 (*Figueira et al.* 2010). Modeling preference in the ELECTRE method is via the comprehensive binary outranking relation S, whose meaning is "at least as good as". In general, $S = P \cup Q \cup I$. Considering two actions $(a, b) \in A \times A$, four cases appear:

 (i) aSb and not bSa, i.e., aPb (a is strictly preferred to b).
 (ii) bSa and not aSb, i.e., bPa (b is strictly preferred to a).
 (iii) aSb and bSa, i.e., aIb (a is indifferent to b).
 (iv) Not aSb and not bSa, i.e., aRb (a is incomparable to b).

Definition 4.7 (*Figueira et al.* 2010). All outranking based methods rely on the concepts of concordance and discordance which represent the reasons for and against an outranking situation:

 (i) Concordance: To validate an outranking aSb, a sufficient majority of criteria in favor of this assertion must occur;
 (ii) Non-discordance: The assertion aSb cannot be validated if a minority of criteria is strongly against this assertion.

4.2.2   The HF-ELECTRE I Method

In this section, we extend the ELECTRE I method to solve the multiple criteria decision making problems under hesitant fuzzy environments.

Now we consider a multiple criteria decision making problem with a discrete set of alternatives $A = \{A_1, A_2, \ldots, A_m\}$ whose assessment information on the criteria set $C = \{C_1, C_2, \ldots, C_n\}$ is represented by $h_{A_i}(C_j) = \{\gamma | \gamma \in h_{A_i}(C_j),\ 0 \le \gamma \le 1\}$, $i = 1, 2, \ldots, m; j = 1, 2, \ldots, n$. $h_{A_i}(C_j)$ represents the possible membership degree of the ith alternative A_i satisfying the jth criterion C_j and can be expressed as a HFE h_{ij}. Given that each criterion has different importance, the weight vector of all criteria is defined as $\omega = (\omega_1, \omega_2, \ldots, \omega_n)^T$, where $0 \le \omega_j \le 1$ and $\sum_{j=1}^{n} \omega_j = 1$ with ω_j denoting the importance degree of the criterion C_j.

The ELECTRE methods are composed of the construction and exploitation of one or several outranking relation(s). The construction is based on the comparison between each pair of actions on the criteria, through which the concordance and discordance indices are obtained and they are further used to analyze the outranking relations among different alternatives. In traditional ELECTRE methods, each criterion over different alternatives can be divided into two different subsets: concordance set and discordance set. The former is composed of all criteria for which A_k is preferred to A_l, and the latter is the complementary subset. Under hesitant fuzzy environments, according to the concepts of score function and deviation function, we can compare different alternatives on the criteria and classify different types of hesitant fuzzy concordance (discordance) sets as the hesitant fuzzy concordance (discordance) set and the weak hesitant fuzzy concordance (discordance) set. A better alternative has the higher score or the lower deviation degree in case the alternatives have the same score.

For a pair of the alternatives A_k and $A_l (k, l = 1, 2, \ldots, m$ and $k \ne l)$, the hesitant fuzzy concordance set $J_{C_{kl}}$ is the sum of all those criteria where the performance of A_k is superior to A_l. It can be formulated as:

$$J_{C_{kl}} = \left\{ j \big| s(h_{kj}) \ge s(h_{lj}) \text{ and } \sigma(h_{kj}) < \sigma(h_{lj}) \right\} \tag{4.15}$$

where $J = \{j | j = 1, 2, \ldots, n\}$ represents a set of subscripts of all criteria. The weak hesitant fuzzy concordance set $J_{C'_{kl}}$ is defined as:

$$J_{C'_{kl}} = \left\{ j \big| s(h_{kj}) \ge s(h_{lj}) \text{ and } \sigma(h_{kj}) \ge \sigma(h_{lj}) \right\} \tag{4.16}$$

The main difference between $J_{C_{kl}}$ and $J_{C'_{kl}}$ lies in the deviation function. The lower deviation values reflect that the opinions of decision makers have a higher consistency degree. So $J_{C_{kl}}$ is more concordant than $J_{C'_{kl}}$.

Similarly, the hesitant fuzzy discordance set $J_{D_{kl}}$, which is composed of all criteria for which A_k is inferior to A_l, can be formulated as:

$$J_{D_{kl}} = \left\{ j \big| s(h_{kj}) < s(h_{lj}) \text{ and } \sigma(h_{kj}) \ge \sigma(h_{lj}) \right\} \tag{4.17}$$

If $s(h_{kj}) < s(h_{lj})$ and the deviation value $\sigma(h_{kj}) < \sigma(h_{lj})$, then we define this circumstance as the weak hesitant fuzzy discordance set $J_{D'_{kl}}$, which is expressed as:

$$J_{D'_{kl}} = \{j | s(h_{kj}) < s(h_{lj}) \text{ and } \sigma(h_{kj}) < \sigma(h_{lj})\} \tag{4.18}$$

Obviously, $J_{D_{kl}}$ is more discordant than $J_{D'_{kl}}$.

Below we construct the corresponding matrices for different types of the hesitant fuzzy concordance sets and the hesitant fuzzy discordance sets.

The hesitant fuzzy concordance index is the ratio of the sum of the weights related to criteria in the hesitant fuzzy concordance sets to that of all criteria. The concordance index c_{kl} of A_k and A_l in the HF-ELECTRE I method is defined as:

$$c_{kl} = \frac{\lambda_C \times \sum\limits_{j \in J_{C_{kl}}} \omega_j + \lambda_{C'} \times \sum\limits_{j \in J_{C'_{kl}}} \omega_j}{\sum\limits_{j=1}^{n} \omega_j} = \lambda_C \times \sum\limits_{j \in J_{C_{kl}}} \omega_j + \lambda_{C'} \times \sum\limits_{j \in J_{C'_{kl}}} \omega_j \tag{4.19}$$

where λ_C and $\lambda_{C'}$ are respectively the weights of the hesitant fuzzy concordance sets and the weak hesitant fuzzy concordance sets, which are depended on the attitudes of the decision makers. The index c_{kl} reflects the relative importance of A_k with respect to A_l. Obviously, $0 \le c_{kl} \le 1$. The big value of c_{kl} indicates that the alternative A_k is superior to the alternative A_l. We can thus construct the asymmetrical hesitant fuzzy concordance matrix C:

$$C = \begin{bmatrix} - & \cdots & c_{1l} & \cdots & c_{1(m-1)} & c_{1m} \\ \vdots & \ddots & \vdots & \ddots & \vdots & \vdots \\ c_{k1} & \cdots & c_{kl} & \cdots & c_{k(m-1)} & c_{km} \\ \vdots & \ddots & \vdots & \ddots & \vdots & \vdots \\ c_{m1} & \cdots & c_{ml} & \cdots & c_{m(m-1)} & - \end{bmatrix} \tag{4.20}$$

Different from the hesitant fuzzy concordance index, the hesitant fuzzy discordance index is reflective of relative difference of A_k with respect to A_l in terms of discordance criteria. The discordance index is defined as:

$$d_{kl} = \frac{\max\limits_{j \in J_{D_{kl}} \cup J_{D'_{kl}}} \{\lambda_D \times d(\omega_j h_{kj}, \omega_j h_{lj}), \lambda_{D'} \times d(\omega_j h_{kj}, \omega_j h_{lj})\}}{\max\limits_{j \in J} d(\omega_j h_{kj}, \omega_j h_{lj})} \tag{4.21}$$

where λ_D and $\lambda_{D'}$ denote the weights of the hesitant fuzzy discordance set and the weak discordance set, respectively, which are depended on the decision makers' attitudes, and $d(\omega_j h_{kj}, \omega_j h_{lj})$ is the distance measure between the HFEs $\omega_j h_{kj}$ and $\omega_j h_{lj}$, defined as Eq. (4.7).

The hesitant fuzzy discordance matrix is established by the hesitant fuzzy discordance index for all pairwise comparisons of alternatives:

$$D = \begin{bmatrix} - & \cdots & d_{1l} & \cdots & d_{1(m-1)} & d_{1m} \\ \vdots & \ddots & \vdots & \ddots & \vdots & \vdots \\ d_{k1} & \cdots & d_{kl} & \cdots & d_{k(m-1)} & d_{km} \\ \vdots & \ddots & \vdots & \ddots & \vdots & \vdots \\ d_{m1} & \cdots & d_{ml} & \cdots & d_{m(m-1)} & - \end{bmatrix} \qquad (4.22)$$

As seen in Eqs. (4.18) and (4.20), the elements of C differ substantially from those of D, making the two matrices have complementary relationship; that is, the matrix C represents the weights resulted from hesitant fuzzy concordance indices, whereas the asymmetrical matrix D reflects the relative difference of $\omega_j h_{ij}$ for all hesitant fuzzy discordance indices. Note that the discordance matrix reflects limited compensation between alternatives. When the difference of two alternatives on a criterion arrives at a certain extent, compensation of the loss on a given criterion by a gain on another one may not be acceptable for the decision makers (Figueira et al. 2005). Due to this reason, the discordance matrix is established differently from the establishment of the concordance matrix.

The hesitant fuzzy concordance dominance matrix can be calculated according to the cut-level of hesitant fuzzy concordance indices. If the hesitant fuzzy concordance index c_{kl} of A_k relative to A_l is over a minimum level, then the superiority degree of A_k to A_l increases. The hesitant fuzzy concordance level can be defined as the average of all hesitant fuzzy concordance indices, which is in mathematical terms of

$$\bar{c} = \sum_{k=1}^{m} \sum_{l=1, l \neq k}^{m} \frac{c_{kl}}{m(m-1)} \qquad (4.23)$$

Based on the concordance level, the concordance dominance matrix F (i.e., a Boolean matrix) can be expressed as:

$$F = \begin{bmatrix} - & \cdots & f_{1l} & \cdots & f_{1(m-1)} & f_{1m} \\ \vdots & \ddots & \vdots & \ddots & \vdots & \vdots \\ f_{k1} & \cdots & f_{kl} & \cdots & f_{k(m-1)} & f_{km} \\ \vdots & \ddots & \vdots & \ddots & \vdots & \vdots \\ f_{m1} & \cdots & f_{ml} & \cdots & f_{m(m-1)} & - \end{bmatrix} \qquad (4.24)$$

whose elements satisfy

$$\begin{cases} f_{kl} = 1, & \text{if } c_{kl} \geq \bar{c} \\ f_{kl} = 0, & \text{if } c_{kl} < \bar{c} \end{cases} \qquad (4.25)$$

The element 1 in the matrix F indicates that the alternative is preferable to the other one.

Likewise, the elements of the hesitant fuzzy discordance matrix are also measured by the discordance level $\bar{d}$, which can be defined as the average of the elements in the hesitant fuzzy discordance matrix:

$$\bar{d} = \sum_{k=1}^{m} \sum_{l=1,l\neq k}^{m} \frac{d_{kl}}{m(m-1)} \tag{4.26}$$

Then, based on the discordance level, the discordance dominance matrix G can be constructed as:

$$G = \begin{bmatrix} - & \cdots & g_{1l} & \cdots & g_{1(m-1)} & g_{1m} \\ \vdots & \ddots & \vdots & \ddots & \vdots & \vdots \\ g_{k1} & \cdots & g_{kl} & \cdots & g_{k(m-1)} & g_{km} \\ \vdots & \ddots & \vdots & \ddots & \vdots & \vdots \\ g_{m1} & \cdots & g_{ml} & \cdots & g_{m(m-1)} & - \end{bmatrix} \tag{4.27}$$

where

$$\begin{cases} g_{kl} = 1, & \text{if } d_{kl} \leq \bar{d} \\ g_{kl} = 0, & \text{if } d_{kl} > \bar{d} \end{cases} \tag{4.28}$$

The elements of the matrix G measure the degree of the discordance. Hence, the discordant statement would be no longer valid if the element value $d_{kl} \leq \bar{d}$. That is to say, the elements of the matrix G, whose values are 1, show the dominant relations among the alternatives.

The aggregated dominance matrix E is constructed from the elements of the matrix F and the matrix G through the following formula:

$$E = F \otimes G \tag{4.29}$$

where each element e_{kl} in E is derived by

$$e_{kl} = f_{kl} \cdot g_{kl} \tag{4.30}$$

Finally, we exploit the outranking relations aiming at elaborating recommendations from the results obtained in previous construction of the outranking relations. By means of binary relations presented in Definition 4.5, we can construct a graph, from which the preferable alternative is selected. Specific details are depicted in Fig. 4.1. If $e_{kl} = 1$, then A_k is strictly preferred to A_l or A_k is weakly preferred to A_l; If $e_{kl} = 1$ and $e_{lk} = 1$, then A_k is indifferent to A_l; If $e_{kl} = 0$ and $e_{lk} = 0$, then A_k is incomparable to A_l.

We summarize the HF-ELECTRE I method in the following steps.

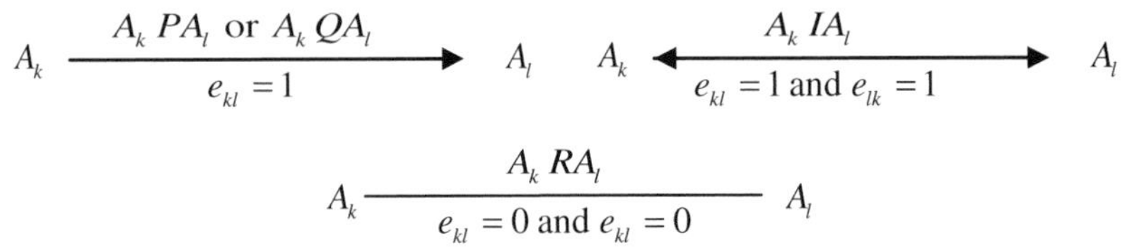

$$A_k \xrightarrow[\substack{e_{kl} = 1}]{A_k \, PA_l \text{ or } A_k \, QA_l} A_l \qquad A_k \xleftrightarrow[\substack{e_{kl} = 1 \text{ and } e_{lk} = 1}]{A_k \, IA_l} A_l$$

$$A_k \xleftrightarrow[\substack{e_{kl} = 0 \text{ and } e_{kl} = 0}]{A_k \, RA_l} A_l$$

Fig. 4.1 A graphical representation of the binary relations

Algorithm 4.2

Step 1. A group of decision makers determine the criteria of the alternatives and give the evaluation information of the alternatives with respect to the criteria in HFEs and thus construct the hesitant fuzzy decision matrix. They also determine the importance vector $\omega = (\omega_1, \omega_2, \ldots, \omega_n)^T$ for the criteria, and the relative weight vector $\lambda = (\lambda_C, \lambda_{C'}, \lambda_D, \lambda_{D'})^T$ of different types of hesitant fuzzy concordance sets and hesitant fuzzy discordance sets.

Step 2. Calculate the score function and the deviation function of each HFEs according to Eqs. (1.17)–(1.20).

Step 3. Construct the hesitant fuzzy concordance sets and the weak hesitant fuzzy concordance sets using Eqs. (4.15) and (4.16).

Step 4. Construct the hesitant fuzzy discordance sets and the weak hesitant fuzzy discordance sets using Eqs. (4.17) and (4.18).

Step 5. Calculate the hesitant fuzzy concordance indexes using Eq. (4.19) and obtain the hesitant fuzzy concordance matrix using Eq. (4.20).

Step 6. Calculate the hesitant fuzzy discordance indexes using Eq. (4.21) and obtain the hesitant fuzzy discordance matrix using Eq. (4.22).

Step 7. Identify the concordance dominance matrix using Eqs. (4.23)–(4.25).

Step 8. Identify the discordance dominance matrix using Eqs. (4.26)–(4.28).

Step 9. Construct the aggregation dominance matrix using Eqs. (4.29) and (4.30).

Step 10. Draw a decision graph and choose the preferable alternative.

Step 11. End.

4.2.3   The HF-ELECTRE II Method

In this section, we introduce the HF-ELECTRE II method to solve the multiple criteria decision making problem under hesitant fuzzy environment.

For each pair of the alternatives A_k and $A_l (k, l = 1, 2, \ldots, m$ and $k \neq l)$, the hesitant fuzzy concordance sets of A_k and A_l are the sum of all those criteria where the performance of A_k is superior to A_l. Here we classify them into three types:

- The hesitant fuzzy strong concordance set $J_{C_{kl}}$:

$$J_{C_{kl}} = \{j | s(h_{kj}) > s(h_{lj}) \text{ and } \sigma(h_{kj}) < \sigma(h_{lj})\} \tag{4.31}$$

- The hesitant fuzzy medium concordance set $J_{C'_{kl}}$:

$$J_{C'_{kl}} = \{j | s(h_{kj}) > s(h_{lj}) \text{ and } \sigma(h_{kj}) \geq \sigma(h_{lj})\} \tag{4.32}$$

- The hesitant fuzzy weak concordance set $J_{C''_{kl}}$:

$$J_{C''_{kl}} = \{j | s(h_{kj}) = s(h_{lj}) \text{ and } \sigma(h_{kj}) < \sigma(h_{lj})\} \tag{4.33}$$

The three types of hesitant fuzzy concordance sets exhibit the different degrees that A_k is superior to A_l. It is the deviation function that reflects the main difference between $J_{C_{kl}}$ and $J_{C'_{kl}}$. Moreover, a lower deviation value shows that the opinions of the decision makers have a higher consistency degree. Thus, $J_{C_{kl}}$ is more concordant than $J_{C'_{kl}}$. Relative to the deviation function, the score function plays a greater role in determining the magnitudes of HFEs. Hence $J_{C'_{kl}}$ having a higher score value is more concordant than $J_{C''_{kl}}$.

In analogous, the hesitant fuzzy discordance sets can be categorized into three types:

- The hesitant fuzzy strong discordance set $J_{D_{kl}}$:

$$J_{D_{kl}} = \{j | s(h_{kj}) < s(h_{lj}) \text{ and } \sigma(h_{kj}) > \sigma(h_{lj})\} \tag{4.34}$$

- The hesitant fuzzy medium discordance set $J_{D'_{kl}}$:

$$J_{D'_{kl}} = \{j | s(h_{kj}) < s(h_{lj}) \text{ and } \sigma(h_{kj}) \leq \sigma(h_{lj})\} \tag{4.35}$$

- The hesitant fuzzy weak discordance set $J_{D''_{kl}}$:

$$J_{D''_{kl}} = \{j | s(h_{kj}) = s(h_{lj}) \text{ and } \sigma(h_{kj}) > \sigma(h_{lj})\} \tag{4.36}$$

Apart from the above mentioned hesitant fuzzy concordance (discordance) sets, if $s(h_{kj}) = s(h_{lj})$ and $\sigma(h_{kj}) = \sigma(h_{lj})$, then we define this case as a hesitant fuzzy indifferent set:

$$J_{kl}^{=} = \{j | s(h_{kj}) = s(h_{lj}) \text{ and } \sigma(h_{kj}) = \sigma(h_{lj})\} \tag{4.37}$$

The concordance index c_{kl} of A_k and A_l in the HF-ELECTRE II method are computed as:

$$c_{kl} = \frac{\lambda_C \times \sum\limits_{j \in J_{C_{kl}}} \omega_j + \lambda_{C'} \times \sum\limits_{j \in J_{C'_{kl}}} \omega_j + \lambda_{C''} \times \sum\limits_{j \in J_{C''_{kl}}} \omega_j + \lambda_{J^=} \times \sum\limits_{j \in J^=_{kl}} \omega_j}{\sum\limits_{j=1}^{n} \omega_j}$$

$$= \lambda_C \times \sum\limits_{j \in J_{C_{kl}}} \omega_j + \lambda_{C'} \times \sum\limits_{j \in J_{C'_{kl}}} \omega_j + \lambda_{C''} \times \sum\limits_{j \in J_{C''_{kl}}} \omega_j + \lambda_{J^=} \times \sum\limits_{j \in J^=_{kl}} \omega_j$$

$$(4.38)$$

Here λ_C, $\lambda_{C'}$, $\lambda_{C''}$ and $\lambda_{J^=}$ are the attitude weights of hesitant fuzzy strong, medium and weak concordance sets and the hesitant fuzzy indifferent sets, which are all depended on the attitudes of the decision makers. c_{kl} shows the relative importance of A_k with respect to A_l and $0 \leq c_{kl} \leq 1$. The hesitant fuzzy concordance matrix C can thus be constructed.

The hesitant fuzzy discordance index reflects the relative difference of A_k with respect to A_l in terms of discordance criteria and is defined as:

$$d_{kl} = \frac{\max_{j \in J_{D_{kl}} \cup J_{D'_{kl}} \cup J_{D''_{kl}}} \left\{ \lambda_D \times d(\omega_j h_{kj}, \omega_j h_{lj}), \lambda_{D'} \times d(\omega_j h_{kj}, \omega_j h_{lj}), \lambda_{D''} \times d(\omega_j h_{kj}, \omega_j h_{lj}) \right\}}{\max_{j \in J} d(\omega_j h_{kj}, \omega_j h_{lj})}$$

$$(4.39)$$

Here λ_D, $\lambda_{D'}$ and $\lambda_{D''}$ are respectively the weights of three types of hesitant fuzzy discordance sets, which depend on the decision makers' attitudes. $d(\omega_j h_{kj}, \omega_j h_{lj})$ is distance measure defined as Eq. (4.7). With the hesitant fuzzy discordance indices for all pairwise comparisons of alternatives, the hesitant fuzzy discordance matrix D can be formulated.

The ELECTRE II method takes into account the notion of two embedded outranking relations. After computing the concordance and discordance indices for each pair of alternatives, two types of outranking relations, a strong relationship S^F and a weak relationship S^f are constructed by comparing these indices with the concordance and discordance levels.

A strong relationship leads to a better discrimination between alternatives and thus yields a more refined and stricter ranking procedure than the weak relationship (Duckstein and Gershon 1983; Hokkanen et al. 1995). To define the two relationships, let c^*, c^0 and c^- be three decreasing levels of concordance, which are denoted by $0 < c^- < c^0 < c^* < 1$. Also, let d^0 and d^* represent two increasing levels of discordance satisfying $0 < d^0 < d^* < 1$. With these specifications, $A_k S^F A_l$ is defined if and only if one or both of the following sets of conditions hold:

$$(I) \quad \begin{cases} C(A_k, A_l) \geq c^* \\ D(A_k, A_l) \leq d^* \\ C(A_k, A_l) \geq C(A_l, A_k) \end{cases}$$

$$\text{(II)} \quad \begin{cases} C(A_k, A_l) \geq c^0 \\ D(A_k, A_l) \leq d^0 \\ C(A_k, A_l) \geq C(A_l, A_k) \end{cases} \tag{4.40}$$

The weak relationship $A_k S^f A_l$ is defined if and only if the following conditions hold:

$$\begin{cases} C(A_k, A_l) \geq c^- \\ D(A_k, A_l) \leq d^* \\ C(A_k, A_l) \geq C(A_l, A_k) \end{cases} \tag{4.41}$$

As a result of the two pairwise outranking relationships, the strong graphs and the weak graphs are respectively constructed for the strong relationship and for the weak relationship. These graphs will be used in an iterative procedure to obtain the desired ranking of the alternatives. Specifically, the ranking procedure consists of a forward ranking v', a reverse ranking v'' and an average ranking $v(A_i)\left[= \frac{v'+v''}{2}(A_i)\right]$. We rank the alternatives according to the values of $v(A_i)$. This process produces the final ranking. For more details, please refer to Duckstein and Gershon (1983) and Hokkanen et al. (1995).

We summarize the proposed HF-ELECTRE II method in Algorithm 4.3.

Algorithm 4.3

Step 1. Determine the matrix H, the weight vector $(\omega_1, \omega_2, \ldots, \omega_n)^T$ of criteria, and the relative attitude weight vector $(\lambda_C, \lambda_{C'}, \lambda_{C''}, \lambda_D, \lambda_{D'}, \lambda_{D''}, \lambda_{J=})^T$ of different types of hesitant fuzzy concordance, discordance and indifferent sets.

Step 2. Calculate $s(h)$ and $\sigma(h)$ of the evaluation value h of each alternative with respect to each criterion according to Eqs. (1.17)–(1.20).

Step 3. Construct the hesitant fuzzy strong, medium and weak concordance sets by Eqs. (4.31)–(4.33), respectively.

Step 4. Construct the hesitant fuzzy strong, medium and weak discordance sets by Eqs. (4.34)–(4.36), respectively.

Step 5. Construct the hesitant fuzzy indifferent set by means of Eq. (4.37).

Step 6. Calculate the hesitant fuzzy concordance index by Eq. (4.38) and obtain the concordance matrix.

Step 7. Calculate the weighted distance between any two alternatives with respect to each criterion by Eq. (4.7).

Step 8. Calculate the hesitant fuzzy discordance index by Eq. (4.39) based on the weighted distance and obtain the discordance matrix.

Step 9. Construct the outranking relations from the given concordance and discordance levels by Eqs. (4.40) and (4.41).

Step 10. Draw the strong and weak outranking graphs and obtain the final ranking of all alternatives.

Step 11. End.

4.2.4   Application of the HF-ELECTRE Methods in Multiple Criteria Decision Making

(1) Application of the HF-ELECTRE I Method

In the following, we use a numerical example to illustrate the details of the HF-ELECTRE I method:

Example 4.3 (Chen et al. 2015) Suppose the board with five directors of an enterprise is planning the development of four large projects (strategy initiatives) $A_i(i = 1, 2, 3, 4)$ in the following 5 years. It is necessary to compare these projects to select the most important one from them as well as order them from the point of view of their importance, taking into account four criteria suggested by the balanced scorecard methodology (Kaplan and Norton 1996): C_1, financial perspective; C_2, the customer satisfaction; C_3, internal business process perspective; and C_4: learning and growth perspective. It should be noted that all of them are of the maximization type. In order to avoid psychic contagion, the decision makers are required to provide their preferences in anonymity. Suppose that the weight vector of the attributes is $\omega = (0.2, 0.3, 0.15, 0.35)^T$, and the hesitant fuzzy decision matrix is presented as:

	C_1	C_2	C_3	C_4
A_1	$\{0.2, 0.4, 0.7\}$	$\{0.2, 0.6, 0.8\}$	$\{0.2, 0.3, 0.6, 0.7, 0.9\}$	$\{0.3, 0.4, 0.5, 0.7, 0.8\}$
A_2	$\{0.2, 0.4, 0.7, 0.9\}$	$\{0.1, 0.2, 0.4, 0.5\}$	$\{0.3, 0.4, 0.6, 0.9\}$	$\{0.5, 0.6, 0.8, 0.9\}$
A_3	$\{0.3, 0.5, 0.6, 0.7\}$	$\{0.2, 0.4, 0.5, 0.6\}$	$\{0.3, 0.5, 0.7, 0.8\}$	$\{0.2, 0.5, 0.6, 0.7\}$
A_4	$\{0.3, 0.5, 0.6\}$	$\{0.2, 0.4\}$	$\{0.5, 0.6, 0.7\}$	$\{0.8, 0.9\}$

Step 1. The hesitant fuzzy decision matrix and the weight vector of the attributes are given above. The decision makers also give the relative weights of the hesitant fuzzy concordance sets, the weak hesitant fuzzy concordance sets, the hesitant fuzzy discordance sets and the weak hesitant fuzzy discordance sets, respectively as $\lambda = (\lambda_C, \lambda_{C'}, \lambda_D, \lambda_{D'})^T = (1, 2/3, 1, 2/3)^T$

Step 2. Calculate the score and the deviation values of each evaluation information of alternatives on the criteria. The results are presented in Tables 4.6 and 4.7.

Table 4.6 Score values obtained by the score function

	C_1	C_2	C_3	C_4
A_1	0.4333	0.5333	0.54	0.54
A_2	0.55	0.3	0.55	0.7
A_3	0.525	0.425	0.575	0.5
A_4	0.4667	0.3	0.6	0.85

Table 4.7 Deviation values obtained by the deviation function

	C_1	C_2	C_3	C_4
A_1	0.2055	0.2494	0.2577	0.1855
A_2	0.2693	0.1581	0.2291	0.1581
A_3	0.1479	0.1479	0.1920	0.1871
A_4	0.1247	0.1	0.0816	0.05

Step 3. Construct the hesitant fuzzy concordance sets and the weak hesitant fuzzy concordance sets:

$$J_C = \begin{bmatrix} - & - & 4 & - \\ 3,4 & - & 4 & - \\ 1,3 & 2,3 & - & - \\ 1,3,4 & 2,3,4 & 3,4 & - \end{bmatrix}, \quad J_{C'} = \begin{bmatrix} - & 2 & 2 & 2 \\ 1 & - & 1 & 1,2 \\ - & - & - & 1,2 \\ - & - & - & - \end{bmatrix}$$

For example, since $s(h_{14}) > s(h_{34})$ and $\sigma(h_{14}) < \sigma(h_{34})$, then $J_{C_{13}} = \{4\}$. In addition, since $s(h_{12}) > s(h_{32})$ and $\sigma(h_{12}) > \sigma(h_{32})$, then $J_{C'_{13}} = \{2\}$.

Step 4. Construct the hesitant fuzzy discordance sets and the weak hesitant fuzzy discordance sets:

$$J_D = \begin{bmatrix} - & 3,4 & 1,3 & 1,3,4 \\ - & - & 2,3 & 3,4 \\ 4 & 4 & - & 3,4 \\ - & - & - & - \end{bmatrix}, \quad J_{D'} = \begin{bmatrix} - & 1 & - & - \\ 2 & - & - & - \\ 2 & 1 & - & - \\ 2 & 1 & 1,2 & - \end{bmatrix}$$

Step 5. Calculate the hesitant fuzzy concordance indices and the hesitant fuzzy concordance matrix:

$$C = (c_{kl})_{4\times4} = \begin{bmatrix} - & 0.2 & 0.55 & 0.2 \\ 0.6333 & - & 0.4833 & 0.3333 \\ 0.35 & 0.45 & - & 0.3333 \\ 0.7 & 0.8 & 0.5 & - \end{bmatrix}$$

Step 6. Calculate the hesitant fuzzy discordance indices and the hesitant fuzzy discordance matrix:

$$D = (d_{kl})_{4\times4} = \begin{bmatrix} - & 0.7778 & 0.3429 & 1 \\ 0.6667 & - & 0.5357 & 1 \\ 0.6667 & 1 & - & 1 \\ 0.3361 & 0.3265 & 0.1143 & - \end{bmatrix}$$

For example,

$$d_{31} = \frac{\max\limits_{j \in J_{D_{31}} \cup J_{D'_{31}}} \left\{ \lambda_D \times d(\omega_j h_{3j}, \omega_j h_{1j}), \lambda_{D'} \times d(\omega_j h_{3j}, \omega_j h_{1j}) \right\}}{\max\limits_{j \in J} d(\omega_j h_{3j}, \omega_j h_{1j})}$$

$$= \frac{\max\left\{ 1 \times 0.028, \frac{2}{3} \times 0.0525 \right\}}{0.0525} = \frac{0.035}{0.0525} = 0.6667$$

where

$$d(\omega_1 h_{31}, \omega_1 h_{11}) = \frac{1}{4} \times 0.2 \times (|0.2 - 0.3| + |0.4 - 0.5| + |0.7 - 0.6| + |0.7 - 0.7|) = 0.015$$

$$d(\omega_2 h_{32}, \omega_2 h_{12}) = 0.0525, d(\omega_3 h_{33}, \omega_3 h_{13}) = 0.018, d(\omega_4 h_{34}, \omega_4 h_{14}) = 0.028.$$

Step 7. Calculate the hesitant fuzzy concordance level and identify the concordance dominance matrix, respectively, which are

$$\bar{c} = \sum_{k=1}^{m} \sum_{l=1, l \neq k}^{m} \frac{c_{kl}}{m(m-1)} = 0.4611, \quad F = \begin{bmatrix} - & 0 & 1 & 0 \\ 1 & - & 1 & 0 \\ 0 & 0 & - & 0 \\ 1 & 1 & 1 & - \end{bmatrix}$$

Step 8. Calculate the hesitant fuzzy discordance level and identify the discordance dominance matrix, respectively, which are

$$\bar{d} = \sum_{k=1}^{m} \sum_{l=1, l \neq k}^{m} \frac{d_{kl}}{m(m-1)} = 0.6472, \quad G = \begin{bmatrix} - & 0 & 1 & 0 \\ 0 & - & 1 & 0 \\ 0 & 0 & - & 0 \\ 1 & 1 & 1 & - \end{bmatrix}$$

Step 9. Construct the aggregation dominance matrix:

$$E = F \otimes G = \begin{bmatrix} - & 0 & 1 & 0 \\ 0 & - & 1 & 0 \\ 0 & 0 & - & 0 \\ 1 & 1 & 1 & - \end{bmatrix}$$

Step 10. As it can be seen from the aggregation dominance matrix, A_1 is preferred to A_3, A_2 is preferred to A_3 and A_4 is preferred to A_1, A_2 and A_3. Hence, A_4 is the best alternative. The results are depicted in Fig. 4.2.

(1) Comparison with the aggregation operator based approach

Example 4.3 was also considered by Xia and Xu (2011a), who used the aggregation operators to fuse the hesitant fuzzy information and made the ranking of projects. We find that the average aggregation operators and the HF-ELECTRE I

Fig. 4.2 Decision graph of
Example 4.3

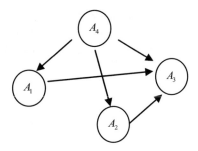

method give a consistent result, which illustrates the validity of the HF-ELECTRE I method. These two approaches are complementary when solving different types of the multiple criteria decision making problems. When the number of criteria in a multiple criteria decision making problem is not larger than 4, the aggregation operator based approach is a suitable tool because of the simple solving processes. However, when the number of criteria exceeds 4 for hesitant fuzzy information, which often appears in some actual multiple criteria decision making problems, the aggregation operator based approach might encounter a barrier in applications because of the need for tremendous computation. For such cases, the HF-ELECTRE I method is particularly useful, which is logically simple and demands less computational efforts. In what follows, we shall illustrate its application for the selection of investments where the number of criteria arrives at 6.

Example 4.4 (Chen et al. 2015). Assume that an enterprise wants to invest money in another country. A group composing of four decision makers in the enterprise considers five possible investments: A_1, invest in the Asian market; A_2, invest in the South American market; A_3, invest in the African market; A_4, invest in the three continents; and A_5, do not invest in any continent. When analyzing the investments, the decision makers considered the following general characteristics: C_1, risks of the investment; C_2, benefits in the short term; C_3, benefits in the midterm; C_4, benefits in the long term; C_5, difficulty of the investment; and C_6, other aspects. After a careful analysis of these characteristics, the decision makers gave the following information in the form of HFEs shown in Table 4.8.

Suppose that the weight vector of the criteria is $(0.25, 0.2, 0.15, 0.1, 0.2, 0.1)^T$, the decision makers also gave the relative weights of the hesitant fuzzy concordance

Table 4.8 Hesitant fuzzy decision matrix for Example 4.4

	A_1	A_2	A_3	A_4	A_5
C_1	{0.5, 0.6, 0.7}	{0.2, 0.3, 0.6, 0.7}	{0.3, 0.4, 0.6}	{0.2, 0.3, 0.6}	{0.4, 0.5, 0.6, 0.7}
C_2	{0.4, 0.6, 0.7, 0.9}	{0.5, 0.6, 0.8}	{0.7, 0.8}	{0.3, 0.4, 0.6, 0.7}	{0.7, 0.9}
C_3	{0.3, 0.6, 0.8, 0.9}	{0.2, 0.4, 0.6}	{0.3, 0.6, 0.7, 0.9}	{0.2, 0.4, 0.5, 0.6}	{0.5, 0.7, 0.8, 0.9}
C_4	{0.4, 0.5, 0.6, 0.8}	{0.3, 0.6, 0.7, 0.8}	{0.2, 0.3, 0.5, 0.6}	{0.1, 0.3, 0.4}	{0.3, 0.5, 0.6}
C_5	{0.2, 0.5, 0.6, 0.7}	{0.3, 0.6, 0.8}	{0.3, 0.4, 0.7}	{0.2, 0.3, 0.7}	{0.4, 0.5}
C_6	{0.2, 0.3, 0.4, 0.6}	{0.1, 0.4}	{0.4, 0.5, 0.7}	{0.2, 0.5, 0.6}	{0.2, 0.3, 0.7}

(weak hesitant fuzzy concordance) sets and the hesitant fuzzy discordance (weak hesitant fuzzy discordance) sets as $(\lambda_C, \lambda_{C'}, \lambda_D, \lambda_{D'})^T = (1, 3/4, 1, 3/4)^T$. Similar to the solving procedure used in Example 4.3, we obtain the aggregation dominance matrix:

$$
E = F \otimes G =
\begin{bmatrix}
- & 1 & 0 & 1 & 0 \\
0 & - & 0 & 1 & 0 \\
0 & 0 & - & 1 & 0 \\
0 & 0 & 0 & - & 0 \\
0 & 1 & 0 & 1 & -
\end{bmatrix}
$$

and a decision graph shown as Fig. 4.3.

From Fig. 4.3, three preference relations are obtained, which are $(1) A_1 \succ A_2 \succ A_4$; $(2) A_5 \succ A_2 \succ A_4$; $(3) A_3 \succ A_4$. In contrast, if adopting the hesitant fuzzy aggregation operators to aggregate the present hesitant information, the amount of data is extremely huge. For example, the number of computed data after aggregating A_1 reaches $3 \times 4^5 = 3072$. If we aggregate all $A_i(i = 1, 2, 3, 4, 5)$, then the corresponding number of computed data is 6677. The number will grow rapidly with the increasing of the number of alternatives and criteria. In the HF-ELECTR I method, the value for the number of calculation is 462. To determine the trends of the computation complexity for the two methods, we generate a great number of $n \times n$ hesitant fuzzy decision matrices $H = (h_{ij})_{n \times n}$ (as an example, here we take the number of values for each h_{ij} to be (4) by the Matlab Optimization Toolbox. The calculation times for the HF-ELECTRE I method and the hesitant fuzzy aggregation operator based method are $(5n^3 + 9n^2 - 10n + 4)/2$ and $n(4^n + 1)$, respectively. To be clearer, we choose the cases of $n = 4, 5, 10, 15, 20$ to demonstrate the trends of the computation complexity with increasing n for these two methods. The results are given in Table 4.9.

(2) Outranking relations for different number of alternatives and criteria

In Example 4.3, only four alternatives are considered. To check possible influence arisen from the number of alternatives, we compare the outranking relation for different number of alternatives (i.e., N = 4, 5, 6 and 7) under the same criteria. For this purpose, we perform a calculation with the Matlab Optimization Toolbox based on the HF-ELECTRE I method.

(i) N = 5

Fig. 4.3 Decision graph of Example 4.3

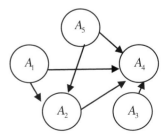

Table 4.9 Calculation times of the HF-ELECTRE I method and the aggregation operator method

Method	$n = 4$	$n = 5$	$n = 10$	$n = 15$	$n = 20$
The HF-ELECTRE I method	214	402	2902	9377	21,702
The aggregation operator method	1028	5125	1.05×10^7	1.61×10^{10}	2.19×10^{13}

Table 4.10 Score and deviation values of A_5

A_5	C_1	C_2	C_3	C_4
Score values	0.65	0.4333	0.5	0.7333
Deviation values	0.1803	0.1247	0.1871	0.1247

We use the same calculation procedure as that in Example 4.3.

Step 1. Give the hesitant fuzzy decision matrix for the case of N = 5. Note that the former four alternatives are the same as those in Example 4.3.

	C_1	C_2	C_3	C_4
A_1	$\{0.2, 0.4, 0.7\}$	$\{0.2, 0.6, 0.8\}$	$\{0.2, 0.3, 0.6, 0.7, 0.9\}$	$\{0.3, 0.4, 0.5, 0.7, 0.8\}$
A_2	$\{0.2, 0.4, 0.7, 0.9\}$	$\{0.1, 0.2, 0.4, 0.5\}$	$\{0.3, 0.4, 0.6, 0.9\}$	$\{0.5, 0.6, 0.8, 0.9\}$
A_3	$\{0.3, 0.5, 0.6, 0.7\}$	$\{0.2, 0.4, 0.5, 0.6\}$	$\{0.3, 0.5, 0.7, 0.8\}$	$\{0.2, 0.5, 0.6, 0.7\}$
A_4	$\{0.3, 0.5, 0.6\}$	$\{0.2, 0.4\}$	$\{0.5, 0.6, 0.7\}$	$\{0.8, 0.9\}$
A_5	$\{0.4, 0.6, 0.7, 0.9\}$	$\{0.3, 0.4, 0.6\}$	$\{0.3, 0.4, 0.5, 0.8\}$	$\{0.6, 0.7, 0.9\}$

Step 2. Calculate the score values and the deviation values of the evaluation information of the alternative A_5 on criteria. The results are listed in Table 4.10.

Step 3. Construct the hesitant fuzzy concordance sets and the weak hesitant fuzzy concordance sets:

$$J_C = \begin{bmatrix} - & - & 4 & - & - \\ 3,4 & - & 4 & - & - \\ 1,3 & 2,3 & - & - & - \\ 1,3,4 & 2,3,4 & 3,4 & - & 3,4 \\ 1,4 & 1,2,4 & 2,4 & - & - \end{bmatrix}, \quad J_{C'} = \begin{bmatrix} - & 2 & 2 & 2 & 2,3 \\ 1 & - & 1 & 1,2 & 3 \\ - & - & - & 1,2 & 3 \\ - & - & - & - & - \\ - & - & 1 & 1,2 & - \end{bmatrix}$$

Step 4. Construct the hesitant fuzzy discordance sets and the weak hesitant fuzzy discordance sets:

$$J_D = \begin{bmatrix} - & 3,4 & 1,3 & 1,3,4 & 1,4 \\ - & - & 2,3 & 3,4 & 1,2,4 \\ 4 & 4 & - & 3,4 & 2,4 \\ - & - & - & - & - \\ - & - & - & 3,4 & - \end{bmatrix}, \quad J_{D'} = \begin{bmatrix} - & 1 & - & - & - \\ 2 & - & - & - & - \\ 2 & 1 & - & - & 1 \\ 2 & 1 & 1,2 & - & 1,2 \\ 2,3 & 3 & 3 & - & - \end{bmatrix}$$

Step 5. Calculate the hesitant fuzzy concordance indices and the hesitant fuzzy concordance matrix:

$$C = (c_{kl})_{5 \times 5} = \begin{bmatrix} - & 0.2 & 0.55 & 0.2 & 0.3 \\ 0.6333 & - & 0.4833 & 0.3333 & 0.1 \\ 0.35 & 0.45 & - & 0.3333 & 0.1 \\ 0.7 & 0.8 & 0.5 & - & 0.5 \\ 0.55 & 0.85 & 0.7833 & 0.3333 & - \end{bmatrix}$$

Step 6. Calculate the hesitant fuzzy discordance indices and the hesitant fuzzy discordance matrix:

$$D = (d_{kl})_{5 \times 5} = \begin{bmatrix} - & 0.7778 & 0.3429 & 1 & 1 \\ 0.6667 & - & 0.5357 & 1 & 1 \\ 0.6667 & 1 & - & 1 & 1 \\ 0.3361 & 0.3265 & 0.1143 & - & 0.4285 \\ 0.3663 & 0.0952 & 0.0779 & 1 & - \end{bmatrix}$$

As it can be seen from Step 3 to Step 6, adding the alternative A_5 does not change the hesitant fuzzy concordance (discordance) sets, the weak hesitant fuzzy concordance (discordance) sets, the hesitant fuzzy concordance (discordance) indices, and the hesitant fuzzy concordance (discordance) matrix of the former four alternatives. Its role is to add the last column and the last line of the resulting corresponding matrices. Similar situations also appear when the number of the alternatives is N = 6 and N = 7, respectively.

Step 7. Calculate the hesitant fuzzy concordance level and identify the concordance dominance matrix, respectively:

$$\bar{c} = \sum_{k=1}^{m} \sum_{l=1, l \neq k}^{m} \frac{c_{kl}}{m(m-1)} = 0.4525, \quad F = \begin{bmatrix} - & 0 & 1 & 0 & 0 \\ 1 & - & 1 & 0 & 0 \\ 0 & 0 & - & 0 & 0 \\ 1 & 1 & 1 & - & 1 \\ 1 & 1 & 1 & 0 & - \end{bmatrix}$$

Step 8. Calculate the hesitant fuzzy discordance level and identify the discordance dominance matrix, respectively:

$$\bar{d} = \sum_{k=1}^{m} \sum_{l=1, l \neq k}^{m} \frac{d_{kl}}{m(m-1)} = 0.6367, \quad G = \begin{bmatrix} - & 0 & 1 & 0 & 0 \\ 0 & - & 1 & 0 & 0 \\ 0 & 0 & - & 0 & 0 \\ 1 & 1 & 1 & - & 1 \\ 1 & 1 & 1 & 0 & - \end{bmatrix}$$

Step 9. Construct the aggregation dominance matrix:

$$E = F \otimes G = \begin{bmatrix} - & 0 & 1 & 0 & 0 \\ 0 & - & 1 & 0 & 0 \\ 0 & 0 & - & 0 & 0 \\ 1 & 1 & 1 & - & 1 \\ 1 & 1 & 1 & 0 & - \end{bmatrix}$$

Step 10. As it can be seen from the aggregation dominance matrix, A_1 is preferred to A_3; A_2 is preferred to A_3; A_4 is preferred to A_1 A_2, A_3 and A_5; A_5 is preferred to A_1, A_2 and A_3. The results are depicted in Fig. 4.4.

(ii) N = 6

Results obtained in Steps 1 and 2 regarding to the alternative A_6 are listed in Tables 4.11 and 4.12, respectively.

When the number of alternatives arrives at 6, the results obtained from Step 3 to Step 6 can be expressed with the 6×6 matrix in which except the data of the 6th line and the 6th column, other data are the same as those in 5×5 matrix when the number of alternatives is 5. Thus, we only list in Table 4.13 the data of the 6th line and the 6th column of the 6×6 matrix, which are the hesitant fuzzy concordance set $J_{C_{kl}}$, the weak hesitant fuzzy concordance set $J_{C'_{kl}}$, the weak hesitant fuzzy discordance set $J_{D_{kl}}$, the hesitant fuzzy discordance set $J_{D'_{kl}}$, the hesitant fuzzy concordance indexes c_{kl}, and the hesitant fuzzy discordance indexes d_{kl}, respectively.

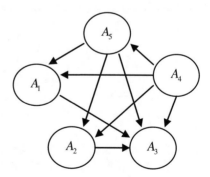

Fig. 4.4 Decision graph of five alternatives

Table 4.11 Hesitant fuzzy decision information for A_6

	C_1	C_2	C_3	C_4
A_6	{0.5, 0.7, 0.9}	{0.4, 0.5, 0.7, 0.9}	{0.5, 0.8}	{0.4, 0.5, 0.8}

Table 4.12 Score values and deviation values of A_6

A_6	C_1	C_2	C_3	C_4
Score values	0.7	0.625	0.65	0.5667
Deviation values	0.1633	0.1920	0.15	0.1700

Table 4.13 Data of $J_{C_{kl}}$, $J_{C'_{kl}}$, $J_{D_{kl}}$, $J_{D'_{kl}}$, c_{kl} and d_{kl} of the 6th line and the 6th column for N = 6

	$J_{C_{kl}}$	$J_{C'_{kl}}$	$J_{D_{kl}}$	$J_{D'_{kl}}$	c_{kl}	d_{kl}
$k = 1, l = 6$	–	–	1, 2, 3, 4	–	0	1
$k = 2, l = 6$	4	–	1, 3	2	0.35	0.6667
$k = 3, l = 6$	–	–	3, 4	1, 2	0	0.7292
$k = 4, l = 6$	4	–	–	1, 2, 3	0.35	0.5238
$k = 5, l = 6$	4	–	1, 3	2	0.35	0.5786
$k = 6, l = 1$	1, 2, 3, 4	–	–	–	1	0
$k = 6, l = 2$	1, 3	2	4	–	0.55	0.2692
$k = 6, l = 3$	3, 4	1, 2	–	–	0.8333	0
$k = 6, l = 4$	–	1, 2, 3	4	–	0.4333	1
$k = 6, l = 5$	1, 3	2	4	–	0.55	1

Steps 7 and 8. Calculate the hesitant fuzzy concordance level and discordance level:

$$\bar{c} = \sum_{k=1}^{m} \sum_{l=1,l\neq k}^{m} \frac{c_{kl}}{m(m-1)} = 0.4489, \quad \bar{d} = \sum_{k=1}^{m} \sum_{l=1,l\neq k}^{m} \frac{d_{kl}}{m(m-1)} = 0.6167$$

Step 9. Construct the aggregation dominance matrix:

$$E = \begin{bmatrix} - & 0 & 1 & 0 & 0 & 0 \\ 0 & - & 1 & 0 & 0 & 0 \\ 0 & 0 & - & 0 & 0 & 0 \\ 1 & 1 & 1 & - & 1 & 0 \\ 1 & 1 & 1 & 0 & - & 0 \\ 1 & 1 & 1 & 0 & 0 & - \end{bmatrix}$$

Step 10. As it can be seen from the aggregation dominance matrix, A_1 is preferred to A_3; A_2 is preferred to A_3; A_4 is preferred to A_1, A_2, A_3 and A_5; A_5 is preferred to A_1, A_2 and A_3; A_6 is preferred to A_1, A_2 and A_3 (see Fig. 4.5).

(iii) N = 7

Analogous to the case of N = 6, main results obtained from Step 1 to Step 6 are summarized in Tables 4.14, 4.15 and 4.16.

Fig. 4.5 Decision graph of six alternatives

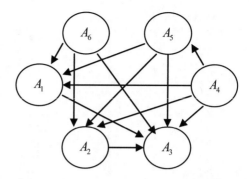

Steps 7 and 8. Calculate the hesitant fuzzy concordance level and the discordance level:

$$\bar{c} = \sum_{k=1}^{m} \sum_{l=1,l\neq k}^{m} \frac{c_{kl}}{m(m-1)} = 0.4258$$

$$\bar{d} = \sum_{k=1}^{m} \sum_{l=1,l\neq k}^{m} \frac{d_{kl}}{m(m-1)} = 0.5520$$

Step 9. Construct the aggregation dominance matrix:

$$E = \begin{bmatrix} - & 0 & 1 & 0 & 0 & 0 & 1 \\ 0 & - & 1 & 0 & 0 & 0 & 1 \\ 0 & 0 & - & 0 & 0 & 0 & 1 \\ 1 & 1 & 1 & - & 1 & 0 & 1 \\ 1 & 1 & 1 & 0 & - & 0 & 1 \\ 1 & 1 & 1 & 0 & 0 & - & 1 \\ 0 & 0 & 0 & 0 & 0 & 0 & - \end{bmatrix}$$

Step 10. As it can be seen from the aggregation dominance matrix, A_1 is preferred to A_3 and A_7; A_2 is preferred to A_3 and A_7; A_3 is preferred to A_7; A_4 is preferred to A_1, A_2, A_3, A_5 and A_7; A_5 is preferred to A_1, A_2, A_3 and A_7; A_6 is preferred to A_1, A_2, A_3 and A_7 (see Fig. 4.6).

Table 4.14 Hesitant fuzzy decision information for A_7

	C_1	C_2	C_3	C_4
A_7	{0.2, 0.3, 0.4}	{0.2, 0.3, 0.4, 0.7}	{0.3, 0.5, 0.6, 0.7}	{0.2, 0.3, 0.4, 0.5}

Table 4.15 Scores and deviations of A_7

A_7	C_1	C_2	C_3	C_4
Score values	0.3	0.4	0.525	0.35
Deviation values	0.0816	0.1871	0.1479	0.1118

Table 4.16 Data of $J_{C_{kl}}$, $J_{C'_{kl}}$, $J_{D_{kl}}$, $J_{D'_{kl}}$, c_{kl} and d_{kl} of the 7th line and the 7th column for N = 7

	$J_{C_{kl}}$	$J_{C'_{kl}}$	$J_{D_{kl}}$	$J_{D'_{kl}}$	c_{kl}	d_{kl}
$k = 1, l = 7$	–	1, 2, 3, 4	–	–	0.6667	0
$k = 2, l = 7$	–	1, 3, 4	–	2	0.4667	0.1633
$k = 3, l = 7$	2	1, 3, 4	–	–	0.7667	0
$k = 4, l = 7$	3,4	1	–	2	0.6333	0.1088
$k = 5, l = 7$	2	1, 4	3	–	0.6667	0.0756
$k = 6, l = 7$	–	1, 2, 3, 4	–	–	0.6667	0
$k = 7, l = 1$	–	–	–	1, 2, 3, 4	0	0.6667
$k = 7, l = 2$	–	2	–	1, 3, 4	0.2	0.6667
$k = 7, l = 3$	–	–	2	1, 3, 4	0	0.6667
$k = 7, l = 4$	–	2	3, 4	1	0.2	1
$k = 7, l = 5$	3	–	2	1, 4	0.15	0.6667
$k = 7, l = 6$	–	–	–	1, 2, 3, 4	0	0.6667

To see the possible influence resulted from the change of numbers of alternatives, the outranking relations obtained for the cases of N = 4, 5, 6 and 7 are listed in Table 4.17.

It can be seen from Table 4.17 that varying the number of alternatives does not change the outranking relations in Example 4.3 and the result shows that A_4 is a non-outranked alternative. It can be explained as follows: Firstly, when the number of alternatives is respectively 5, 6 and 7, the hesitant fuzzy concordance (discordance) indices given in Example 4.3 are not changed. Secondly, although a variation in the number of alternatives slightly modifies the hesitant fuzzy concordance level $\bar{c}$ and discordance level $\bar{d}$ (see Table 4.17), but the changes in $\bar{c}$ and $\bar{d}$ are still within the sensitivity range of Example 4.3 in which the parameter changes will not affect the set of the non-outranked alternatives. Specifically, in Example 4.3, $\bar{c} = 0.4611$ and $\bar{d} = 0.6472$. A decrease (increase) in $\bar{c}$ $(\bar{d})$ cannot bring about a change in the set of alternatives that are not outranked by other alternatives. Thus, $\bar{c}$ could

Fig. 4.6 Decision graph of 7 alternatives

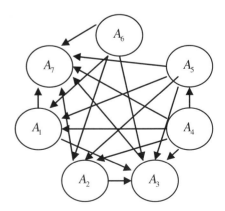

Table 4.17 Comparison of the outranking relations for different number of alternatives

Number of alternatives	$\bar{c}$	$\bar{d}$	Results
N = 4	0.4611	0.6472	$A_1 \succ A_3$; $A_2 \succ A_3$; $A_4 \succ A_1, A_2, A_3$
N = 5	0.4525	0.6367	$A_1 \succ A_3$; $A_2 \succ A_3$; $A_4 \succ A_1, A_2, A_3, A_5$; $A_5 \succ A_1, A_2, A_3$
N = 6	0.4489	0.6167	$A_1 \succ A_3$; $A_2 \succ A_3$; $A_4 \succ A_1, A_2, A_3, A_5$; $A_5 \succ A_1, A_2, A_3$; $A_6 \succ A_1, A_2, A_3$
N = 7	0.4258	0.5520	$A_1 \succ A_3, A_7; A_2 \succ A_3, A_7; A_3 \succ A_7$; $A_4 \succ A_1, A_2, A_3, A_5, A_7$; $A_5 \succ A_1, A_2, A_3, A_7; A_6 \succ A_1, A_2, A_3, A_7$

be lowered to zero and $\bar{d}$ could be raised to 1. Moreover, we note from the hesitant fuzzy concordance (discordance) matrices (given in Step 5 (Step 6) in Example 4.3) that when $\bar{c}$ is increased above 0.55, the alternative A_3 is no longer outranked by any other alternative. For the alternative $A_1(A_2)$, the corresponding value is 0.7 (0.8), as for $\bar{c} < 0.7(\bar{c} < 0.8)$ it remains outranked by the alternative A_4. The first change in the set of the non-outranked alternatives occurs when $\bar{c}$ is increased to 0.55. A similar analysis can be performed for the discordance level. When $\bar{d}$ is lowered below 0.3361, A_1 is no longer outranked. The corresponding value for $A_2(A_3)$ is 0.3265 (0.1143), so the lower bound for $\bar{d}$ is thus given by 0.3361. When the parameters satisfy $0 < \bar{c} < 0.55$ and $0.3361 < \bar{d} < 1$, the set of the non-outranked alternatives will not be affected.

In the following, we survey outranking relations for different numbers of criteria. To this end, we increase the numbers of criteria by 5 and 6 respectively (see Table 4.18).

Because the weights of all criteria satisfy the normalization constraint $\sum_{j=1}^{n} \omega_j = 1$, varying the number of criteria inevitably changes the original weights of criteria. This will affect the outranking relations. As an illustration, we list in Table 4.19 the corresponding results for outranking relations within the HF-ELECTRE I framework and the comparison with the case that the number of criteria is 4. As it can be seen in Table 4.19, the results of outranking relations for different numbers of criteria calculated with the HF-ELECTRE I method are consistent with general expectation.

In order to further evaluate the HF-ELECTRE I method, a simulation with randomly generated cases is made in a direct and transparent way. Random data are generated to form multiple criteria decision making problems with all possible combinations of 4, 6, 8, 10 alternatives and 4, 6, 8, 10 criteria. So, 16 different instances are examined in this study. We find that the preferred choice of alternatives in each instance is the alternative A_1 within the framework of HF-ELECTRE I (see below for details).

In the following, the cases corresponding to the combinations of 4 alternatives (i.e., $A = \{A_1, A_2, A_3, A_4\}$) with 4, 6, 8, 10 criteria are employed to illustrate our simulation process. Under the environment of group decision making, the

Table 4.18 Hesitant fuzzy information of the 5th and 6th criteria

	A_1	A_2	A_3	A_4
C_5	{0.3, 0.6, 0.7}	{0.2, 0.4, 0.5, 0.6}	{0.3, 0.7, 0.8}	{0.4, 0.5, 0.7, 0.9}
C_6	{0.4, 0.5, 0.7, 0.8}	{0.5, 0.7, 0.9}	{0.3, 0.4, 0.6, 0.9}	{0.5, 0.6, 0.8}

Table 4.19 Comparison of outranking relations for different numbers of criteria

Number of criteria	Weight	Results
4	$\omega = (0.2, 0.3, 0.15, 0.35)^T$	$A_1 \succ A_3; \quad A_2 \succ A_3; \quad A_4 \succ A_1, A_2, A_3$
5	$\omega = (0.2, 0.3, 0.15, 0.2, 0.15)^T$	$A_2 \succ A_1; \quad A_3 \succ A_1; \quad A_4 \succ A_1, A_2, A_3$
6	$\omega = (0.2, 0.15, 0.15, 0.2, 0.15, 0.15)^T$	$A_1 \succ A_3; A_2 \succ A_1, A_3; A_3 \succ A_1; A_4 \succ A_1, A_2, A_3$

performance of each alternative on each criterion can be considered as a HFE, h_{ij}, which indicates the possible satisfaction degrees of the criterion C_j to the alternative A_i. The characteristic of satisfaction degrees assigned to these 4 alternatives is as follows: For $\gamma \in h_{1j}$, $\gamma \sim U(0.8, 1)$; for $\gamma \in h_{2j}, \gamma \sim U(0.5, 0.8)$; for $\gamma \in h_{3j}, \gamma \sim U(0.3, 0.5)$; for $\gamma \in h_{4j}$, $\gamma \sim U(0, 0.3)$, where $U(a, b)$ means the uniform distribution on the interval $[a, b]$. In this way, we simulate 100 times corresponding to each instance with the number of criteria being 4, 6, 8 and 10, respectively. In each simulation, the number of γ in each h_{ij} and the weights of criteria are also randomly generated. Consequently, all quantities calculated by the HF-ELECTRTE I fluctuate in each simulation. Figure 4.7 displays the scores of h_{ij} of 4 alternatives on a criterion, which shows an obvious variation for different simulations.

Following the steps of the HF-ELECTRE I method outlined previously, our simulation results demonstrate that the outranking relations are consistent with the expectation. That is, the alternative A_1 is the best choice due to a larger role of the performance of alternatives in influencing the outranking relations as compared to the weights and threshold values. This consistency further indicates the validity of the HF-ELECTRE I method.

(3) Comparison with the ELECTRE III and ELECTRE IV methods

There exist several other types of ELECTRE methods, such as ELECTRE III (Figueira et al. 2005; Buchanan and Vanderpooten 2007) and ELECTRE IV (Roy and Hugonnard 1982). It is interesting to compare the results obtained by using these different methods to see the ranking differences among them. To facilitate the comparison, the same example (Example 4.3) used to illustrate the HF-ELECTRE I method is also considered for the ELECTRE III and ELECTRE IV methods, which is composed of the construction and the exploitation of the outranking relations.

Firstly, the ELECTRE III method is described briefly as follows (for more details see Figueira et al. (2005) and Buchanan and Vanderpooten (2007)):

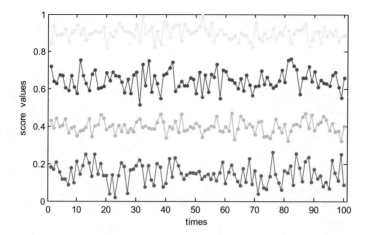

Fig. 4.7 Score values of h_{ij} in 100 times simulations. From *top* to *bottom*, they correspond to A_1 (*yellow line*), A_2 (*blue line*), A_3 (*green line*) and A_4 (*red line*), respectively

Let two alternatives A_k and A_l belong to a given set of actions, and $y_j(A_k)$ and $y_j(A_l)$ be the performances of A_k and A_l in terms of the criterion C_j. We denote indifference, preference and veto thresholds on the criterion C_j determined by the decision makers as q_j, p_j and v_j, respectively, where $j = 1, 2, \ldots, n$. The partial concordance index $c_j(A_k, A_l)$ for each criterion C_j is defined as:

$$c_j(A_k, A_l) = \begin{cases} 1 & \text{if } y_j(A_k) + q_j \geq y_j(A_l), \\ 0 & \text{if } y_j(A_k) + p_j \leq y_j(A_l), \quad j = 1, \ldots, n \\ (p_j + y_j(A_k) - y_j(A_l))/(p_j - q_j) & \text{otherwise.} \end{cases}$$

(4.42)

Let ω_j be the weight of the criterion C_j. The overall concordance index $C(A_k, A_l)$ is defined as:

$$C(A_k, A_l) = \sum_{j=1}^{n} \omega_j c_j(A_k, A_l)$$

(4.43)

The discordance index $d_j(A_k, A_l)$ for the criterion C_j is calculated as:

$$d_j(A_k, A_l) = \begin{cases} 0 & \text{if } y_j(A_k) + p_j \geq y_j(A_l) \\ 1 & \text{if } y_j(A_k) + v_j \leq y_j(A_l) \\ (y_j(A_l) - y_j(A_k) - p_j) & \text{otherwise} \end{cases}$$

(4.44)

The credibility degree $S(A_k, A_l)$ for the pair (A_k, A_l) is defined as:

$$S(A_k, A_l) = \begin{cases} C(A_k, A_l) & \text{if } d_j(A_k, A_l) \leq C(A_k, A_l), \forall j & \text{otherwise} \\ C(A_k, A_l) \cdot \prod\limits_{\substack{j \in J: \\ d_j(A_k, A_l) > C(A_k, A_l)}} \frac{1 - d_j(A_k, A_l)}{1 - C(A_k, A_l)} & \end{cases} \quad (4.45)$$

The ELECTRE III model is "exploited" to produce a ranking of alternatives from the credibility matrix. In what follows, we use the ELECTRE III method to tackle the problem in Example 4.3.

Step 1. Construct the fuzzy group decision matrix. To derive the group decision matrix, we aggregate the decision makers' individual decision information by the averaging operator, which is defined as $\tilde{h}_{ij} = \frac{1}{l_{h_{ij}}} \sum_{\gamma \in h_{ij}} \gamma$. As it can be easily seen that the results obtained from the formula are just the score function values of HFEs, which have been given in Table 4.6. So, the fuzzy group decision matrix $\tilde{H} = (\tilde{h}_{ij})_{4 \times 4}$ is obtained.

Step 2. Indifference, preference and veto threshold values on the criterion C_j are introduced by the decision makers, as shown in Table 4.20.

Step 3. Calculate the partial concordance index $c_j(A_k, A_l)$ with Eq. (4.42). The results are given in Table 4.21.

Step 4. Calculate the overall concordance index $C(A_k, A_l)$ with Eq. (4.43). The results are set out in Table 4.22.

Step 5. Calculate the discordance index $d_j(A_k, A_l)$ with Eq. (4.44). The results are given in Table 4.23.

Step 6. Calculate the credibility degree $S(A_k, A_l)$ with Eq. (4.45). The results are set out in Table 4.24.

Step 7. Perform the exploiting procedure with the credibility matrix. Two pre-orders are obtained via a descending and ascending distillation process respectively (see Fig. 4.8). Based on these two pre-orders, the result of outranking is $A_4 \succ A_2 \succ A_1 \succ A_3$.

We next analyze the calculations with the ELECTRE IV method. Let $y_j(A_k)$ and $y_j(A_l)$ be the performance of two alternatives A_k and A_l on the criterion C_j. $J^+(A_k, A_l)$ and $J^-(A_k, A_l)$ represent respectively the sums of all those criteria on which the performance of A_k is superior and inferior to A_l:

$$J^+(A_k, A_l) = \{j \in J | j : y_j(A_l) + q_j(y_j(A_l)) < y_j(A_k)\} \quad (4.46)$$

$$J^-(A_k, A_l) = \{j \in J | j : y_j(A_k) + q_j(y_j(A_k)) < y_j(A_l)\} \quad (4.47)$$

Table 4.20 Three types of threshold values

Threshold values	C_1	C_2	C_3	C_4
q_j	0.02	0.05	0.02	0.02
p_j	0.1	0.1	0.05	0.1
v_j	0.3	0.3	0.2	0.3

Table 4.21 Partial concordance index for each criterion

		A_1	A_2	A_3	A_4
$c_1(A_k, A_l)$	A_1	1	0	0.1038	0.8325
	A_2	1	1	1	1
	A_3	1	0.9375	1	1
	A_4	1	0.2088	0.5213	1
$c_2(A_k, A_l)$	A_1	1	1	1	1
	A_2	0	1	0	1
	A_3	0	1	1	1
	A_4	0	1	0	1
$c_3(A_k, A_l)$	A_1	1	1	0.5	0
	A_2	1	1	0.8333	0
	A_3	1	1	1	0.8333
	A_4	1	1	1	1
$c_4(A_k, A_l)$	A_1	1	0	1	0
	A_2	1	1	1	0
	A_3	0.75	0	1	0
	A_4	1	1	1	1

Table 4.22 Concordance matrix

	A_1	A_2	A_3	A_4
A_1	1	0.45	0.7458	0.4665
A_2	0.7	1	0.6750	0.5
A_3	0.6125	0.6375	1	0.6250
A_4	0.7	0.8418	0.6043	1

Table 4.23 Discordance index for each criterion

		A_1	A_2	A_3	A_4
$d_1(A_k, A_l)$	A_1	0	0.0835	0	0
	A_2	0	0	0	0
	A_3	0	0	0	0
	A_4	0	0	0	0
$d_2(A_k, A_l)$	A_1	0	0	0	0
	A_2	0.6665	0	0.125	0
	A_3	0.0415	0	0	0
	A_4	0.6665	0	0.125	0
$d_3(A_k, A_l)$	A_1	0	0	0	0.0667
	A_2	0	0	0	0
	A_3	0	0	0	0
	A_4	0	0	0	0
$d_4(A_k, A_l)$	A_1	0	0.3	0	1
	A_2	0	0	0	0.25
	A_3	0	0.5	0	1
	A_4	0	0	0	0

Table 4.24 Credibility matrix

	A_1	A_2	A_3	A_4
A_1	1	0.45	0.7458	0
A_2	0.7	1	0.6750	0.5
A_3	0.6125	0.6375	1	0
A_4	0.7	0.8418	0.6043	1

(a)   **(b)**

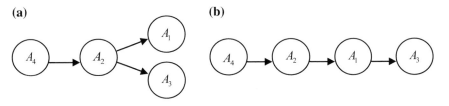

Fig. 4.8 a Descending distillation. **b** Ascending distillation

The ELECTRE IV method contains two levels of outranking relations, i.e., the strong outranking relation (S^F) and the weak outranking relation (S^f), which are defined as:

$$A_k S^F A_l \Leftrightarrow \forall j, y_j(A_k) + p_j(y_j(A_k)) \geq y_j(A_l) \quad \text{and} \quad \|J^+(A_k, A_l)\| > \|J^-(A_k, A_l)\| \tag{4.48}$$

$$A_k S^f A_l \Leftrightarrow \forall j, y_j(A_k) + p_j(y_j(A_k)) \geq y_j(A_l)$$
$$\text{or} \begin{cases} \exists l, y_l(A_k) + v_l(y_l(A_k)) \geq y_l(A_l) > y_l(A_k) + p_l(y_l(A_k)) \\ \text{and} \|j : y_j(A_k) - y_j(A_l) > p_j(y_j(A_k))\| \geq n/2 \end{cases} \tag{4.49}$$

where $\|J\|$ denotes the number of elements in the set J.

The ELECTRE IV exploiting procedure is the same as that of the ELECTRE III. The detailed steps of using the ELECTRE IV method to deal with Example 4.3 are given as follows:

Step 1. The process is the same as that in ELECTRE III method.
Step 2. The indifference, preference and veto threshold values on the criterion j are the same as those given in Table 4.20.
Step 3. Calculate $J^+(A_k, A_l)$ and $J^-(A_k, A_l)$ with Eq. (4.46) and (4.47). The results are summarized in Table 4.25.
Step 4. Calculate the outranking relation S between alternatives with Eqs. (4.48) and (4.49). The results are also displayed in Table 4.25. From Table 4.25, we know $A_1 \succ_{S^f} A_3$, $A_2 \succ_{S^f} A_1$, $A_4 \succ_{S^f} A_1$, and $A_4 \succ_{S^F} A_2$.

Table 4.25 $J^+(A_k, A_l), J^-(A_k, A_l)$ and the outrankings

	A₁			A₂			A₃			A₄		
	J^+	J^-	S	J^+	J^-	S	J^+	J^-	S	J^+	J^-	S
A₁				2	1, 4		2, 4	1, 3	S^f	2	1, 3, 4	
A₂	1, 4	2	S^f				1, 4	2, 3		1	3, 4	
A₃	1, 3	2, 4		2, 3	1, 4					1, 2	3, 4	
A₄	1, 3, 4	2	S^f	3, 4	1	S^F	3, 4	1, 2				

Step 5. Draw the strong and weak outranking graphs, shown as Fig. 4.9. With the exploiting procedure, the final outranking is $A_4 \succ A_2 \succ A_1 \succ A_3$.

It is observed that the conclusions obtained from the ELECTRE III and ELECTRE IV methods are partially consistent with that derived with the HF-ELECTRE I method, whereas a difference in the outranking relations is also noticed. In the ELECTRE III and ELECTRE IV group outranking methods, only information involving the average opinions of all decision makers is considered, whereas in the HF-ELECTRE I method, in addition to that consideration, the deviation degrees which reflect the difference in opinions between the individual decision makers and their averages are also accounted for. It should be pointed out that the ELECTRE III and ELECTRE IV methods can give the ranking of all alternatives, which is different from the HF-ELECTRE I method that gives the partial outranking relations. Thus, these two kinds of group decision making approaches are complementary.

(2) Application of the HF-ELECTRE II Method

Below, we use some numerical examples to illustrate the details of the HF-ELECTRE II method.

Example 4.5 (Chen and Xu 2015). A battery industry involved in the recycling process desires to select a suitable 3rd-party reverse logistics provider (3PRLP) to perform the reverse logistics activities. A committee of three experts has been formed to select the most suitable 3PRLP. They evaluate the performance of each 3PRLP from seven aspects: (1) C_1: quality; (2) C_2: delivery; (3) C_3: reverse

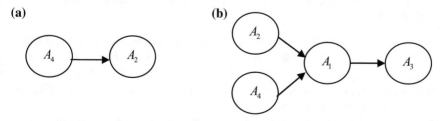

Fig. 4.9 **a** Strong outranking graph. **b** Weak outranking graph

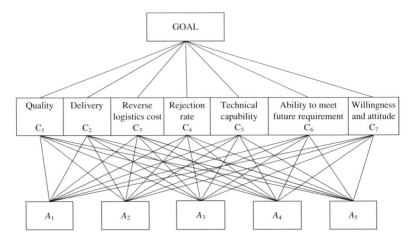

Fig. 4.10 Hierarchical structure of decision problem

Table 4.26 Hesitant fuzzy decision matrix

	A_1	A_2	A_3	A_4	A_5
C_1	{0.7, 0.8, 0.9}	{0.7, 0.8}	{0.7, 0.9}	{0.6, 0.8}	{0.9, 1}
C_2	{0.6, 0.7}	{0.2, 0.3, 0.6}	{0.6, 0.7, 0.8}	{0.3, 0.4, 0.5}	{0.8, 0.9}
C_3	{0.2, 0.3}	{0.2, 0.3}	{0.1, 0.3, 0.4}	{0.3, 0.5}	{0.2, 0.5, 0.6}
C_4	{0.5, 0.6, 0.7}	{0.4, 0.5}	{0.7, 0.8, 1}	{0.5, 0.6, 0.7}	{0.7}
C_5	{0.7, 0.8, 1}	{0.5, 0.7}	{0.5, 0.7}	{0.6, 0.8, 0.9}	{0.8, 0.9, 1}
C_6	{0.2, 0.3}	{0.3, 0.4, 0.6}	{0.3, 0.5}	{0.3, 0.5}	{0.4, 0.6}
C_7	{0.6, 0.9}	{0.2, 0.3}	{0.4, 0.6, 0.7}	{0.4, 0.5, 0.6}	{0.7, 0.8, 0.9}

logistics cost; (4) C_4: rejection rate; (5) C_5: technical capability; (6) C_6: ability to meet future requirement; (7) C_7: willingness and attitude. The hierarchical structure of the decision making problem is depicted in Fig. 4.10.

To avoid psychic contagion, the experts are required to provide their preferences in anonymity. Suppose that the weight vector of the attributes is $\omega = (0.2, 0.2, 0.15, 0.1, 0.1, 0.15, 0.1)^T$ and the hesitant fuzzy decision matrix is presented in Table 4.26.

Step 1. The decision makers assign the relative attitude weights of strong, medium and weak hesitant fuzzy concordance, discordance and indifferent sets as:
$\lambda = (\lambda_C, \lambda_{C'}, \lambda_{C''}, \lambda_D, \lambda_{D'}, \lambda_{D''}, \lambda_{J=})^T = (1, 0.9, 0.8, 1, 0.9, 0.8, 0.7)^T$.

Step 2. Calculate the score value and the deviation value of each HFE (see Tables 4.27 and 4.28).

Step 3. Construct the hesitant fuzzy strong, medium and weak concordance sets:

Table 4.27 Score values obtained by the score function

	C_1	C_2	C_3	C_4	C_5	C_6	C_7
A_1	0.8	0.65	0.25	0.6	0.8333	0.25	0.75
A_2	0.75	0.3667	0.25	0.45	0.6	0.4333	0.25
A_3	0.8	0.7	0.2667	0.8333	0.6	0.4	0.5667
A_4	0.7	0.4	0.4	0.6	0.7667	0.4	0.5
A_5	0.95	0.85	0.4333	0.7	0.9	0.5	0.8

Table 4.28 Deviation values obtained by the deviation function

	C_1	C_2	C_3	C_4	C_5	C_6	C_7
A_1	0.0816	0.05	0.05	0.0816	0.1247	0.05	0.15
A_2	0.05	0.17	0.05	0.05	0.1	0.1247	0.05
A_3	0.1	0.0816	0.1247	0.1247	0.1	0.1	0.1247
A_4	0.1	0.0816	0.1	0.0816	0.1247	0.1	0.0816
A_5	0.05	0.05	0.17	0	0.0816	0.1	0.0816

$$
J_C = \begin{bmatrix}
- & 2 & - & 1,2 & - \\
- & - & - & 1 & - \\
- & 2 & - & - & - \\
- & 2 & 3 & - & - \\
1,4,5,7 & 2,4,5,6 & 1,2,5,7 & 1,2,4,5 & -
\end{bmatrix}
$$

$$
J_{C'} = \begin{bmatrix}
- & 1,4,5,7 & 5,7 & 5,7 & - \\
6 & - & 6 & 6 & - \\
2,3,4,6 & 1,3,4,7 & - & 1,2,4,7 & 4 \\
3,6 & 3,4,5,7 & 5 & - & - \\
2,3,6 & 1,3,7 & 3,6 & 3,6,7 & -
\end{bmatrix}
$$

$$
J_{C''} = \begin{bmatrix}
- & - & 1 & - & - \\
- & - & - & - & - \\
- & - & - & - & - \\
- & - & - & - & - \\
- & - & - & - & -
\end{bmatrix}
$$

For instance, since $s(h_{12}) > s(h_{22})$ and $\sigma(h_{12}) < \sigma(h_{22})$, then $J_{C_{12}} = \{2\}$.
Step 4. Construct the hesitant fuzzy strong, medium and weak discordance sets:

$$J_D = \begin{bmatrix} - & - & - & - & 1,4,5,7 \\ 2 & - & 2 & 2 & 2,4,5,6 \\ - & - & - & 3 & 1,2,5,7 \\ 1,2 & 1 & - & - & 1,2,4,5 \\ - & - & - & - & - \end{bmatrix}$$

$$J_{D'} = \begin{bmatrix} - & 6 & 2,3,4,6 & 3,6 & 2,3,6 \\ 1,4,5,7 & - & 1,3,4,7 & 3,4,5,7 & 1,3,7 \\ 5,7 & 6 & - & 5 & 3,6 \\ 5,7 & 6 & 1,2,4,7 & - & 3,6,7 \\ - & - & 4 & - & - \end{bmatrix}$$

$$J_{D''} = \begin{bmatrix} - & - & - & - & - \\ - & - & - & - & - \\ 1 & - & - & - & - \\ - & - & - & - & - \\ - & - & - & - & - \end{bmatrix}$$

For example, since $s(h_{22}) < s(h_{32}), \sigma(h_{22}) > \sigma(h_{32})$, then $J_{D_{23}} = \{2\}$.

Step 5. Construct the hesitant fuzzy indifferent set by using Eq. (4.37):

$$J^= = \begin{bmatrix} - & 3 & - & 4 & - \\ 3 & - & 5 & - & - \\ - & 5 & - & 6 & - \\ 4 & - & 6 & - & - \\ - & - & - & - & - \end{bmatrix}$$

For example, since $s(h_{13}) = s(h_{23}), \sigma(h_{13}) = \sigma(h_{23})$, then $J^=_{12} = \{3\}$.

Step 6. Calculate the hesitant fuzzy concordance index and construct concordance matrix:

$$C = (c_{kl})_{5 \times 5} = \begin{bmatrix} - & 0.755 & 0.34 & 0.65 & 0 \\ 0.24 & - & 0.205 & 0.335 & 0 \\ 0.54 & 0.765 & - & 0.645 & 0.09 \\ 0.34 & 0.605 & 0.345 & - & 0 \\ 0.95 & 0.955 & 0.87 & 0.96 & - \end{bmatrix}$$

For example, $c_{12} = 0.2 + 0.5 \times 0.9 + 0.15 \times 0.7 = 0.755$.

Step 7. Calculate the weighted distance between any two alternatives with respect to each criterion by using Eq. (4.7), see Table 4.29.

Step 8. Calculate the hesitant fuzzy discordance index and construct the discordance matrix:

Table 4.29 Weighted distances

	h_{11}	h_{21}	h_{31}	h_{41}	h_{51}	h_{12}	h_{22}	h_{32}	h_{42}	h_{52}
h_{11}	–	0.0067	0.0067	0.0133	0.0333	–	0.06	0.0067	0.0533	0.04
h_{21}	–	–	0.01	0.01	0.04	–	–	0.0667	0.02	0.1
h_{31}	–	–	–	0.02	0.03	–	–	–	0.06	0.0333
h_{41}	–	–	–	–	0.05	–	–	–	–	0.0933
h_{51}	–	–	–	–	–	–	–	–	–	–

	h_{13}	h_{23}	h_{33}	h_{43}	h_{53}	h_{14}	h_{24}	h_{34}	h_{44}	h_{54}
h_{13}	–	0	0.01	0.0225	0.025	–	0.0133	0.0233	0	0.01
h_{23}	–	–	0.01	0.0225	0.025	–	–	0.0367	0.0133	0.025
h_{33}	–	–	–	0.025	0.025	–	–	–	0.0233	0.0133
h_{43}	–	–	–	–	0.01	–	–	–	–	0.01
h_{53}	–	–	–	–	–	–	–	–	–	–

	h_{15}	h_{25}	h_{35}	h_{45}	h_{55}	h_{16}	h_{26}	h_{36}	h_{46}	h_{56}
h_{15}	–	0.02	0.02	0.0067	0.0067	–	0.025	0.0225	0.0225	0.0375
h_{25}	–	–	0	0.0133	0.0267	–	–	0.01	0.01	0.015
h_{35}	–	–	–	0.0133	0.0267	–	–	–	0	0.015
h_{45}	–	–	–	–	0.0133	–	–	–	–	0.015
h_{55}	–	–	–	–	–	–	–	–	–	–

	h_{17}	h_{27}	h_{37}	h_{47}	h_{57}
h_{17}	–	0.05	0.0233	0.03	0.0067
h_{27}	–	–	0.03	0.0233	0.0533
h_{37}	–	–	–	0.0067	0.0233
h_{47}	–	–	–	–	0.03
h_{57}	–	–	–	–	–

$$D = (d_{kl})_{5\times 5} = \begin{bmatrix} - & 0.375 & 0.9 & 0.3799 & 0.9 \\ 1 & - & 1 & 0.9 & 1 \\ 0.9 & 0.1349 & - & 0.4167 & 1 \\ 1 & 0.4292 & 0.9 & - & 1 \\ 0 & 0 & 0.3595 & 0 & - \end{bmatrix}$$

For instance,

$$d_{21} = \frac{\max\limits_{j\in J_{D_{21}} \cup J_{D'_{21}} \cup J_{D''_{21}}} \left\{ \lambda_D \times d(\omega_j h_{2j}, \omega_j h_{1j}), \lambda_{D'} \times d(\omega_j h_{2j}, \omega_j h_{1j}), \lambda_{D''} \times d(\omega_j h_{2j}, \omega_j h_{1j}) \right\}}{\max\limits_{j\in J} d(\omega_j h_{2j}, \omega_j h_{1j})}$$

$$= \frac{\max\{1 \times 0.06, 0.9 \times 0.05\}}{0.06} = \frac{0.06}{0.06} = 1.$$

Step 9. Construct the outranking relations following the given concordance and
discordance levels. Here the concordance and discordance levels for the
strong and weak outranking relations are chosen by the experts as
$(c^-, c^0, c^*) = (0.5, 0.6, 0.75)$ and $d^0 = 0.4$, $d^* = 0.45$. According to
Eqs. (4.40) and (4.41), the outranking relations are derived as shown in
Table 4.30.

Step 10. Draw the strong and weak outranking graphs in Fig. 4.11.

Table 4.30 Outranking relations

	A_1	A_2	A_3	A_4	A_5
A_1	–	S^F		S^F	
A_2		–			
A_3		S^F	–	S^f	
A_4		S^f		–	
A_5	S^F	S^F	S^F	S^F	–

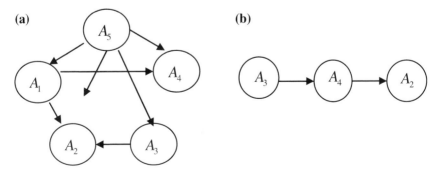

Fig. 4.11 a Strong outranking graph. **b** Weak outranking graph

Table 4.31 Results of ranking

	A_1	A_2	A_3	A_4	A_5
Forward ranking v'	3	5	2	4	1
Reverse ranking v''	2	5	3	4	1
Average ranking $\bar{v}$	2.5	5	2.5	4	1

Now we consider the strong and weak outranking graphs plotted in Fig. 4.11. The forward ranking v', the reverse ranking v'' and the average ranking $\bar{v}$ are deduced and summarized in Table 4.31. From Table 4.31, the ranking of the alternatives is $A_5 \succ A_3 \sim A_1 \succ A_4 \succ A_2$.

To survey the performance of the algorithm introduced for the HF-ELECTRE II approach for a larger group, we extend the number of decision makers from three assumed in Example 4.5 to ten, and other initial conditions are the same as those in Example 4.5. Table 4.32 lists the hesitant fuzzy decision matrix constructed for the larger group with ten experts.

For the larger group, we still use the same steps displayed in Example 4.5. Specifically speaking, after the score values and the deviations are calculated, the hesitant fuzzy strong, medium and weak concordance sets and the discordance sets as well as indifferent sets are constructed. The concordance and discordance matrices are then obtained as:

Table 4.32 Hesitant fuzzy decision matrix

	A_1	A_2	A_3
C_1	{0.3, 0.35, 0.45, 0.5, 0.55, 0.6, 0.7, 0.8}	{0.55, 0.6, 0.75, 0.85, 0.9}	{0.25, 0.45, 0.65, 0.7, 0.75, 0.85}
C_2	{0.1, 0.2, 0.3, 0.45}	{0.1, 0.15, 0.2, 0.3, 0.35, 0.4, 0.45, 0.6, 0.7, 0.75}	{0.25, 0.3, 0.4, 0.6, 0.65, 0.7, 0.8}
C_3	{0.15, 0.2, 0.3, 0.35, 0.4, 0.6}	{0.6, 0.8, 0.9}	{0.1, 0.2, 0.25, 0.4, 0.55, 0.6, 0.65, 0.7, 0.95, 1}
C_4	{0.2, 0.3, 0.35, 0.45, 0.5, 0.55, 0.7, 0.85}	{0.3, 0.45, 0.5, 0.55, 0.65, 0.7, 0.75, 0.8, 0.95, 1}	{0.15, 0.2, 0.4}
C_5	{0.3, 0.35, 0.4, 0.45}	{0.45, 0.5, 0.6, 0.7, 0.9}	{0.1, 0.35, 0.4, 0.5, 0.55, 0.75, 0.8, 0.85}
C_6	{0.4, 0.55, 0.65, 0.85, 0.9, 0.95}	{0.1, 0.2, 0.25, 0.3, 0.35, 0.45, 0.5, 0.6, 0.65, 0.75}	{0.2, 0.4, 0.6}
C_7	{0.35, 0.4, 0.5, 0.65, 0.7, 0.75, 0.8, 0.9, 0.95, 1}	{0.1, 0.2, 0.4, 0.45}	{0.45, 0.5, 0.65, 0.75, 0.9, 0.95}

	A_4	A_5
C_1	{0.1, 0.25, 0.3, 0.4, 0.5, 0.65, 0.7, 0.85, 0.95, 1}	{0.45, 0.7, 0.8}
C_2	{0.3, 0.5, 0.6, 0.75, 0.9}	{0.5, 0.6, 0.65, 0.7, 0.85, 0.9, 0.95}
C_3	{0.2, 0.3, 0.35, 0.5}	{0.15, 0.25, 0.3, 0.35, 0.4, 0.5, 0.55, 0.7}
C_4	{0.45, 0.6, 0.75, 0.8, 0.85}	{0.2, 0.45, 0.5, 0.55, 0.7}
C_5	{0.15, 0.25, 0.3, 0.45, 0.5, 0.65, 0.7, 0.75, 0.8, 0.9}	{0.25, 0.3, 0.45, 0.65, 0.7, 0.75, 0.8}
C_6	{0.6, 0.7, 0.85, 0.95}	{0.4, 0.6, 0.65, 0.7, 0.8, 0.85, 0.95}
C_7	{0.3, 0.4, 0.45, 0.65, 0.7, 0.75, 0.95}	{0.35, 0.65, 0.7, 0.75, 0.85, 0.9}

$$C = (c_{kl})_{5 \times 5} = \begin{bmatrix} - & 0.24 & 0.225 & 0.09 & 0.225 \\ 0.71 & - & 0.675 & 0.435 & 0.54 \\ 0.665 & 0.29 & - & 0.435 & 0.135 \\ 0.85 & 0.54 & 0.51 & - & 0.25 \\ 0.685 & 0.44 & 0.805 & 0.735 & - \end{bmatrix}$$

$$D = (d_{kl})_{5 \times 5} = \begin{bmatrix} - & 1 & 0.9 & 0.9 & 0.9 \\ 0.79 & - & 0.8 & 1 & 1 \\ 0.7884 & 1 & - & 1 & 1 \\ 0.0508 & 0.8797 & 0.7590 & - & 1 \\ 0.9 & 0.8438 & 0.3587 & 0.4038 & - \end{bmatrix}$$

Based on these matrices, the strong and weak outranking relations are constructed with $(c^-, c^0, c^*) = (0.5, 0.6, 0.7)$ and $d^0 = 0.5, d^* = 0.8$, as shown in Table 4.33.

With the exploration procedure, we deduce the forward ranking v', the reverse ranking v'' and the average ranking $\bar{v}$ and present them in Table 4.34. From Table 4.34, we know that the final ranking of the five alternatives is $A_5 \succ A_2 \succ A_4 \succ A_3 \succ A_1$.

In the following, one real case is considered to further illustrate our method:

Example 4.6 (Chen and Xu 2015). Owing to the great convenience provided by online shopping, Electronic Commence (e-Commerce) has attracted great attention. To enable the online shopping to make success, it is critical to help customers to reduce their time spent on searching for products that are suitable for their requirements. In this context, a recommendation system is required. The system attempts to find some interesting sets of products from individual evaluation of multiple features characterizing the products. Multiple criteria decision making models can therefore be effectively utilized because they provide a method for evaluating decision alternatives with a finite number of features or criteria. When

Table 4.33 Outranking relations

	A_1	A_2	A_3	A_4	A_5
A_1	–				
A_2	S^F	–	S^f		
A_3	S^f		–		
A_4	S^F		S^f	–	
A_5			S^F	S^F	–

Table 4.34 Results of ranking

	A_1	A_2	A_3	A_4	A_5
Forward ranking v'	4	1	3	2	1
Reverse ranking v''	4	2	3	2	1
Average ranking $\bar{v}$	4	1.5	3	2	1

one buys a laptop, in terms of his/her requirements (including price, quality, appearance, running speed and customer service) for this product, the recommendation system will recommend a kind of laptop that is most suitable for the customer according to evaluation information provided by those customers who had ever bought similar products. In general, these different customers could give different evaluations for a certain feature of the product. HFEs are particularly suitable for denoting the difference.

Suppose that the recommendation system suggests a suitable laptop to the customer from six candidate products (alternatives) $A_i(i = 1, 2, 3, 4, 5, 6)$ and each of these alternatives is evaluated by considering a set of five criteria $C = \{C_1, C_2, \ldots, C_5\}$ representing price (C_1), quality (C_2), appearance (C_3), running speed (C_4) and customer service (C_5) whose weight vector is $\omega = (0.3, 0.2, 0.15, 0.15, 0.2)^T$. The evaluation information is denoted by a hesitant fuzzy matrix as shown in Table 4.35.

Step 1.   This step is given as Table 4.35. We start our calculation from Step 2 directly.

Steps 2–3. Give score and deviation values of hesitant fuzzy decision matrix in Tables 4.36 and 4.37.

Steps 3–6. Construct the concordance matrix based on hesitant fuzzy strong, medium and weak concordance sets:

Table 4.35 Hesitant fuzzy decision matrix for Example 4.6

	C_1	C_2	C_3	C_4	C_5
A_1	{0.2, 0.5, 0.6, 0.7}	{0.4, 0.5, 0.6, 0.8, 0.9}	{0.3, 0.4, 0.6, 0.8, 0.9}	{0.4, 0.6, 0.7, 0.9}	{0.5, 0.6, 0.7, 0.8}
A_2	{0.3, 0.6, 0.7, 0.8, 0.9}	{0.3, 0.6, 0.8, 0.9}	{0.2, 0.4, 0.6, 0.7}	{0.5, 0.6, 0.8, 0.9}	{0.2, 0.3, 0.6, 0.7, 0.8}
A_3	{0.3, 0.4, 0.7}	{0.2, 0.3, 0.5, 0.6, 0.7}	{0.4, 0.6, 0.7, 0.8, 0.9}	{0.7, 0.8, 0.9, 1}	{0.3, 0.4, 0.6, 0.7}
A_4	{0.2, 0.3, 0.4, 0.6, 0.7}	{0.1, 0.2, 0.3, 0.4}	{0.2, 0.4, 0.5, 0.6}	{0.3, 0.4, 0.6, 0.7}	{0.2, 0.3, 0.5, 0.6, 0.7}
A_5	{0.4, 0.5, 0.7, 0.8, 0.9}	{0.3, 0.5, 0.6, 0.7}	{0.5, 0.6, 0.7, 0.8, 0.9}	{0.7, 0.8, 0.9}	{0.4, 0.5, 0.6, 0.8}
A_6	{0.2, 0.3, 0.4, 0.6}	{0.1, 0.3, 0.4}	{0.4, 0.5, 0.7}	{0.2, 0.5, 0.6, 0.7, 0.9}	{0.2, 0.3, 0.7, 0.8, 0.9}

Table 4.36 Score values obtained by the score function

	A_1	A_2	A_3	A_4	A_5	A_6
C_1	0.5	0.66	0.4667	0.44	0.66	0.375
C_2	0.64	0.65	0.46	0.25	0.525	0.2667
C_3	0.6	0.475	0.68	0.425	0.7	0.5333
C_4	0.65	0.7	0.85	0.5	0.8	0.58
C_5	0.65	0.52	0.5	0.46	0.575	0.58

Table 4.37 Deviation values obtained by the deviation function

	A_1	A_2	A_3	A_4	A_5	A_6
C_1	0.1871	0.2059	0.1700	0.1855	0.1855	0.1479
C_2	0.1855	0.2291	0.1855	0.1118	0.1479	0.1247
C_3	0.2280	0.1920	0.1720	0.1479	0.1414	0.1247
C_4	0.1803	0.1581	0.1118	0.1581	0.0816	0.2315
C_5	0.1118	0.2315	0.1581	0.1855	0.1479	0.2786

$$C = (c_{kl})_{5 \times 5} = \begin{bmatrix} - & 0.335 & 0.65 & 0.92 & 0.38 & 0.935 \\ 0.6 & - & 0.63 & 0.9 & 0.18 & 0.6 \\ 0.3 & 0.3 & - & 0.965 & 0.135 & 0.735 \\ 0 & 0 & 0 & - & 0 & 0.27 \\ 0.6 & 0.74 & 0.82 & 0.95 & - & 0.735 \\ 0 & 0.33 & 0.18 & 0.645 & 0.18 & - \end{bmatrix}$$

Steps 7–8. Construct the discordance matrix based on the weighted distance that is formulated by using the corresponding discordance sets:

$$D = (d_{kl})_{5 \times 5} = \begin{bmatrix} - & 0.9 & 0.8333 & 0 & 1 & 0 \\ 0.8889 & - & 0.5 & 0 & 1 & 0.2083 \\ 0.9 & 0.9 & - & 0 & 0.9 & 0.48 \\ 0.9 & 0.9 & 1 & - & 0.9 & 0.9375 \\ 0.4167 & 0.8333 & 0.1125 & 0 & - & 0.3 \\ 0.9 & 0.9 & 1 & 0.25 & 0.9 & - \end{bmatrix}$$

Step 9. Construct the strong and weak outranking relations with $(c^-, c^0, c^*) = (0.6, 0.7, 0.9)$ and $d^0 = 0.5$, $d^* = 0.6$. The outranking relations are given in Table 4.38.

Step 10. Draw the strong and weak outranking graphs in Fig. 4.12.

The forward ranking v', the reverse ranking v'' and the average ranking $\bar{v}$ are derived and shown in Table 4.39.

From Table 4.39, we obtain the final ranking as $A_2 \sim A_5 \succ A_1 \sim A_3 \succ A_6 \succ A_4$.

Table 4.38 The outranking relations

	A_1	A_2	A_3	A_4	A_5	A_6
A_1	–			S^F		S^F
A_2		–	S^f	S^F		S^f
A_3			–	S^F		S^F
A_4				–		
A_5	S^f		S^F	S^F	–	S^F
A_6				S^f		–

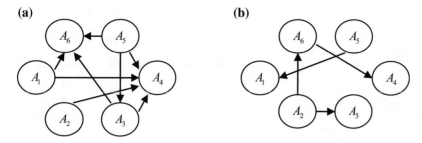

Fig. 4.12 a Strong outranking graph. **b** Weak outranking graph

Table 4.39 Results of ranking

	A_1	A_2	A_3	A_4	A_5	A_6
Forward ranking v'	2	1	2	4	1	3
Reverse ranking v''	2	1	2	4	1	3
Average ranking $\bar{v}$	2	1	2	4	1	3

References

Buchanan J, Vanderpooten D (2007) Ranking projects for an electricity utility using ELECTRE III. Int Trans Oper Res 14:309–323

Chen N, Xu ZS (2015) Hesitant fuzzy ELECTRE II approach: A new way to handle multi-criteria decision making problems. Inf Sci 292:175–197

Chen N, Xu ZS, Xia MM (2015) The ELECTRE I multi-criteria decision making method based on hesitant fuzzy sets. Int J Inf Technol Decis Making 14(3):621–657

Dias LC, Clímaco JN (2005) Dealing with imprecise information in group multicriteria decisions: a methodology and a GDSS architecture. Eur J Oper Res 160:291–307

Du Y, Liu P (2011) Extended fuzzy VIKOR method with intuitionistic trapezoidal fuzzy numbers. Inf Int Interdisc J 14:2575–2583

Duckstein L, Gershon M (1983) Multicriterion analysis of a vegetation management problem using ELECTRE II. Appl Math Model 7:254–261

Duckstein L, Opricovic S (1980) Multiobjective optimization in river basin development. Water Resour Res 16:14–20

Fernandez E, Olmedo R (2005) An agent model based on ideas of concordance and discordance for group ranking problems. Decis Support Syst 39:429–443

Figueira JR, Mousseau V, Roy B (2005) ELECTRE methods. In: Figueira J, Greco S, Ehrgott M (eds) Multiple Criteria Decision Analysis: The State of the Art Surveys. Springer Science, Business Media Inc., New York, pp 133–162

Figueira JR, Greco S, Roy B, Słowiński R (2010) ELECTRE methods: Main features and recent developments. In: Zopounidis C, Pardalos PM (eds) Handbook of Multicriteria Analysis, Applied Optimization, Springer-Verlag, Berlin, Heidelberg, pp 51–89

Hatami-Marbini A, Tavana M (2011) An extension of the Electre I method for group decision-making under a fuzzy environment. Omega 39:373–386

Hatami-Marbini A, Tavana M, Moradi M, Kangi F (2013) A fuzzy group Electre method for safety and health assessment in hazardous waste recycling facilities. Saf Sci 51:414–426

Hokkanen J, Salminen P, Rossi E, Ettala M (1995) The choice of a solid waste management system using the ELECTRE II decision-aid method. Waste Manage Res 13:175–193

Hwang CL, Yoon K (1981) Multiple attribute decision making. Lecture Notes in Economics and Mathematical Systems. Springer, Berlin

Kaplan RS, Norton D (1996) The Balanced Scorecard: Translating Strategy into Action. Harvard Business School Press, Boston

Kaya T, Kahraman C (2011) Fuzzy multiple criteria forestry decision making based on an integrated VIKOR and AHP approach. Expert Syst Appl 38:7326–7333

Leyva JC, Fernandez E (2003) A new method for group decision support based on ELECTRE III methodology. Eur J Oper Res 148:14–27

Liao HC, Xu ZS (2013) A VIKOR-based method for hesitant fuzzy multi-criteria decision making. Fuzzy Optim Decis Making 12(4):373–392

Liao HC, Xu ZS (2014a) Multi-criteria decision making with intuitionistic fuzzy PROMETHEE. J Intell Fuzzy Syst 27(4):1703–1717

Liao HC, Xu ZS (2014b) Intuitionistic fuzzy hybrid weighted aggregation operators. Int J Intell Syst 29(11):971–993

Liao HC, Xu ZS (2015) Extended hesitant fuzzy hybrid weighted aggregation operators and their application in decision making. Soft Comput 19(9):2551–2564

Liou J, Tsai CY, Lin RH, Tzeng GH (2011) A modified VIKOR multiple-criteria decision method for improving domestic airlines service quality. J Air Transp Manage 17:57–61

Mousseau V, Dias LC (2004) Valued outranking relations in ELECTRE providing manageable disaggregation procedures. Eur J Oper Res 156:467–482

Opricovic S (1998) Multicriteria optimization of civil engineering systems. Fac Civil Eng, Belgrade

Opricovic S (2002) Multicriteria planning of post-earthquake sustainable reconstruction. Comput. Aided Civil Infrastruct Eng 17:211–220

Opricovic S, Tzeng GH (2004) Compromise solution by MCDM methods: A comparative analysis of VIKOR and TOPSIS. Eur J Oper Res 156:445–455

Opricovic S, Tzeng GH (2007) Extended VIKOR method in comparison with outranking methods. Eur J Oper Res 178:514–529

Pareto V (1986) Cours d'economie politique. Droz, Geneva

Park JH, Cho HJ, Kwun YC (2011) Extension of VIKOR method for group decision making with interval-valued intuitionistic fuzzy information. Fuzzy Optim Decis Making 10:233–253

Ribeiro RA (1996) Fuzzy multiple attribute decision making: a review and new preference elicitation techniques. Fuzzy Sets Syst 78:155–181

Roy B (1968) Classement et choix en présence de points de vue multiples (la méthode ELECTRE), RIRO, 857–875

Roy B, Hugonnard JC (1982) Ranking of Suburban line extension projects on the Paris metro system by a multicriteria method. Transp Res 16:301–312

Roy B, Présent M, Silhol D (1986) A programming method for determining which Paris metro stations should be renovated. Eur J Oper Res 24:318–334

Saaty TL (1980) The Analytic Hierarchy Process. McGraw-Hill, New York

Salminen P, Hokkanen J, Lahdelma R (1998) Comparing multicriteria methods in the context of environmental problems. Eur J Oper Res 104:485–496

Vahdani B, Jabbari AHK, Roshanaei V, Zandieh M (2010) Extension of the ELECTRE method for decision-making problems with interval weights and data. The International Journal of Advanced Manufacturing Technology 50:793–800

Vahdani B, Mousavi SM, Moghaddama RT, Hashemi H (2013) A new design of the elimination and choice translating reality method for multi-criteria group decision-making in an intuitionistic fuzzy environment. Appl Math Model 37:1781–1799

Wang XT, Triantaphyllou E (2008) Ranking irregularities when evaluating alternatives by using some ELECTRE methods. Omega 36:45–63

Wu MC, Chen TY (2011) The ELECTRE multicriteria analysis approach based on Atanassov's intuitionistic fuzzy sets. Expert Syst Appl 38:12318–12327

Xu ZS, Xia MM (2011) Distance and similarity measures for hesitant fuzzy sets. Inf Sci 181: 2128–2138

Xu ZS, Yager RR (2009) Intuitionistic and interval-valued intuitionistic fuzzy preference relations and their measures of similarity for the evaluation of agreement within a group. Fuzzy Optim Decis Making 8:123–139

Yu PL (1973) A class of solutions for group decision problems. Manage Sci 19:936–946

Chapter 5
Hesitant Fuzzy Multiple Criteria Decision Making Methods with Incomplete Weight Information

In Chap. 4, we discuss the multiple criteria decision making problems in which the evaluation values of alternatives over different criteria are given as the hesitant fuzzy decision matrix H (shown as Eq. (5.1)) and the weight information of the criteria are completely known. However, in many cases, the weight information is not completely known. It is important for us to develop some procedures to handle hesitant fuzzy multiple criteria decision making problems with incomplete weights.

$$H = \begin{bmatrix} h_{11} & h_{12} & \cdots & h_{1n} \\ h_{21} & h_{22} & \cdots & h_{2n} \\ \vdots & \vdots & \ddots & \vdots \\ h_{m1} & h_{m2} & \cdots & h_{mn} \end{bmatrix} \tag{5.1}$$

To simplify the representation, let Ω be the set of all known weight information provided by the decision makers. Once the weight information in Ω is contradictory, Ω becomes an empty set and it should be reconsidered by the decision maker until there is no contradiction. Without loss of generality, the structure of the weights in Ω would be in the following forms (Liao and Xu 2014a):

(1) A weak ranking: $\{\omega_i \geq \omega_j\}$;
(2) A strict ranking: $\{\omega_i - \omega_j > \alpha_i\}$, where $\{\alpha_i\}$ are non-negative constants;
(3) A ranking of differences: $\{\omega_i - \omega_j \geq \omega_k - \omega_l\}$, for $j \neq k \neq l$;
(4) A ranking with multiples: $\{\omega_i \geq \alpha_i \omega_j\}$, where $\{\alpha_i\}$ are non-negative constants;
(5) An interval form: $\{\alpha_i \leq \omega_i \leq \alpha_i + \varepsilon_i\}$, where $\{\alpha_i\}$ and $\{\varepsilon_i\}$ are non-negative constants.

In the context of multiple criteria decision making with incomplete weight information, the first thing we shall do is to determine the weights of criteria under the restricted domain Ω. Some scholars have noticed this and carried out a series of

© Springer Nature Singapore Pte Ltd. 2017
H. Liao and Z. Xu, *Hesitant Fuzzy Decision Making Methodologies and Applications*, Uncertainty and Operations Research,
DOI 10.1007/978-981-10-3265-3_5

studies on it. Xia et al. (2013) proposed the hesitant fuzzy quasi-arithmetic means and utilized the Choquet integral to obtain the weights of criteria. Xu and Zhang (2013) constructed a programming model to get the weights of criteria. Observing that in the course of decision making, the decision makers sometimes want to interact with the analysts by modifying their preference information gradually, in the first part of this chapter, we introduce a satisfaction degree based interactive decision making method for hesitant fuzzy multiple criteria decision making with incomplete weight information. We propose the hesitant fuzzy positive ideal solution and the hesitant fuzzy negative ideal solution. Then, we definite the satisfaction degree of an alternative, based on which several optimization models are developed to determinate the weights of criteria. Subsequently, in order to make a reasonable decision, we introduce an interactive method based on some optimization models for multiple criteria decision making with hesitant fuzzy information. A practical example on evaluating the service quality of airlines is provided to illustrate the models and method.

Note that the methods in Xia et al. (2013) and Xu and Zhang (2013) are based on the objective information of alternatives over criteria. However, in some cases, the decision makers would like to provide some subjective preferences to the alternatives in advance. Consequently, when determining the weights of criteria, we should consider both the objective information (manifested as criteria values) and subjective preferences provided by the decision makers. In the second part of this chapter, we consider this scenario and propose a method to derive the weights of criteria by minimizing the deviations between the subjective and objective preferences. We also extend the method to IVHFSs. A numerical example on energy policy selection is conducted to demonstrate the effectiveness of the models and methods.

The third part of this chapter discusses the multiple stages multiple criteria decision making problems where the judgments are collected from different stages and represented in HFEs. The key to handle the multiple stages multiple criteria decision making problem is to determine the dynamic weight vector of different stages. After assigning the weights to different stages, it is easy to aggregate the dynamic hesitant information by the developed operators and then rank the alternatives. Therefore, after introducing the definition of hesitant fuzzy variable and giving some basic operational laws and aggregation operators, we introduce some novel methods to determine the weights of hesitant fuzzy variables from different stages.

5.1 Satisfaction Degree Based Interactive Decision Making Methods for Hesitant Fuzzy Multiple Criteria Decision Making with Incomplete Weight Information

5.1.1 A Satisfaction Degree Based Method for Hesitant Fuzzy Multiple Criteria Decision Making with Incomplete Weight Information

According to the aggregation operators presented in Chap. 3, it is easy to calculate the overall values $g_i(i = 1, 2, \ldots, m)$ of the alternatives $A_i(i = 1, 2, \ldots, m)$ over the criteria $C_j(j = 1, 2, \ldots, n)$. Taking the AHFWA operator (see Definition 3.4) as an illustration, the overall value of the alternative A_i with the weight vector ω can be obtained as:

$$g_i(\omega) = \text{AHFWA}(h_{i1}, h_{i2}, \ldots, h_{in}) = \overset{n}{\underset{j=1}{\oplus}} (\omega_j h_{ij})$$

$$= \left\{ 1 - \prod_{j=1}^{n} (1 - h_{ij}^{\sigma(t)})^{\omega_j} \,\middle|\, t = 1, 2, \ldots, l; l = \max\{l_{h_{i1}}, l_{h_{i2}}, \ldots, l_{h_{in}}\} \right\} \qquad (5.2)$$

It is observed that the alternative whose ratings over criteria are all full sets is ideal and desirable, and similarly the alternative whose evaluation values over criteria are all empty sets is negative and undesirable. If all ratings of the alternative A_i over the criteria $C_j(j = 1, 2, \ldots, n)$ are full sets, then the overall value of A_i can be calculated as:

$$g^+ = \text{AHFWA}(\{1\}, \{1\}, \ldots, \{1\}) = \overset{n}{\underset{j=1}{\oplus}} (\omega_j \{1\})$$

$$= \left\{ 1 - \prod_{j=1}^{n} (1 - 1)^{\omega_j} \,\middle|\, t = 1, 2, \ldots, l \right\} = \{1\} \qquad (5.3)$$

Analogously, if all evaluation values of the alternative A_i over the criteria $C_j(j = 1, 2, \ldots, n)$ are empty sets, then the overall value of A_i can be calculated as:

$$g^- = \text{AHFWA}(\{0\}, \{0\}, \ldots, \{0\}) = \overset{n}{\underset{j=1}{\oplus}} (\omega_j \{0\})$$

$$= \left\{ 1 - \prod_{j=1}^{n} (1 - 0)^{\omega_j} \,\middle|\, t = 1, 2, \ldots, l \right\} = \{0\} \qquad (5.4)$$

Definition 5.1 (*Liao and Xu* 2014b). $G^+ = (\{1\}, \{1\}, \ldots, \{1\})$ is called the hesitant fuzzy positive ideal solution and $G^- = (\{0\}, \{0\}, \ldots, \{0\})$ is called the hesitant fuzzy negative ideal solution.

For the alternative A_i with the ratings $h_i = (h_{i1}, h_{i2}, \ldots, h_{in})$, based on the distance measure of HFEs (taking the hesitant normalized Manhattan distance shown as Eq. (4.7) as an illustration), we can calculate the distance between the alternative A_i and the hesitant fuzzy positive ideal solution G^+ and also the distance between the alternative A_i and the hesitant fuzzy negative ideal solution G^-, respectively.

$$d_{hnm}(h_i, G^+) = \frac{1}{l} \sum_{t=1}^{l} \left| \left(1 - \prod_{j=1}^{n} (1 - h_{ij}^{\sigma(t)})^{\omega_j} \right) - 1 \right| = \frac{1}{l} \sum_{t=1}^{l} \prod_{j=1}^{n} (1 - h_{ij}^{\sigma(t)})^{\omega_j}$$

(5.5)

$$d_{hnm}(h_i, G^-) = \frac{1}{l} \sum_{t=1}^{l} \left| \left(1 - \prod_{j=1}^{n} (1 - h_{ij}^{\sigma(t)})^{\omega_j} \right) - 0 \right|$$
$$= \frac{1}{l} \sum_{t=1}^{l} \left(1 - \prod_{j=1}^{n} (1 - h_{ij}^{\sigma(t)})^{\omega_j} \right)$$

(5.6)

where $l = \max\{l_{h_{i1}}, l_{h_{i2}}, \ldots, l_{h_{in}}\}$.

Since $g_i(\omega) = \left\{ 1 - \prod_{j=1}^{n} (1 - h_{ij}^{\sigma(t)})^{\omega_j} \Big| t = 1, 2, \ldots, l; \ l = \max\{l_{h_{i1}}, l_{h_{i2}}, \ldots, l_{h_{in}}\} \right\}$, then we can see that $d_{hnm}(h_i, G^-) = s(g_i)$, i.e., the distance between h_i and G^- equals to the score function value of g_i, which is just a coincidence.

Ostensibly, all of these derivations seem to be quite easy. However, we may ignore an importance precondition, i.e., the weight information is partially known and we cannot get the crisp weights corresponding to different criteria. Consequently, it is hard or impossible to calculate the distance between the alternatives $A_i(i = 1, 2, \ldots, m)$ and the hesitant fuzzy positive ideal solution G^+ and the distance between the alternatives $A_i(i = 1, 2, \ldots, m)$ and the hesitant fuzzy negative ideal solution G^- by Eqs. (5.5) and (5.6), respectively. To determinate the weights is a difficult or insurmountable question due to the fact that the unknown parameters $\omega_j(j = 1, 2, \ldots, n)$ are in the exponential term. Hence, we need to find a novel way to solve this issue.

Below we reconsider the main idea of what we have done above. We first fuse the values $h_{ij}(j = 1, 2, \ldots, n)$ of the alternative A_i over the criteria $C_j(j = 1, 2, \ldots, n)$ into an overall value g_i, and then calculate the distance between the derived overall value g_i and G^+ or G^-. What about changing the order of these two steps? If we first calculate the distance between each rating and the hesitant fuzzy positive ideal point or the hesitant fuzzy negative ideal point, and then fuse the distances with respect to the criteria $C_j(j = 1, 2, \ldots, n)$, the computational complexity will be changed significantly, which makes the problem easy to handle.

Since

$$d_{hnm}(h_{ij}, h_{ij}^+) = \frac{1}{l}\sum_{t=1}^{l}\left|h_{ij}^{\sigma(t)} - 1\right| = \frac{1}{l}\sum_{t=1}^{l}\left(1 - h_{ij}^{\sigma(t)}\right) = 1 - \frac{1}{l}\sum_{t=1}^{l}h_{ij}^{\sigma(t)} = 1 - s(h_{ij})$$

(5.7)

$$d_{hnm}(h_{ij}, h_{ij}^-) = \frac{1}{l}\sum_{t=1}^{l}\left|h_{ij}^{\sigma(t)} - 0\right| = \frac{1}{l}\sum_{t=1}^{l}h_{ij}^{\sigma(t)} = s(h_{ij})$$

(5.8)

then we can fuse the distances with respect to different criteria by some developed operators. These operators are not limited by the hesitant fuzzy aggregation operators but all the classical operators because the distance $d_{hnm}(h_{ij}, h_{ij}^+)$ and $d_{hnm}(h_{ij}, h_{ij}^-)$ are all crisp values. Taking the weighted averaging (WA) operator as an example, the overall distance between the alternative A_i and the hesitant fuzzy positive ideal solution G^+ and also the distance between the alternative A_i and the hesitant fuzzy negative ideal solution G^- can be derived respectively as:

$$d'_{hnm}(h_i, G^+) = \sum_{j=1}^{n}\omega_j\left(1 - s(h_{ij})\right) = 1 - \sum_{j=1}^{n}\omega_j s(h_{ij})$$

(5.9)

$$d'_{hnm}(h_i, G^-) = \sum_{j=1}^{n}\omega_j s(h_{ij})$$

(5.10)

Intuitively, the smaller the distance $d'_{hnm}(h_i, G^+)$ is, the better the alternative should be; while the larger the distance $d'_{hnm}(h_i, G^-)$ is, the better the alternative should be. Motivated by the TOPSIS method, we shall take both the distance $d'_{hnm}(h_i, G^+)$ and the distance $d'_{hnm}(h_i, G^-)$ into consideration rather than consider them individually. Then we can derive the definition of satisfaction degree naturally.

Definition 5.2 (*Liao and Xu* 2014b). A satisfaction degree of the alternative A_i over the criteria $C_j(j = 1, 2, \ldots, n)$ with the weight vector $\omega = (\omega_1, \omega_2, \ldots, \omega_n)^T \in \Omega$ is defined as:

$$\sigma(h_i(\omega)) = \frac{d'_{hnm}(h_i, G^-)}{d'_{hnm}(h_i, G^+) + d'_{hnm}(h_i, G^-)}$$

(5.11)

where $0 \le \omega_j \le 1$, $j = 1, 2, \ldots, n$, and $\sum_{j=1}^{n}\omega_j = 1$.

Combining Eqs. (5.9), (5.10) and (5.11), we have

$$\sigma(h_i(\omega)) = \frac{d'_{hnm}(h_i, G^-)}{d'_{hnm}(h_i, G^+) + d'_{hnm}(h_i, G^-)} = \frac{\sum_{j=1}^n \omega_j s(h_{ij})}{1 - \sum_{j=1}^n \omega_j s(h_{ij}) + \sum_{j=1}^n \omega_j s(h_{ij})}$$
$$= \sum_{j=1}^n \omega_j s(h_{ij})$$

$$(5.12)$$

From Eqs. (5.10) and (5.12), we can see that the satisfaction degree reduces to the distance between the alternative A_i and the hesitant fuzzy negative ideal solution G^-, which is a coincidence. In order not to lose much information and make our method more applicable, we introduce a parameter θ, which denotes the risk preference of the decision maker: $\theta > 0.5$ means the decision maker is pessimist and the further the distance between the alternative and the positive ideal solution, the better the choice; while $\theta < 0.5$ means the opposite. The value of the parameter θ is provided by the decision maker in advance. Consequently, the satisfaction degree becomes

$$\sigma(h_i(\omega)) = \frac{(1 - \theta) \sum_{j=1}^n \omega_j s(h_{ij})}{\theta \left(1 - \sum_{j=1}^n \omega_j s(h_{ij})\right) + (1 - \theta) \sum_{j=1}^n \omega_j s(h_{ij})}$$

$$(5.13)$$

It is obvious that $0 \leq \sigma(h_i(\omega)) \leq 1$, for any $\theta \in [0, 1]$, $i = 1, 2, \ldots, m$. As our purpose is to select the alternative with the highest satisfaction degree, the following multiple objective optimization model can be generated naturally.

Model 5.1

$$\max \ \sigma(h_1(\omega)), \sigma(h_2(\omega)), \ldots, \sigma(h_m(\omega))$$
$$s.\,t.\,\omega = (\omega_1, \omega_2, \ldots, \omega_n) \in \Omega$$
$$0 \leq \omega_j \leq 1, j = 1, 2, \ldots, n$$
$$\sum_{j=1}^n \omega_j = 1$$

We change Model 5.1 into a single objective optimization model by using the equal weighted summation method (French et al. 1983).

Model 5.2

$$\max \sum_{i=1}^m \sigma(h_i(\omega))$$
$$s.\,t.\,\omega = (\omega_1, \omega_2, \ldots, \omega_n) \in \Omega$$
$$0 \leq \omega_j \leq 1, j = 1, 2, \ldots, n$$
$$\sum_{j=1}^n \omega_j = 1$$

Combining Eq. (5.13) and Model 5.2, we obtain

Model 5.3

$$\max \sum_{i=1}^{m} \frac{(1 - \theta) \sum_{j=1}^{n} \omega_j s(h_{ij})}{\theta \left(1 - \sum_{j=1}^{n} \omega_j s(h_{ij})\right) + (1 - \theta) \sum_{j=1}^{n} \omega_j s(h_{ij})}$$

$$s.\,t.\,\omega = (\omega_1, \omega_2, \ldots, \omega_n) \in \Omega$$

$$0 \leq \omega_j \leq 1, j = 1, 2, \ldots, n$$

$$\sum_{j=1}^{n} \omega_j = 1$$

This model can be solved by using many efficient algorithms (Terlaky 1996) or using the MATLAB or the Lingo mathematic software package. Suppose that the optimal solution of Model 5.3 is $\omega^* = (\omega_1^*, \omega_2^*, \ldots, \omega_n^*)^T$, then we can calculate the overall value g_i of each alternative A_i according to Eq. (5.2). Subsequently, the ranking order of alternatives can be derived by the comparison method in Sect. 1.1.3.

Let us reconsider the satisfaction based method again deeply. In general, in the process of decision making with incomplete weight information of criteria, the basic thing we should do is to find the weights that are as adequate as possible to the opinions of decision makers. Does our proposed model reflect the decision makers' opinions? The answer is "yes". Actually, starting from calculating the overall distance between the alternative A_i and the hesitant fuzzy positive ideal solution G^+ and the distance between the alternative A_i and the hesitant fuzzy negative ideal solution G^- by Eqs. (5.9) and (5.10), respectively, the decision makers' opinions have been taken into account. The aim of introducing the satisfaction degree is also to model the decision makers' opinions more comprehensive since it includes both of the above two distances. In addition, the parameter θ, which denotes the risk preferences of the decision makers, is also used to enhance the reflection of the decision makers' ideas. Since the unknown weight information cannot be obtained directly, maximizing those satisfaction degrees simultaneously is a good choice to find a solution which does not show any discrimination to certain alternative(s), and meanwhile reflects the decision makers' opinions comprehensively. Certainly we can also minimize those satisfaction degrees simultaneously if we want to select the worst alternative(s).

5.1.2 An Interactive Method for Hesitant Fuzzy Multiple Criteria Decision Making with Incomplete Weight Information

Section 5.1.1 presents a satisfaction degree based method to handle multiple criteria decision making problem whose weight information is partially known. However, by using this method, the satisfaction degrees of certain alternatives are sometimes too high and others are simultaneously too low. Satisfaction degrees with a wide range may match with some decision makers' requirement, but, in many cases, the decision makers may want to modify their satisfaction degrees slightly in order to provide new preference information or modify the previous preference information. Interacting with the decision makers gradually is an acceptable and applicable way for doing so in reality. In the following, we propose an interactive method for multiple criteria decision making with hesitant fuzzy information.

The main idea of this method can be clarified easily. Firstly, the decision makers give the lower bounds of the satisfaction degrees with respect to each alternative, and then according to these lower bounds, we can establish the weights of different criteria. Once we have determined the different weights, the satisfaction degrees of different alternatives can be calculated and the analysts then ask the decision makers whether they want to reconsider the satisfaction degrees or not. If the decision makers are not satisfied with the derived satisfaction degrees, then the analysts shall inform the decision makers to reconsider their lower bounds of the satisfaction degrees and then go to do iteration till acceptable.

In order to help the decision makers establish the lower bounds of the alternatives, motivated by the max-min operator developed by Zimmermann and Zysno (1980), we derive Model 5.4.

Model 5.4

$$\max \tau$$

$$s.t.\ \sigma\big(h_i(\omega)\big) \geq \tau$$

$$\omega = (\omega_1, \omega_2, \ldots, \omega_n)^T \in \Omega$$

$$0 \leq \omega_j \leq 1,\ j = 1, 2, \ldots, n,$$

$$\sum_{j=1}^{n} \omega_j = 1$$

Solving Model 5.4, we obtain the initial optimal weight vector $\omega^{(0)} = (\omega_1^{(0)}, \omega_2^{(0)}, \ldots, \omega_n^{(0)})^T$ and the initial satisfaction degrees $\sigma(h_i(\omega^{(0)}))(i = 1, 2, \ldots, m)$ of the alternatives $A_i(i = 1, 2, \ldots, m)$. Then the decision makers can provide the lower bounds $\tau_i^{(0)}(i = 1, 2, \ldots, m)$ of the satisfaction degrees of the alternatives $A_i(i = 1, 2, \ldots, m)$ according to $\sigma(h_i(\omega^{(0)}))\ (i = 1, 2, \ldots, m)$. Once we obtain the lower bounds, the criteria weights can be re-established by Model 5.5.

Model 5.5

$$\max \sum_{i=1}^{m} \tau_i$$

$$s.\,t.\,\sigma(h_i(\omega)) \geq \tau_i \geq \tau_i^{(0)}, \quad i = 1, 2, \ldots, m$$

$$\omega = (\omega_1, \omega_2, \ldots, \omega_n) \in \Omega$$

$$0 \leq \omega_j \leq 1, j = 1, 2, \ldots, n$$

$$\sum_{j=1}^{n} \omega_j = 1$$

Solving Model 5.5 we can get a new weight vector $\omega^{(1)} = (\omega_1^{(1)}, \omega_2^{(1)}, \ldots, \omega_n^{(1)})^T$. If Model 5.5 has no optimal solution, this means that some lower bounds are greater than the corresponding initial satisfaction degrees. Hence it needs to be reconsidered till the optimal solution is obtained.

For the convenience of application, the procedure of the interactive method for multiple criteria decision making under hesitant fuzzy environment with incomplete weight information can be described as follows:

Algorithm 5.1

Step 1. Construct the hesitant fuzzy decision matrix, and then go to the next step.

Step 2. Use Model 5.4 to determinate the initial weight vector $\omega^{(0)} = (\omega_1^{(0)}, \omega_2^{(0)}, \ldots, \omega_n^{(0)})$ and the initial satisfaction degrees $\sigma(h_i(\omega^{(0)}))(i = 1, 2, \ldots, m)$ of the alternatives $A_i(i = 1, 2, \ldots, m)$, and then go to the next step.

Step 3. The decision makers provide the lower bounds $\tau_i^{(t)}(i = 1, 2, \ldots, m)$ of the satisfaction degrees of the alternatives $A_i(i = 1, 2, \ldots, m)$ according to $\sigma(h_i(\omega^{(0)}))$ $(i = 1, 2, \ldots, m)$. Let $t = t + 1$, then go to the next step.

Step 4. Solve Model 5.5 to determinate the weight vector $\omega^{(t)} = (\omega_1^{(t)}, \omega_2^{(t)}, \ldots, \omega_n^{(t)})^T$ and the satisfaction degrees $\sigma(h_i(\omega^{(t)}))(i = 1, 2, \ldots, m)$ of the alternatives $A_i(i = 1, 2, \ldots, m)$, and then go to the next step.

Step 5. If the model has an optimal solution, then go to Step 6; Otherwise go to Step 3.

Step 6. Calculate the overall values $g_i(\omega^{(t)})(i = 1, 2, \ldots, m)$ of the alternatives $A_i(i = 1, 2, \ldots, m)$ and rank the alternatives according to the comparison law, and then choose the best one(s). Go to the next step.

Step 7. End.

5.1.3   Application of the Approaches in Domestic Airline Service Quality Evaluation

We now consider the same example concerning the domestic airline service quality evaluation as shown in Example 4.2 to illustrate the satisfaction based method.

Example 5.1 (Liao and Xu 2014b). In Example 4.2, the weight vector of the criteria is given directly as $\omega = (0.1, 0.2, 0.4, 0.3)^T$. This is somehow subjective. Here we suppose that the weight information is incomplete. All the information regarding to the weights of criteria is $\Omega = \{\omega_3 \geq \omega_4 \geq \omega_2 \geq \omega_1 \geq 0.1, 0.5 \geq \omega_3 \geq 0.4, \omega_2 \leq 2\omega_1, \sum_{j=1}^{4} \omega_j = 1\}$. Below we use the satisfaction based method to solve this problem:

Firstly, we calculate the distance between each rating and the hesitant fuzzy positive ideal solution and the hesitant fuzzy negative ideal solution by using Eqs. (5.7) and (5.8). To do so, we only need to compute the score function value matrix S:

$$S = \begin{bmatrix} 0.7333 & 0.7 & 0.6 & 0.6 \\ 0.8 & 0.7333 & 0.6 & 0.6 \\ 0.6333 & 0.7333 & 0.5 & 0.6 \\ 0.75 & 0.8 & 0.4333 & 0.45 \end{bmatrix}$$

Take $\theta = 0.4$ as an illustration. Based on the score function value matrix S and the partially known weight information Ω, motivated by Model 5.3, we can construct Model 5.6:

Model 5.6

$$\max \quad \frac{0.6(0.7333\omega_1 + 0.7\omega_2 + 0.6\omega_3 + 0.6\omega_4)}{0.4(0.2667\omega_1 + 0.3\omega_2 + 0.4\omega_3 + 0.4\omega_4) + 0.6(0.7333\omega_1 + 0.7\omega_2 + 0.6\omega_3 + 0.6\omega_4)}$$
$$+ \frac{0.6(0.8\omega_1 + 0.7333\omega_2 + 0.6\omega_3 + 0.6\omega_4)}{0.4(0.2\omega_1 + 0.2667\omega_2 + 0.4\omega_3 + 0.4\omega_4) + 0.6(0.8\omega_1 + 0.7333\omega_2 + 0.6\omega_3 + 0.6\omega_4)}$$
$$+ \frac{0.6(0.6333\omega_1 + 0.7333\omega_2 + 0.5\omega_3 + 0.6\omega_4)}{0.4(0.3667\omega_1 + 0.2667\omega_2 + 0.5\omega_3 + 0.4\omega_4) + 0.6(0.6333\omega_1 + 0.7333\omega_2 + 0.5\omega_3 + 0.6\omega_4)}$$
$$+ \frac{0.6(0.75\omega_1 + 0.8\omega_2 + 0.4333\omega_3 + 0.45\omega_4)}{0.4(0.25\omega_1 + 0.2\omega_2 + 0.5667\omega_3 + 0.55\omega_4) + 0.6(0.75\omega_1 + 0.8\omega_2 + 0.4333\omega_3 + 0.45\omega_4)}$$

s.t. $\omega_3 \geq \omega_4 \geq \omega_2 \geq \omega_1 \geq 0.1$
$0.5 \geq \omega_3 \geq 0.4; \ \omega_2 \leq 2\omega_1$
$\omega_1 + \omega_2 + \omega_3 + \omega_4 = 1$

The objective function can be simplified as:

$$\max \quad \frac{0.44\omega_1 + 0.42\omega_2 + 0.36\omega_3 + 0.36\omega_4}{0.5467\omega_1 + 0.54\omega_2 + 0.52\omega_3 + 0.52\omega_4} + \frac{0.48\omega_1 + 0.44\omega_2 + 0.36\omega_3 + 0.36\omega_4}{0.56\omega_1 + 0.5467\omega_2 + 0.52\omega_3 + 0.52\omega_4}$$
$$+ \frac{0.38\omega_1 + 0.44\omega_2 + 0.3\omega_3 + 0.36\omega_4}{0.5267\omega_1 + 0.5467\omega_2 + 0.5\omega_3 + 0.2\omega_4} + \frac{0.45\omega_1 + 0.48\omega_2 + 0.26\omega_3 + 0.27\omega_4}{0.55\omega_1 + 0.56\omega_2 + 0.4867\omega_3 + 0.49\omega_4}$$

Solving this model, we get the optimal solution: $\omega^* = (0.12, 0.24, 0.4, 0.24)^T$. According to Eq. (5.2), the overall value of each domestic airline can be obtained as:

$$g_1(\omega^*) = \{0.4485, 0.5293, 0.8819\}; \ g_2(\omega^*) = \{0.4942, 0.6335, 0.8282\}$$
$$g_3(\omega^*) = \{0.4578, 0.5694, 0.7805\}; \ g_4(\omega^*) = \{0.4571, 0.5161, 0.7716\}$$

Hence, the score function values of the overall values $g_i(\omega^*)(i = 1, 2, 3, 4)$ are

$$s(g_1(\omega^*)) = 0.6199, \ s(g_2(\omega^*)) = 0.6520, \ s(g_3(\omega^*)) = 0.6026,$$
$$s(g_4(\omega^*)) = 0.5816$$

Since $s(g_2(\omega^*)) > s(g_1(\omega^*)) > s(g_3(\omega^*)) > s(g_4(\omega^*))$, then we can rank these four domestic airlines in descending order as $g_2(\omega^*) \succ g_1(\omega^*) \succ g_3(\omega^*) \succ g_4(\omega^*)$, where "$\succ$" denotes "be prior to". That is to say, the service quality of Transasia is the best among the service quality of domestic airline in Taiwan.

To show the applicability of the interactive method and its corresponding Algorithm 5.1, let us use the same example in Example 5.1 as an illustration:

Example 5.2 (Liao and Xu 2014b). In this example, we also take $\theta = 0.4$ as an illustration. Based on Model 5.4, we can establish Model 5.7:

Model 5.7

$$\max \quad \tau$$
$$s.t. \ \frac{0.44\omega_1 + 0.42\omega_2 + 0.36\omega_3 + 0.36\omega_4}{0.5467\omega_1 + 0.54\omega_2 + 0.52\omega_3 + 0.52\omega_4} \geq \tau$$
$$\frac{0.48\omega_1 + 0.44\omega_2 + 0.36\omega_3 + 0.36\omega_4}{0.56\omega_1 + 0.5467\omega_2 + 0.52\omega_3 + 0.52\omega_4} \geq \tau$$
$$\frac{0.38\omega_1 + 0.44\omega_2 + 0.3\omega_3 + 0.36\omega_4}{0.5267\omega_1 + 0.5467\omega_2 + 0.5\omega_3 + 0.2\omega_4} \geq \tau$$
$$\frac{0.45\omega_1 + 0.48\omega_2 + 0.26\omega_3 + 0.27\omega_4}{0.55\omega_1 + 0.56\omega_2 + 0.4867\omega_3 + 0.49\omega_4} \geq \tau$$
$$\omega_3 \geq \omega_4 \geq \omega_2 \geq \omega_1 \geq 0.1$$
$$0.5 \geq \omega_3 \geq 0.4; \ \omega_2 \leq 2\omega_1$$
$$\omega_1 + \omega_2 + \omega_3 + \omega_4 = 1$$

Solving this model, we get the initial weight vector $\omega^{(0)} = (\omega_1^{(0)}, \omega_2^{(0)}, \ldots, \omega_n^{(0)})^T = (0.2, 0.2, 0.4, 0.2)^T$ and the initial satisfaction degrees $\sigma^{(0)} = \left(\sigma^{(0)}(h_1), \sigma^{(0)}(h_2), \sigma^{(0)}(h_3), \sigma^{(0)}(h_4)\right)^T = (73.3\%, 75\%, 78.3\%, 66.84\%)^T$ on these four domestic airlines.

Suppose that the decision makers provide the lower bounds of the satisfaction degrees on these four domestic airlines as $\tau^{(0)} = (\tau_1^{(0)}, \tau_2^{(0)}, \tau_3^{(0)}, \tau_4^{(0)})^T = (70\,\%,$ $75\,\%, 80\,\%, 70\,\%)^T$. Model 5.5 becomes the following optimal programming problem:

Model 5.8

$$
\begin{aligned}
&\max \quad (\tau_1 + \tau_2 + \tau_3 + \tau_4) \\
&s.t. \quad \frac{0.44\omega_1 + 0.42\omega_2 + 0.36\omega_3 + 0.36\omega_4}{0.5467\omega_1 + 0.54\omega_2 + 0.52\omega_3 + 0.52\omega_4} \geq \tau_1 \geq 0.7 \\
&\qquad \frac{0.48\omega_1 + 0.44\omega_2 + 0.36\omega_3 + 0.36\omega_4}{0.56\omega_1 + 0.5467\omega_2 + 0.52\omega_3 + 0.52\omega_4} \geq \tau_2 \geq 0.75 \\
&\qquad \frac{0.38\omega_1 + 0.44\omega_2 + 0.3\omega_3 + 0.36\omega_4}{0.5267\omega_1 + 0.5467\omega_2 + 0.5\omega_3 + 0.2\omega_4} \geq \tau_3 \geq 0.8 \\
&\qquad \frac{0.45\omega_1 + 0.48\omega_2 + 0.26\omega_3 + 0.27\omega_4}{0.55\omega_1 + 0.56\omega_2 + 0.4867\omega_3 + 0.49\omega_4} \geq \tau_4 \geq 0.7 \\
&\qquad \omega_3 \geq \omega_4 \geq \omega_2 \geq \omega_1 \geq 0.1 \\
&\qquad 0.5 \geq \omega_3 \geq 0.4; \ \omega_2 \leq 2\omega_1 \\
&\qquad \omega_1 + \omega_2 + \omega_3 + \omega_4 = 1
\end{aligned}
$$

Solving Model 5.8, we find that there is no feasible solution. This may result from the fact that some of the lower bounds of the satisfaction degrees given by the decision makers are too high. We appeal to the decision makers and then they modify some of their lower bounds of the satisfaction degrees referring to the initial satisfaction degrees. Suppose that the modified lower bounds are $\tau^{(0)} = (\tau_1^{(0)}, \tau_2^{(0)}, \tau_3^{(0)}, \tau_4^{(0)})^T = (70\,\%, 72\,\%, 80\,\%, 65\,\%)^T$, then we modify Model 5.8 to Model 5.9.

Model 5.9

$$
\begin{aligned}
&\max \quad (\tau_1 + \tau_2 + \tau_3 + \tau_4) \\
&s.t. \quad \frac{0.44\omega_1 + 0.42\omega_2 + 0.36\omega_3 + 0.36\omega_4}{0.5467\omega_1 + 0.54\omega_2 + 0.52\omega_3 + 0.52\omega_4} \geq \tau_1 \geq 0.7 \\
&\qquad \frac{0.48\omega_1 + 0.44\omega_2 + 0.36\omega_3 + 0.36\omega_4}{0.56\omega_1 + 0.5467\omega_2 + 0.52\omega_3 + 0.52\omega_4} \geq \tau_2 \geq 0.72 \\
&\qquad \frac{0.38\omega_1 + 0.44\omega_2 + 0.3\omega_3 + 0.36\omega_4}{0.5267\omega_1 + 0.5467\omega_2 + 0.5\omega_3 + 0.2\omega_4} \geq \tau_3 \geq 0.8 \\
&\qquad \frac{0.45\omega_1 + 0.48\omega_2 + 0.26\omega_3 + 0.27\omega_4}{0.55\omega_1 + 0.56\omega_2 + 0.4867\omega_3 + 0.49\omega_4} \geq \tau_4 \geq 0.65 \\
&\qquad \omega_3 \geq \omega_4 \geq \omega_2 \geq \omega_1 \geq 0.1 \\
&\qquad 0.5 \geq \omega_3 \geq 0.4; \ \omega_2 \leq 2\omega_1 \\
&\qquad \omega_1 + \omega_2 + \omega_3 + \omega_4 = 1
\end{aligned}
$$

Solving Model 5.9, we get $\omega^{(1)} = (\omega_1^{(1)}, \omega_2^{(1)}, \omega_3^{(1)}, \omega_4^{(1)})^T = (0.12, 0.24,$ $0.4, 0.24)^T$ and the satisfaction degrees $\sigma^{(1)} = (\sigma^{(1)}(h_1), \sigma^{(1)}(h_2), \sigma^{(1)}(h_3), \sigma^{(1)}$ $(h_4))^T = (72.73\%, 74.1\%, 80.83\%, 65.93\%)^T$ on these four domestic airlines. If the decision makes are not satisfied with these results, they can further modify the lower bounds of the satisfaction degrees. Suppose that the decision makes are satisfied with these results, then we go to the next step.

We further calculate the overall values $g_i(\omega^{(1)})(i = 1, 2, 3, 4)$ of these four domestic airlines and rank them according to the comparison law in Sect. 1.1.3. In analogy to Example 5.1, we can obtain $g_2(\omega^{(1)}) \succ g_1(\omega^{(1)}) \succ g_3(\omega^{(1)}) \succ g_4(\omega^{(1)})$, where "$\succ$" denotes "be prior to", i.e., the service quality of Transasia is best among the service quality of domestic airline in Taiwan. But from the satisfaction degrees, we can see $\sigma_3(\omega^{(1)}) \succ \sigma_2(\omega^{(1)}) \succ \sigma_1(\omega^{(1)}) \succ \sigma_4(\omega^{(1)})$. The reason for this is that when deriving the satisfaction degree, in order not to decrease the computational complexity, we use Eqs. (5.9) and (5.10) to substitute Eqs. (5.5) and (5.6), respectively. So, after we obtain the weights of the criteria, in order not to lose much information, we shall calculate the satisfaction degree by Eq. (5.11) instead of Eq. (5.12).

5.2 Minimum Deviation Methods for Hesitant Fuzzy Multiple Criteria Decision Making with Incomplete Weight Information

5.2.1 Minimum Deviation Methods for Hesitant Fuzzy Multiple Criteria Decision Making with Incomplete Weight Information

In practical decision making problems, people are often difficult to give the explicit weight information. Sometimes even there is an extreme case that the weights are completely unknown. Meanwhile, the decision maker often has particular subjective preference to the alternatives. How to solve this kind of decision making problem becomes a necessary and interesting thing. In view of this situation, Xu (2004) proposed a method based on the minimum deviation between the subjective and objective preferences. But this method is useful for triangular fuzzy information. In the following, we introduce some minimum deviation methods for hesitant fuzzy multiple criteria decision making in which the weights of criteria are not really sure and the decision maker has preferences over all the alternatives.

1. **Method Based on Hesitant Fuzzy Expected Values and Minimum Deviations**

In the following, we first define the concept of hesitant fuzzy expected value, which includes the risk preference of the decision maker:

Definition 5.3 (*Zhao et al.* 2016). Let h be a HFE and $h = \{\gamma_1, \gamma_2, \ldots \gamma_l\}$, then the expected value of h is

$$h^{(T)} = \frac{1}{l-1}\left[(1-T)\gamma_{\sigma(l)} + \gamma_{\sigma(l-1)} + \cdots + \gamma_{\sigma(2)} + T\gamma_{\sigma(1)}\right] \qquad (5.14)$$

where $\gamma_{\sigma(t)}$ is the tth largest number of γ_k ($k = 1, 2, \ldots, l$), T is a real number lying between 0 and 1.

The choice of T depends on the risk attitude of the decision maker. $T > 0.5$ shows that the decision maker prefers to risk; $T = 0.5$ implies that the decision maker is risk neutral; $T < 0.5$ means the decision maker is risk-averse. In fact, $\gamma_{\sigma(1)}$ and $\gamma_{\sigma(l)}$ in Eq. (5.14) reveal the most optimistic attitude and the most pessimistic attitude of the decision maker. Thus, by Eq. (5.14), a medium value between $\gamma_{\sigma(1)}$ and $\gamma_{\sigma(l)}$ can be derived. Obviously, Definition 5.3 is just a special case of the expected value of probability theory but reflects the risk preference clearly.

We can use Eq. (5.14) to calculate the expected values $h_{ij}^{(T)}$ of all evaluation values h_{ij} ($i = 1, 2, \ldots, m$; $j = 1, 2, \ldots, n$) and get the hesitant fuzzy expected value decision matrix $H^{(T)} = \left(h_{ij}^{(T)}\right)_{m \times n}$. If the subjective preference s_i to the ith alternative A_i given by the decision maker is a HFE, we can also compute its expected value $s_i^{(T)}$.

In the following, we introduce a hesitant fuzzy multiple criteria decision making method based on the hesitant fuzzy expected value and deviations, where the weights of criteria are completely unknown or incompletely known and the decision maker has subjective preferences over the alternatives.

(1) Case with completely unknown weight information on criteria

Due to various constraints, there usually is a certain deviation between the subjective and objective preferences. Suppose that the deviation between the expected values $h_{ij}^{(T)}$ and the subjective preference values $s_i^{(T)}$ given by the decision maker is denoted as $\sigma_{ij} = h_{ij}^{(T)} - s_i^{(T)}$, then $\sigma_{ij}^2 = \left(h_{ij}^{(T)} - s_i^{(T)}\right)^2$. Thus, the deviations between all the expected values $h_{ij}^{(T)}$ ($j = 1, 2, \ldots, n$) of the ith alternative A_i and the ith subjective preference value $s_i^{(T)}$ can be expressed by $\sigma_i^2 = \sum_{j=1}^{n}\left(\sigma_{ij}\omega_j\right)^2$. In order to make the decision result more scientific and reasonable, the weights of criteria should minimize the total deviation between the subjective preferences and the objective ones. Therefore, we construct the following single objective optimization model:

Model 5.10

$$\min \ \sigma(\omega) = \sum_{i=1}^{m} \sum_{j=1}^{n} \sigma_{ij}^2 \omega_j^2$$

$$s.t. \ \ \omega_j \geq 0, \ \sum_{j=1}^{n} \omega_j = 1$$

To solve the model, we construct the Lagrange function as:

$$\sigma(\omega, \lambda) = \sum_{i=1}^{m} \sum_{j=1}^{n} \sigma_{ij}^2 \omega_j^2 + 2\lambda (\sum_{j=1}^{n} \omega_j - 1) \tag{5.15}$$

Computing the partial derivative and let

$$\frac{\partial \sigma}{\partial \omega_j} = 2 \sum_{i=1}^{m} \sigma_{ij}^2 \omega_j + 2\lambda = 0, \ j = 1, 2, \dots, n \tag{5.16}$$

$$\frac{\partial \sigma}{\partial \lambda} = \sum_{j=1}^{n} \omega_j - 1 = 0 \tag{5.17}$$

From Eq. (5.16), we can get

$$\omega_j = \frac{-\lambda}{\sum_{i=1}^{m} \sigma_{ij}^2}, \ j = 1, 2, \dots, n \tag{5.18}$$

Substituting Eq. (5.18) into Eq. (5.17), we obtain

$$\lambda = -1 / \sum_{j=1}^{n} 1 / \sum_{i=1}^{m} \sigma_{ij}^2 \tag{5.19}$$

Then we get

$$\omega_j = \frac{1}{\sum_{j=1}^{n} 1 / \sum_{i=1}^{m} \sigma_{ij}^2} \bigg/ \sum_{i=1}^{m} \sigma_{ij}^2, \ j = 1, 2, \dots, n \tag{5.20}$$

Using the weight vector $\omega = (\omega_1, \omega_2 \cdots, \omega_n)^T$ and the weighted average method:

$$h_i^{(T)} = \sum_{j=1}^{n} h_{ij}^{(T)} \omega_j, \ \ i = 1, 2, \dots, m \tag{5.21}$$

we can compute the overall expected value $h_i^{(T)}$ of all the alternatives A_i $(i = 1, 2, \ldots, m)$. Then we can sort the alternatives and choose the best one(s) according to the value of $h_i^{(T)}$ $(i = 1, 2, \ldots, m)$.

(2) Case with partly known weight information on criteria

Sometimes people can provide partly weight information when making decisions. If the criterion weight vector $\omega = (\omega_1, \omega_2 \cdots, \omega_n)^T$ satisfies the constraints $0 \le a_j \le \omega_j \le b_j, j = 1, 2, \ldots, n$, where a_j and b_j are the upper and lower bounds of ω_j, respectively. In this situation, we give another minimum deviation method to determine the criterion weight vector.

Model 5.11

$$\min \quad \sigma(\omega) = \sum_{i=1}^{m} \sum_{j=1}^{n} \sigma_{ij}^2 \omega_j$$

$$s.t. \ \ 0 \le a_j \le \omega_j \le b_j, \ j = 1, 2, \ldots, n$$

$$\sum_{j=1}^{n} \omega_j = 1$$

Using the MATLAB or LINGO mathematic software package, we can solve this model and get the optimal criterion weight vector. Next, we use Eq. (5.21) to obtain the overall criterion expected values of all the alternatives and give their rankings.

Taking into account the above two cases, we propose the following hesitant fuzzy decision method:

Algorithm 5.2

Step 1. Construct the hesitant fuzzy decision matrix H, and then go to the next step.

Step 2. Assume that the decision maker has subjective preference over the alternative A_i $(i = 1, 2, \ldots, m)$ and all the preference values s_i $(i = 1, 2, \ldots, m)$ are HFEs. Then we utilize Eq. (5.14) to calculate the expected values $s_i^{(T)}$ $(i = 1, 2, \ldots, m)$ of all the subjective preference values s_i $(i = 1, 2, \ldots, m)$ and the expected values $h_{ij}^{(T)}$ $i = 1, 2, \ldots, m$; $j = 1, 2, \ldots, n$ of $h_{ij}(i = 1, 2, \ldots, m; \ j = 1, 2, \ldots, n)$. After that, we construct the hesitant fuzzy expected value decision matrix $H^{(T)} = \left(h_{ij}^{(T)} \right)_{m \times n}$ and go to the next step.

Step 3. If the information of the criterion weights is completely unknown, then we use Eq. (5.20) to obtain the optimal weight vector $\omega = (\omega_1, \omega_2 \cdots, \omega_n)^T$; otherwise, go to Step 4.

Step 4 If we know the information of the criterion weights in part, then we can solve Model 5.11 to get the criterion weight vector $\omega = (\omega_1, \omega_2 \cdots, \omega_n)^T$, and go to the next step.

Step 5. Use Eq. (5.21) to get the overall expected values $h_i^{(T)}(i = 1, 2, \ldots, m)$ of the alternatives $A_i(i = 1, 2, \ldots, m)$, then we can obtain the ranking of the alternatives, and go to the next step.

Step 6. End.

(2) **Method Based on Distances and Minimum Deviations**

Algorithm 5.2 uses the expected values to characterize the deviations between the subjective and objective preferences. It is simple and clear. When the demand of precision is not very high, it is a good method. Moreover, it can reflect the attitudes of the decision makers by the parameter T. However, this algorithm needs to change the HFEs into real numbers first. This conversion process may lose some information. To reduce the loss of the information as much as possible, in the following, we introduce another method in which we directly use the hesitant fuzzy distance between the subjective and objective preferences to represent the deviations between them.

1) Case with completely unknown weight information on criteria

Here we use the distance to characterize the subjective and objective deviations and then construct a goal programming model to determine the optimal weight vector of criteria in hesitant fuzzy environment. If we use the hesitant normalized hamming distance $d_{ij} = \frac{1}{l} \sum_{k=1}^{l} \left| h_{ij}^{\sigma(k)} - s_i^{\sigma(k)} \right|$ to express the deviation between the values h_{ij} and s_i, then the deviations between the ith alternative and the subjective preference values s_i is

$$d_i(\omega) = \sum_{j=1}^{n} \omega_j d_{ij}, \ i = 1, 2, \ldots, m \tag{5.22}$$

The total deviation of all the alternatives to all the subjective preference values is

$$d(\omega) = \sum_{i=1}^{m} \sum_{j=1}^{n} \omega_j d_{ij} \tag{5.23}$$

To make the decision result reasonable, the total deviation should be minimal. Motivated by this idea, we construct the following single objective programming model:

Model 5.12

$$\min \ d(\omega) = \sum_{i=1}^{m} \sum_{j=1}^{n} \omega_j d_{ij}$$

$$s.t. \ \ \omega_j \geq 0, \ j = 1, 2, \ldots, n; \ \sum_{j=1}^{n} \omega_j^2 = 1$$

Note that in Model 5.12, we use the unification condition of a vector, i.e., $\sum_{j=1}^{n} \omega_j^2 = 1$ instead of the above mentioned normalization condition $\sum_{j=1}^{n} \omega_j = 1$ so that Model 5.12 can be solved easily by the Lagrange method.

To find the optimal solution of Model 5.12, we construct the following Lagrange function:

$$L(\omega, \lambda) = \sum_{i=1}^{m} \sum_{j=1}^{n} \omega_j d_{ij} + \frac{\lambda}{2} \left(\sum_{j=1}^{n} \omega_j^2 - 1 \right) \tag{5.24}$$

where λ is a real number, called Lagrange multiplier variable. Computing the partial derivatives of the function $L(\omega, \lambda)$, we get

$$\frac{\partial L}{\partial \omega_j} = \sum_{i=1}^{m} d_{ij} + \lambda \omega_j = 0, \quad j = 1, 2, \ldots, n \tag{5.25}$$

$$\frac{\partial L}{\partial \lambda} = \frac{1}{2} \left(\sum_{j=1}^{n} \omega_j^2 - 1 \right) = 0 \tag{5.26}$$

It follows from Eq. (5.25) that

$$\omega_j = \frac{-\sum_{i=1}^{m} d_{ij}}{\lambda}, \quad j = 1, 2, \ldots, n \tag{5.27}$$

Taking Eq. (5.27) into Eq. (5.25), we have

$$\lambda = -\sqrt{\sum_{j=1}^{n} \left(\sum_{i=1}^{m} d_{ij} \right)^2} \tag{5.28}$$

Therefore, we obtain

$$\omega_j = \frac{\sum_{i=1}^{m} d_{ij}}{\sqrt{\sum_{j=1}^{n} \left(\sum_{i=1}^{m} d_{ij} \right)^2}}, \quad j = 1, 2, \ldots, n \tag{5.29}$$

Because the criterion weights should satisfy the normalization condition, we get the weights of the criteria as:

$$\omega_j^* = \frac{\omega_j}{\sum_{j=1}^{n} \omega_j}, j = 1, 2, \ldots, n \qquad (5.30)$$

We can see that Models 5.10 and 5.12 consider the deviations between subjective and objective information. But there are some differences. Model 5.10 uses the hesitant fuzzy expected values to construct the objective function, while Model 5.12 computes the deviations by the distances measures. Thus, Model 5.10 may lead to the loss of information. However, it can reflect the risk preference of the decision maker according to the parameter T.

2) Case with partly known weight information on criteria

If the weights of criteria satisfy the constraint conditions $0 \le a_j \le \omega_j \le b_j$, $j = 1, 2, \ldots, n$, and $\sum_{j=1}^{n} \omega_j = 1$, where a_j and b_j are the upper and lower bounds of ω_j, respectively, then we construct the following model to get the optimal weight vector of criteria:

Model 5.13

$$\min \quad d(\omega) = \sum_{i=1}^{m} \sum_{j=1}^{n} \omega_j d_{ij}$$

$$s.t. \quad 0 \le a_j \le \omega_j \le b_j, \ j = 1, 2, \ldots, n$$

$$\sum_{j=1}^{n} \omega_j = 1$$

Using the MATLAB or LINGO mathematic software package, we can get the optimal weight vector $\omega = (\omega_1, \omega_2 \cdots, \omega_n)^T$ of criteria.

Using Models 5.12 and 5.13, we can easily obtain the criterion weights no matter the weight information is completely unknown or partially known. Next we can use the hesitant fuzzy aggregating operators in Chap. 3 to aggregate the decision information and gain the overall value of each alternative, and then select the best one(s). Based on these analysis, we introduce Algorithm 5.3 for multiple criteria decision making problems where the criterion values (objective preference values) and the subjective preference values of the decision maker are all HFEs and the criterion weight information is completely unknown or incompletely known.

Algorithm 5.3

Step 1. Construct the hesitant fuzzy decision matrix H. Meanwhile, the decision makers give the subjective preference values of the ith alternative by s_i, $i = 1, 2, \ldots, m$, which are also HFEs, and then go to the next step.

Step 2. If we do not know the weight information of the criteria completely, then the optimal criteria weights can be obtained by Eq. (5.30) and we can turn to Step 4; Otherwise, go to the next step.

Step 3. If we know the possible range of criteria weights, then the best criterion weights could be obtained by solving Model 5.13, and go to the next step.

Step 4 Use Eq. (3.2) to compute the overall value h_i of each alternative A_i, and then go to the next step.

Step 5. Rank the alternatives according to their overall values h_i $(i = 1, 2, \ldots, m)$ and select the best one(s), then go to the next step.

Step 6. End.

5.2.2 Minimum Deviation Methods for Interval-Valued Hesitant Fuzzy Multiple Criteria Decision Making with Incomplete Weight Information

In practical decision making problems, the decision makers may have difficulty in giving the precise assessments and thus provide the value ranges of membership degrees. For this reason, Chen et al. (2013) introduced the notation of IVHFS, in which the membership degrees of an element to a given set are expressed by intervals. If the evaluation values of alternatives with respect to criteria are represented as IVHFEs, then we can construct the interval valued hesitant fuzzy decision matrix $\widetilde{H}$ shown as:

$$
\widetilde{H} = \begin{bmatrix}
\tilde{h}_{11} & \tilde{h}_{12} & \cdots & \tilde{h}_{1n} \\
\tilde{h}_{21} & \tilde{h}_{22} & \cdots & \tilde{h}_{2n} \\
\vdots & \vdots & \ddots & \vdots \\
\tilde{h}_{m1} & \tilde{h}_{m2} & \cdots & \tilde{h}_{mn}
\end{bmatrix}
\tag{5.31}
$$

Similar to the previous analysis, we shall give two interval-valued hesitant fuzzy multiple criteria decision making methods based on the deviations of subjective and objective preferences.

(1) **Method Based on Interval-Valued Hesitant Fuzzy Expected Values and Minimum Deviations**

We first introduce the concepts of interval-valued hesitant fuzzy expected value and interval-valued hesitant fuzzy expected value decision matrix.

Definition 5.4 (*Zhao et al.* 2016). Let $\tilde{h} = \{\tilde{\gamma}_1, \tilde{\gamma}_2, \ldots \tilde{\gamma}_l\}$ be an IVHFE, where $\tilde{\gamma}_k = \left[\tilde{\gamma}_k^L, \tilde{\gamma}_k^U\right]$ $(k = 1, 2, \ldots, l)$. The interval-valued hesitant fuzzy expected value of $\tilde{h}$ is defined as:

$$\tilde{h}^{(T)} = \frac{1}{l-1}\left[(1-T)\bar{\gamma}_{\sigma(l)} + \bar{\gamma}_{\sigma(l-1)} + \ldots + \bar{\gamma}_{\sigma(2)} + T\bar{\gamma}_{\sigma(1)}\right] \tag{5.32}$$

where $\bar{\gamma}_{\sigma(t)}$ is the tth largest number of $\bar{\gamma}_k$ $(k = 1, 2, \ldots, l)$ and $\bar{\gamma}_k = \frac{\tilde{\gamma}_k^L + \tilde{\gamma}_k^U}{2}$; T is a real number lying between 0 and 1. When $\tilde{\gamma}_k^L = \tilde{\gamma}_k^U$, for all $k = 1, 2, \ldots, l$, Eq. (5.32) turns into Eq. (5.14). Similarly, the value of T reveals the risk attitude of the decision maker. If the decision maker is risk pursuing, then $T > 0.5$; if the decision maker is risk neutral, then $T = 0.5$; if the decision maker is risk averse, then $T < 0.5$.

If the decision maker's subjective preference $\tilde{s}_i$ to the ith alternative A_i is an IVHFE, then we can use Eq. (5.32) to compute the expected values $\tilde{s}_i^{(T)}$ of $\tilde{s}_i$, $i = 1, 2, \ldots, m$. We also can calculate the expected values of the criterion values $\tilde{h}_{ij}$ $(i = 1, 2, \ldots, m; j = 1, 2, \ldots, n)$ in the interval-valued hesitant fuzzy decision matrix $\tilde{H}$. Thus, interval-valued hesitant fuzzy expected value decision matrix $\tilde{H}^{(T)} = \left(\tilde{h}_{ij}^{(T)}\right)_{n \times m}$ can be determined.

1) Case with completely unknown weight information on criteria

As the decision maker's subjective preference $\tilde{s}_i$ and objective preference $\tilde{h}_{ij}$ are all IVHFEs, we first compute the interval-valued hesitant fuzzy expected values $\tilde{h}_{ij}^{(T)}(i = 1, 2, \ldots, m; j = 1, 2, \ldots, n)$ of the criterion values $\tilde{h}_{ij}$ $(i = 1, 2, \ldots, m; j = 1, 2, \ldots, n)$ and the expected values $\tilde{s}_i^{(T)}$ $(i = 1, 2, \ldots, m)$ of the subjective preferences $\tilde{s}_i (i = 1, 2, \ldots, m)$ by Eq. (5.32). Then we can calculate the deviation between the objective preference value $\tilde{h}_{ij}^{(T)}$ and the subjective preference value $\tilde{s}_i^{(T)}$, denoted as $\tilde{\sigma}_{ij} = \tilde{h}_{ij}^{(T)} - \tilde{s}_i^{(T)}$. The overall deviation between the subjective and the objective preference should be minimal. For this purpose, we construct the following optimization model:

Model 5.14

$$\min \tilde{\sigma}(\omega) = \sum_{i=1}^{m}\sum_{j=1}^{n} \tilde{\sigma}_{ij}^2 \omega_j^2$$

$$s.t. \quad \omega_j \geq 0, \quad \sum_{j=1}^{n} \omega_j = 1$$

We first establish the Lagrange function:

$$\tilde{\sigma}(\omega, \lambda) = \sum_{i=1}^{m}\sum_{j=1}^{n} \tilde{\sigma}_{ij}^2 \omega_j^2 + 2\lambda(\sum_{j=1}^{n} \omega_j - 1) \tag{5.33}$$

and then compute its partial derivatives and let

$$\begin{cases} \frac{\partial \tilde{\sigma}}{\partial \omega_j} = 2 \sum_{i=1}^{m} \tilde{\sigma}_{ij}^2 \omega_j + 2\lambda = 0, \ j = 1, 2, \ldots, n \\ \frac{\partial \tilde{\sigma}}{\partial \lambda} = \sum_{j=1}^{n} \omega_j - 1 = 0 \end{cases} \tag{5.34}$$

Solving these equations, we obtain

$$\omega_j = \frac{1}{\sum_{j=1}^{n} 1 \Big/ \sum_{i=1}^{m} \tilde{\sigma}_{ij}^2} \Big/ \sum_{i=1}^{m} \tilde{\sigma}_{ij}^2, \ j = 1, 2, \ldots, n \tag{5.35}$$

Then, the overall expected values of all alternatives are calculated by

$$\tilde{z}_i^{(T)} = \sum_{j=1}^{n} \tilde{h}_{ij}^{(T)} \omega_j, \ i = 1, 2, \ldots, m \tag{5.36}$$

Thus, we can sort the alternatives by the values of $\tilde{z}_i^{(T)}$ ($i = 1, 2, \ldots, m$) and then select the best one(s).

2) Case with partly known weight information on criteria

If the decision makers provide the ranges of the criterion weights as $0 \le a_j \le \omega_j \le b_j$ with a_j and b_j being the upper and lower bounds of ω_j, $j = 1, 2, \ldots, n$, then we construct the following model to obtain the criterion weight vector ω:

Model 5.15

$$\min \tilde{\sigma}(\omega) = \sum_{i=1}^{m} \sum_{j=1}^{n} \tilde{\sigma}_{ij}^2 \omega_j$$

$$s.t. \ 0 \le a_j \le \omega_j \le b_j, \ j = 1, 2, \ldots, n$$

$$\sum_{j=1}^{n} \omega_j = 1$$

Solving Model 5.15 with the MATLAB or LINGO mathematical software package, we can get the optimal criterion weight vector ω. After that, we compute the overall expected values of all alternatives by Eq. (5.36) and choose the best alternative(s).

Based on the above analyses, we introduce the following minimum deviation based algorithm to solve the interval-valued hesitant fuzzy decision making problems:

Algorithm 5.4

Step 1. Construct the interval-valued hesitant fuzzy decision matrix $\widetilde{H}$. Meanwhile, we ask the decision maker to give the subjective preference $\tilde{s}_i$ as IVHFE to each alternative, and then go to the next step.

Step 2. Use Eq. (5.32) to calculate the interval-valued hesitant fuzzy expected values $\tilde{s}_i^{(T)}$, $\tilde{h}_{ij}^{(T)}$ of $\tilde{s}_i$, $\tilde{h}_{ij}$, respectively, and get the interval-valued hesitant fuzzy expected decision matrix $\widetilde{H}^{(T)} = \left(\tilde{h}_{ij}^{(T)}\right)_{m \times n}$, then go to the next step.

Step 3. If the weight information of the criteria is completely unknown, then we use Eq. (5.35) to obtain the optimal weight vector $\omega = (\omega_1, \omega_2, \ldots, \omega_n)^T$ and go to Step 5; Otherwise, go to the next step.

Step 4. If the weight information of the criteria is partly known as $0 \leq a_j \leq \omega_j \leq b_j$, $j = 1, 2, \ldots, n$, then we solve Model 5.15 and get the weight vector $\omega = (\omega_1, \omega_2, \ldots, \omega_n)^T$, and go to the next step.

Step 5. Calculate the overall expected values $\tilde{z}_i^{(T)}$ $(i = 1, 2, \ldots, m)$ by Eq. (5.36) and then sort the alternatives by the values of $\tilde{z}_i^{(T)}$ $(i = 1, 2, \ldots, m)$, and go to the next step.

Step 6. End.

(2) Method Based on Interval-Valued Distances and Minimum Deviations

Chen et al. (2013) introduced a family of distance measures between two IVHFEs $\tilde{h}_1$ and $\tilde{h}_2$. Taking the interval-valued hesitant normalized Hamming distance as an example, it is in mathematical term of

$$\tilde{d}(\tilde{h}_1, \tilde{h}_2) = \frac{1}{2l} \sum_{k=1}^{l} \left(\left| h_1^{\sigma(k)L} - h_2^{\sigma(k)L} \right| + \left| h_1^{\sigma(k)U} - h_2^{\sigma(k)U} \right| \right) \tag{5.37}$$

where $l = \max\{l(\tilde{h}_1), l(\tilde{h}_2)\}$, $h_p^{\sigma(k)}$ is the kth largest interval in $\tilde{h}_p$ $(p = 1, 2)$, and $h_p^{\sigma(k)} = \left[h_p^{\sigma(k)L}, h_p^{\sigma(k)U}\right]$ $(p = 1, 2; \ k = 1, 2, \ldots, l)$.

Below we introduce an interval-valued hesitant fuzzy multiple criteria decision making method based on the distance and the minimum deviation.

1) Case with completely unknown weight information on criteria

Since the decision makers have subjective preference $\tilde{s}_i$ over each alternative, we can build an optimization model by minimizing the deviations between the subjective and objective preferences to obtain the optimal criterion weight vector. Firstly, we use the interval-valued hesitant normalized Hamming distance to calculate the deviation between the assessment $\tilde{h}_{ij}$ and the subjective preference $\tilde{s}_i$, which is in form of

$$\tilde{d}_{ij} = \frac{1}{2l} \sum_{k=1}^{l} \left(\left| \tilde{h}_{ij}^{\sigma(k)L} - \tilde{s}_{i}^{\sigma(k)L} \right| + \left| \tilde{h}_{ij}^{\sigma(k)U} - \tilde{s}_{i}^{\sigma(k)U} \right| \right) \qquad (5.38)$$

where $\tilde{h}_{ij}^{\sigma(k)}$ and $\tilde{s}_{i}^{\sigma(k)}$ are the kth intervals in $\tilde{h}_{ij}$ and $\tilde{s}_i$, respectively, $\tilde{h}_{ij}^{\sigma(k)} = \left[\tilde{h}_{ij}^{\sigma(k)L}, \tilde{h}_{ij}^{\sigma(k)U} \right], \tilde{s}_{i}^{\sigma(k)} = \left[\tilde{s}_{i}^{\sigma(k)L}, \tilde{s}_{i}^{\sigma(k)U} \right]$.

Then, the overall deviation between the alternative A_i and the subjective preference value $\tilde{s}_i$ is

$$\tilde{d}_i(\omega) = \sum_{j=1}^{n} \omega_j \tilde{d}_{ij}, \ i = 1, 2, \ldots, m \qquad (5.39)$$

Furthermore, the total deviation between the subjective and objective preferences of all the alternatives is yielded as:

$$\tilde{d}(\omega) = \sum_{i=1}^{m} \sum_{j=1}^{n} \omega_j \tilde{d}_{ij} \qquad (5.40)$$

and the total deviation should be minimal. Thus, we establish the following programming model:

Model 5.16

$$\min \tilde{d}(\omega) = \sum_{i=1}^{m} \sum_{j=1}^{n} \omega_j \tilde{d}_{ij}$$

$$s.t. \quad \omega_j \geq 0, \ j = 1, 2, \ldots, n$$

$$\sum_{j=1}^{n} \omega_j^2 = 1$$

Solving this model, we get the optimal solution:

$$\omega_j = \frac{\sum_{i=1}^{m} \tilde{d}_{ij}}{\sqrt{\sum_{j=1}^{n} \left(\sum_{i=1}^{m} \tilde{d}_{ij} \right)^2}}, j = 1, 2, \ldots, n \qquad (5.41)$$

As $\omega_j (j = 1, 2, \ldots, n)$ satisfy the constrained conditions in Model 5.16, we normalize the weights of criteria as:

$$\omega_j^* = \frac{\omega_j}{\sum_{j=1}^{n} \omega_j}, j = 1, 2, \ldots, n \qquad (5.42)$$

2) Case with partly known weight information on criteria

Consider that sometimes the decision makers may give the value ranges of the criterion weights. Thus, we need to build another model to acquire the criterion weights. Suppose that the criterion weights satisfy $0 \le a_j \le \omega_j \le b_j$ with a_j and b_j being the upper and lower bounds of ω_j, $j = 1, 2, \ldots, n$. Then we construct the following model to obtain the criterion weight vector ω:

Model 5.17

$$\min \tilde{d}(\omega) = \sum_{i=1}^{m} \sum_{j=1}^{n} \omega_j \tilde{d}_{ij}$$

$$s.t. \quad 0 \le a_j \le \omega_j \le b_j, \ j = 1, 2, \ldots, n$$

$$\sum_{j=1}^{n} \omega_j = 1$$

The optimal criterion weight vector $\omega = (\omega_1, \omega_2 \cdots, \omega_n)^T$ can be derived by solving Model 5.17. Then we can integrate the interval-valued hesitant fuzzy decision information for each alternative by the interval-valued hesitant fuzzy aggregating operators (Chen et al. 2013). Finally, the best alternative(s) is selected by the integrated values.

According to the above discussion, the following algorithm is given to tackle the interval-valued hesitant fuzzy multiple criteria decision making problems in which the evaluation values and the subjective preferences to the alternatives are all IVHFEs and the information about the criterion weights is incomplete.

Algorithm 5.5

Step 1. Construct the interval-valued hesitant fuzzy decision matrix $\widetilde{H}$. Meanwhile, we ask the decision maker to give the subjective preference $\tilde{s}_i$ for each alternative as IVHFE, then go to the next step.

Step 2. If there is no any information about the criterion weights, then we calculate the optimal criterion weights by Eq. (5.42), and go to Step 4; Otherwise, go to the next step.

Step 3. If the weight information of criteria is partly known as $0 \le a_j \le \omega_j \le b_j$ $(j = 1, 2, \ldots, n)$, then we solve Model 5.17 to obtain the optimal criterion weights, and go to the next step.

Step 4. Use the interval-valued hesitant fuzzy aggregating operator, such as the IVHFWA operator shown as Eq. (5.43), to compute the overall values $\tilde{h}_i$ $(i = 1, 2, \ldots, m)$ of the alternatives A_i $(i = 1, 2, \ldots, m)$, and then go to the next step.

$$\tilde{h}_i = \text{IVHFWA}\left(\tilde{h}_{i1}, \tilde{h}_{i2}, \ldots, \tilde{h}_{in}\right) = \bigoplus_{j=1}^{n} \left(\omega_j \tilde{h}_{ij}\right)$$

$$= \left\{\left[1 - \prod_{j=1}^{n}(1 - \tilde{\gamma}_j^L)^{\omega_j}, 1 - \prod_{j=1}^{n}(1 - \tilde{\gamma}_j^U)^{\omega_j}\right] \middle| \tilde{\gamma}_1 \in \tilde{h}_1, \tilde{\gamma}_2 \in \tilde{h}_2, \ldots, \tilde{\gamma}_n \in \tilde{h}_n\right\}$$

$$(5.43)$$

Step 5. Rank of the alternatives according to the overall values $\tilde{h}_i$ $(i = 1, 2, \ldots, m)$ and select the best one(s), then go to the next step.

Step 6. End.

5.2.3   Application of the Minimum Deviation Methods in Energy Policy Selection

In this section, we present an illustrative example concerning the selection of optimal energy policies. Energy is always an essential factor to the economic and social development of society. A good energy policy will affect the economic development and environment. How to choose an optimal energy policy is a major concern for the government. In the following, we use the example adapted from Xu and Xia (2011) to demonstrate the effectiveness of our methods. After pre-selection, five energy projects $A_i(i = 1, 2, 3, 4, 5)$ are evaluated, and four criteria involved in are C_1: technological, C_2: environmental, C_3: socio-political and C_4: economic. Suppose that the experts have subjective preferences over each alternative and the preference values are represented as HFEs. We rank the five energy projects under four different scenarios.

Example 5.3 (Zhao et al. 2016) Assume that the evaluation values of the alternatives under the criteria are provided as a hesitant fuzzy decision matrix (see Table 5.1), and the decision maker has subjective preferences over all the alternatives, which are: $s_1 = \{0.6, 0.5, 0.2\}$, $s_2 = \{0.5, 0.4\}$, $s_3 = \{0.4, 0.3, 0.2\}$, $s_4 = \{0.5, 0.3\}$ and $s_5 = \{0.9, 0.5\}$, respectively.

Next, we use Algorithm 5.2 to select the best alternative(s).

(1) If the weight information of the criteria is completely unknown, then the following steps are conducted to choose the best alternative(s):

Assuming $T = 0.5$, we use Eq. (5.14) to compute the expected values of the subjective preference values s_i $(i = 1, 2, 3, 4, 5)$ and get $s_1^{(T)} = 0.45$, $s_2^{(T)} = 0.45$, $s_3^{(T)} = 0.3$, $s_4^{(T)} = 0.4$, $s_5^{(T)} = 0.7$. Then we calculate the hesitant fuzzy expected values of each element in Table 5.1 and obtain the hesitant fuzzy expected value decision matrix $H^{(T)}$ as shown in Table 5.2.

Table 5.1 Hesitant fuzzy decision matrix

	C_1	C_2	C_3	C_4
A_1	{0.5, 0.4}	{0.7}	{0.5, 0.4, 0.2}	{0.6, 0.5}
A_2	{0.5, 0.3}	{0.7, 0.6, 0.5}	{0.5, 0.1}	{0.4}
A_3	{0.7, 0.6}	{0.9, 0.6}	{0.5, 0.3}	{0.6, 0.4}
A_4	{0.7, 0.4}	{0.7, 0.4, 0.2}	{0.8, 0.1}	{0.8}
A_5	{0.6, 0.3, 0.1}	{0.6, 0.4}	{0.8, 0.7}	{0.6}

Table 5.2 Hesitant fuzzy expected value decision matrix

	C_1	C_2	C_3	C_4
A_1	0.45	0.7	0.375	0.55
A_2	0.4	0.6	0.3	0.4
A_3	0.7	0.75	0.4	0.5
A_4	0.55	0.425	0.45	0.8
A_5	0.325	0.5	0.75	0.6

Using Eq. (5.20), we obtain the optimal weight vector $\omega = (0.0933, 0.0925, 0.6775, 0.1367)^T$. Then we calculate the overall expected values of the alternatives by Eq. (5.21) and obtain $z_1^{(T)} = 0.4361$, $z_2^{(T)} = 0.3508$, $z_3^{(T)} = 0.4741$, $z_4^{(T)} = 0.5049$, and $z_5^{(T)} = 0.6667$. Thus, the ranking of the alternatives is $A_5 \succ A_4 \succ A_3 \succ A_1 \succ A_2$. So A_5 is the best one.

(2) If the value ranges of the criterion weights are given as $0.2 \leq \omega_1 \leq 0.3$, $0.3 \leq \omega_2 \leq 0.4$, $0.2 \leq \omega_3 \leq 0.3$, $0.2 \leq \omega_4 \leq 0.3$, then according to Model 5.11, we construct the following linear programming model to solve the optimal criteria weights:

Model 5.18

$$\min \sigma'(\omega) = 0.327\omega_1 + 0.3296\omega_2 + 0.045\omega_3 + 0.223\omega_4$$
$$\text{s.t.} \, 0.2 \leq \omega_1 \leq 0.3$$
$$0.3 \leq \omega_2 \leq 0.4$$
$$0.2 \leq \omega_3 \leq 0.3$$
$$0.2 \leq \omega_4 \leq 0.3$$
$$\omega_1 + \omega_2 + \omega_3 + \omega_4 = 1$$

By the LINGO mathematics software package, we get the optimal criterion weight vector $\omega = (0.2, 0.3, 0.3, 0.2)^T$. With Eq. (5.21), the overall expected value of each alternative are obtained as $h_1^{(T)} = 0.5225$, $h_2^{(T)} = 0.43$, $h_3^{(T)} = 0.585$, $h_4^{(T)} = 0.5325$, and $h_5^{(T)} = 0.56$. This implies that the ranking of the alternatives is $A_3 \succ A_5 \succ A_4 \succ A_1 \succ A_2$.

Below we use Algorithm 5.3 to solve the problem in Example 5.3:

We first compute the hesitant normalized hamming distances between the values h_{ij} ($i = 1, 2, 3, 4, 5; j = 1, 2, 3, 4$) and s_i ($i = 1, 2, 3, 4, 5$). The results are set out in Table 5.3. To get the best criterion weight vector $\omega = (\omega_1, \omega_2, \omega_3, \omega_4)^T$, the total deviation between the subjective and objective preferences should be minimal.

(1) If the weight information of the criteria is completely unknown, then by Eq. (5.30), we get the normalized criterion weights as $w_1^* = 0.2589$, $w_2^* = 0.3125$, $w_3^* = 0.1831$, and $w_4^* = 0.2455$. Then, we calculate the overall values of the alternatives by Eq. (3.2) and the results are as follows:

Table 5.3 Hesitant Hamming distances between the subjective and objective preferences

	h_{1j}	h_{2j}	h_{3j}	h_{4j}
s_1	0.1333	0.2667	0.0667	0.1
s_2	0.05	0.1667	0.15	0.05
s_3	0.3333	0.4	0.0667	0.1667
s_4	0.15	0.1333	0.25	0.4
s_5	0.3	0.2	0.15	0.2

$$h_1 = \begin{pmatrix} 0.5130, 0.5355, 0.5380, 0.5390, 0.5532, 0.5593, 0.5602, 0.5626, 0.5738, 0.5770, \\ 0.5828, 0.5965 \end{pmatrix}$$

$$h_2 = \begin{pmatrix} 0.3647, 0.4075, 0.4177, 0.4295, 0.4569, 0.4584, 0.4679, 0.4771, 0.5036, 0.5123, \\ 0.5137, 0.5543 \end{pmatrix}$$

$$h_3 = \begin{pmatrix} 0.5105, 0.5397, 0.5456, 0.5568, 0.5727, 0.5833, 0.5886, 0.6132, 0.6826, 0.7015, \\ 0.7054, 0.7126, 0.7230, 0.7300, 0.7333, 0.7492 \end{pmatrix}$$

$$h_4 = \begin{pmatrix} 0.4601, 0.5065, 0.5488, 0.5876, 0.5901, 0.6026, 0.6253, 0.6574, 0.6679, 0.6869, \\ 0.6983, 0.7479 \end{pmatrix}$$

$$h_5 = \begin{pmatrix} 0.4686, 0.5021, 0.5067, 0.5319, 0.5377, 0.5614, 0.5654, 0.5693, 0.5927, 0.6001, \\ 0.6205, 0.6477 \end{pmatrix}$$

By Eq. (1.17), the score values of h_i $(i = 1, 2, 3, 4, 5)$ are calculated as $s(h_1) = 0.5576$, $s(h_2) = 0.4636$, $s(h_3) = 0.6405$, $s(h_4) = 0.6150$, and $s(h_5) = 0.5587$, respectively. Thus, the ranking of the alternatives is $A_3 \succ A_4 \succ A_5 \succ A_1 \succ A_2$, which implies A_3 is the best energy policy.

(2) Given the weight information of the criteria is partly known as $0.2 \le \omega_1 \le 0.3$, $0.3 \le \omega_2 \le 0.4$, $0.2 \le \omega_3 \le 0.3$, and $0.2 \le \omega_4 \le 0.3$, to get optimal weight vector, we construct the following linear programming model according to Model 5.13:

Model 5.19

$$\min d(\omega) = 0.9666\omega_1 + 1.1667\omega_2 + 0.6834\omega_3 + 0.9167\omega_4$$
$$\text{s.t.}\, 0.2 \le \omega_1 \le 0.3$$
$$0.3 \le \omega_2 \le 0.4$$
$$0.2 \le \omega_3 \le 0.3$$
$$0.2 \le \omega_4 \le 0.3$$
$$\omega_1 + \omega_2 + \omega_3 + \omega_4 = 1$$

Solving Model 5.19, we get the optimal weights as: $\omega_1 = 0.2$, $\omega_2 = 0.3$, $\omega_3 = 0.3$, and $\omega_4 = 0.2$. Then we use Eq. (3.2) to obtain the overall value of each alternative. The results are set out as follows:

$$h_1 = \begin{pmatrix} 0.4877, 0.5061, 0.5101, 0.5276, 0.5301, 0.5469, 0.5506, 0.5551, 0.5667, 0.5710, \\ 0.5745, 0.5898 \end{pmatrix}$$

$$h_2 = \begin{pmatrix} 0.3384, 0.3812, 0.3814, 0.4215, 0.4324, 0.4453, 0.4693, 0.4812, 0.4814, 0.5150, \\ 0.5241, 0.5551 \end{pmatrix}$$

$$h_3 = \begin{pmatrix} 0.4869, 0.5156, 0.5269, 0.5362, 0.5533, 0.5621, 0.5723, 0.5962, 0.6615, 0.6804, \\ 0.6879, 0.6940, 0.7053, 0.7111, 0.7178, 0.7336 \end{pmatrix}$$

$$h_4 = \begin{pmatrix} 0.4070, 0.4561, 0.4838, 0.5265, 0.5582, 0.6154, 0.6224, 0.6536, 0.6712, 0.6984, \\ 0.7186, 0.7551 \end{pmatrix}$$

$$h_5 = \begin{pmatrix} 5126, 0.5365, 0.5685, 0.5685, 0.5856, 0.5896, 0.5896, 0.6179, 0.6331, 0.6331, \\ 0.6366, 0.6751 \end{pmatrix}$$

After that, we calculate the score values of the overall values with Eq. (1.17) and obtain $s(h_1) = 0.5430$, $s(h_2) = 0.4522$, $s(h_3) = 0.6213$, $s(h_4) = 0.5972$, and $s(h_5) = 0.5956$. Thus, the ranking of these alternatives is $A_3 \succ A_4 \succ A_5 \succ A_1 \succ A_2$.

Example 5.4 (Zhao et al. 2016). In Example 5.3, if the decision makers give their subjective preferences over the alternatives $A_i (i = 1, 2, 3, 4, 5)$ by the following IVHFEs:

$$\tilde{s}_1 = \{[0.5, 0.7], [0.5, 0.6], [0.2, 0.3]\}, \tilde{s}_2 = \{[0.3, 0.5], [0.3, 0.4]\}$$
$$\tilde{s}_3 = \{[0.3, 0.5], [0.3, 0.4], [0.2, 0.3]\}, \tilde{s}_4 = \{[0.5, 0.6], [0.3, 0.4]\}$$
$$\tilde{s}_5 = \{[0.8, 0.9], [0.4, 0.5]\}$$

and they give their evaluation values for each alternative with respect to each criterion by IVHFE, then we can construct an interval-valued hesitant fuzzy decision matrix as shown in Table 5.4:
In the following, we use Algorithm 5.4 to choose the best alternative(s).

(1)   If the weight information of the criteria is completely unknown, then we let $T = 0.5$ and use Eq. (5.32) to compute the interval-valued hesitant fuzzy expected values of the subjective preference $\tilde{s}_i$ and get: $\tilde{s}_1^{(T)} = 0.4875$, $\tilde{s}_2^{(T)} = 0.375$, $\tilde{s}_3^{(T)} = 0.3375$, $\tilde{s}_4^{(T)} = 0.45$, and $\tilde{s}_5^{(T)} = 0.65$. The expected values of the evaluation values $\tilde{h}_{ij} (i = 1, 2, 3, 4, 5; j = 1, 2, 3, 4)$ can also be calculated, which are listed in Table 5.5.

Using Eq. (5.35), we get the optimal criterion weights as $\omega_1 = 0.1272$, $\omega_2 = 0.1072$, $\omega_3 = 0.5893$, and $\omega_4 = 0.1763$. Afterwards, by Eq. (5.36), we get the overall expected values of all the alternatives, which are $\tilde{z}_1^{(T)} = 0.3853$,

Table 5.4 Interval-valued hesitant fuzzy decision matrix

	C_1	C_2	C_3	C_4
A_1	{[0.3, 0.5], [0.3, 0.4]}	{[0.6, 0.7]}	{[0.3, 0.5], [0.3, 0.4], [0.1, 0.2]}	{[0.4, 0.6], [0.4, 0.5]}
A_2	{[0.2, 0.5], [0.2, 0.3]}	{[0.6, 0.7], [0.4, 0.6], [0.4, 0.5]}	{[0.4, 0.5], [0.1, 0.3]}	{[0.3, 0.4]}
A_3	{[0.5, 0.7], [0.5, 0.6]}	{[0.7, 0.9], [0.5, 0.6]}	{[0.4, 0.5], [0.2, 0.3]}	{[0.5, 0.6], [0.3, 0.4]}
A_4	{[0.6, 0.7], [0.3, 0.4]}	{[0.6, 0.7], [0.3, 0.4], [0.1, 0.2]}	{[0.7, 0.8], [0.1, 0.3]}	{[0.7, 0.8]}
A_5	{[0.5, 0.6], [0.2, 0.4], [0.1, 0.3]}	{[0.5, 0.6], [0.3, 0.4]}	{[0.7, 0.8], [0.6, 0.7]}	{[0.5, 0.6]}

Table 5.5 Interval-valued hesitant fuzzy expected value decision matrix

	C_1	C_2	C_3	C_4
A_1	0.375	0.65	0.3125	0.475
A_2	0.3	0.525	0.325	0.35
A_3	0.575	0.675	0.35	0.45
A_4	0.5	0.375	0.475	0.75
A_5	0.3375	0.45	0.7	0.55

$\tilde{z}_2^{(T)} = 0.3477$, $\tilde{z}_3^{(T)} = 0.4329$, $\tilde{z}_4^{(T)} = 0.5159$, and $\tilde{z}_5^{(T)} = 0.6006$. Therefore, the ranking of these alternatives is $A_5 \succ A_4 \succ A_3 \succ A_1 \succ A_2$.

(2) If the decision makers partly give the information about the weights of criteria as $0.2 \leq \omega_1 \leq 0.3$, $0.3 \leq \omega_2 \leq 0.4$, $0.2 \leq \omega_3 \leq 0.3$, and $0.2 \leq \omega_4 \leq 0.3$, then according to Model 5.15, we can construct the following mathematical model:

Model 5.20

$$\min \tilde{\sigma}'(\omega) = 0.176\omega_1 + 0.209\omega_2 + 0.038\omega_3 + 0.127\omega_4$$
$$\text{s.t.} \, 0.2 \leq \omega_1 \leq 0.3$$
$$0.3 \leq \omega_2 \leq 0.4$$
$$0.2 \leq \omega_3 \leq 0.3$$
$$0.2 \leq \omega_4 \leq 0.3$$
$$\omega_1 + \omega_2 + \omega_3 + \omega_4 = 1$$

Utilizing the LINGO mathematical software package, we can get the optimal criterion weight vector as: $\omega_1 = 0.2$, $\omega_2 = 0.3$, $\omega_3 = 0.3$, and $\omega_4 = 0.2$. Utilizing Eq. (5.36), we get the overall expected criteria values of all the alternatives, which are $\tilde{z}_1^{(T)} = 0.4588$, $\tilde{z}_2^{(T)} = 0.385$, $\tilde{z}_3^{(T)} = 0.5125$, $\tilde{z}_4^{(T)} = 0.505$, and $\tilde{z}_5^{(T)} = 0.5225$. This implies that the ranking of the alternative is $A_5 \succ A_3 \succ A_4 \succ A_1 \succ A_2$.

Table 5.6 Interval-valued hesitant Hamming distances between the subjective and objective preferences

	$\tilde{h}_{1j}$	$\tilde{h}_{2j}$	$\tilde{h}_{3j}$	$\tilde{h}_{4j}$
$\tilde{s}_1$	0.1667	0.1833	0.1667	0.1333
$\tilde{s}_2$	0.075	0.1667	0.1	0.025
$\tilde{s}_3$	0.2333	0.3	0.05	0.0833
$\tilde{s}_4$	0.05	0.1	0.175	0.3
$\tilde{s}_5$	0.2333	0.2	0.15	0.2

We now use Algorithm 5.5 to solve the problem in Example 5.4.

First of all, we compute the interval-valued hesitant hamming distances between each criterion values $\tilde{h}_{ij}$ and each subjective values $\tilde{s}_i$ by Eq. (5.38). The results are listed in Table 5.6.

(1) If there is no any information about the weights of criteria, then by Eq. (5.42), we obtain the criteria weights as $\omega_1^* = 0.2453$, $\omega_2^* = 0.3073$, $\omega_3^* = 0.2075$, and $\omega_4^* = 0.2399$. Thus, by Eq. (5.43), we can calculate the overall values $\tilde{h}_j$ ($j = 1, 2, 3, 4, 5$). Then the score values of the alternatives are calculated by Eq. (1.24), which are $s(\tilde{h}_1) = [0.4219, 0.5150]$, $s(\tilde{h}_2) = [0.3298, 0.4753]$, $s(\tilde{h}_3) = [0.4828, 0.6357]$, $s(\tilde{h}_4) = [0.5073, 0.6243]$, and $s(\tilde{h}_5) = [0.4660, 0.5820]$. By the probability degree function shown in Eq. (1.23), we compare the score values and get $A_4 \succ A_3 \succ A_5 \succ A_1 \succ A_2$.

(2) Suppose that the weight information of criteria is partly known as $0.2 \leq \omega_1 \leq 0.3$, $0.3 \leq \omega_2 \leq 0.4$, $0.2 \leq \omega_3 \leq 0.3$, and $0.2 \leq \omega_4 \leq 0.3$. Based on Model 5.17, we establish the following linear programming model:

Model 5.21

$$\min \tilde{d}(\omega) = 0.7583\omega_1 + 0.95\omega_2 + 0.6417\omega_3 + 0.7416\omega_4$$
$$\text{s.t. } 0.2 \leq \omega_1 \leq 0.3$$
$$0.3 \leq \omega_2 \leq 0.4$$
$$0.2 \leq \omega_3 \leq 0.3$$
$$0.2 \leq \omega_4 \leq 0.3$$
$$\omega_1 + \omega_2 + \omega_3 + \omega_4 = 1$$

Solving the above model by the LINGO mathematics software package, we gain the optimal criterion weight vector as $\omega = (0.2, 0.3, 0.3, 0.2)^T$. Then we use Eq. (5.43) to obtain the overall value of each alternative and calculate the score values of them as $s(\tilde{h}_1) = [0.4112, 0.5098]$, $s(\tilde{h}_2) = [0.3288, 0.4738]$, $s(\tilde{h}_3) = [0.4695, 0.6213]$, $s(\tilde{h}_4) = [0.4945, 0.6151]$, and $s(\tilde{h}_5) = [0.4928, 0.6068]$. Using the probability degree function shown in Eq. (1.23), we get $A_4 \succ A_3 \succ A_5 \succ A_1 \succ A_2$.

5.3   Hesitant Fuzzy Multiple Stages Multiple Criteria Decision Making with Incomplete Weight Information

In the previous chapters, we discussed the problems where all the hesitant fuzzy decision information is collected at the same period or stage. However, in many cases, such as multiple stages investment, medical diagnosis, personal dynamic examination, and ecosystem efficiency dynamic evaluation, etc., we often need to make decision with multiple stages preference information. It is reasonable and realistic that the assessment values provided by the decision makers are changed regularly regarding dynamic external factors such as knowledge or meteorological condition. The dynamic decision making environment is now more and more common in our daily life. Particularly, the spread of e-services and wireless or mobile devices has increased accessibility to data, which, in turn, influences the way in which the users make decisions (Perez et al. 2011). Users even can make real-time decisions based on the most up-to-date data accessed via wireless devices, such as portable computer, mobile phone or other electronic products. Considering the powerfulness of HFS in representing vagueness and uncertainty, it is necessary to develop some methods to handle multiple stages decision making problems where the judgments are collected from different stages and represented in HFEs. Hence, in this chapter, we shall develop some methods for hesitant fuzzy multiple stages multiple criteria decision making (MSMCDM) problems.

As to the MSMCDM problems where the information is given in fuzzy set or its extended forms such as IFS, many scholars have developed some decision making methods. As to the MSMCDM problem with fuzzy set, Xu (2008) defined the concept of dynamic weighted averaging (DWA) operator, and introduced some methods, such as the arithmetic series based method, the geometric series based method and the normal distribution based method, to obtain the weights of the DWA operator. As for the decision information expressed in multiplicative linguistic labels at different stages, Xu (2009) introduced the dynamic linguistic weighted geometric (DLWG) operator and used the minimum variance model to derive the time series weights of the DLWG operator, based on which, they then developed an approach to MSMCDM with linguistic assessments. Considering the MSMCDM problem with intuitionistic fuzzy information, Xu and Yager (2008) proposed some methods, including the basic unit-interval monotonic (BUM) function based method, the normal distribution based method, the exponential distribution based method and the average age method, to determine the weight vector of the developed operators. Based on the definition of distance measure between triangular intuitionistic fuzzy numbers, Chen and Li (2011) used the entropy method to determine the weight vector, which is a new way to generate the weight vector. After reviewing the literature, we can find that the key point in handling the MSMCDM problem is to determine the dynamic weight vector. After assigning the weights to different stages, it is easy to aggregate the dynamic hesitant fuzzy information by the developed operators and then rank the alternatives. Hence, in this section, we propose some novel methods to determine the weights of hesitant fuzzy variables at different stages.

5.3.1 Hesitant Fuzzy Variable and Dynamic Hesitant Fuzzy Aggregation Operators

In the literature, scholars have paid great attention to the problems where all hesitant fuzzy data were collected at a fixed stage. However, in many practical problems, the evaluation information is usually collected over different stages. For example, considering a scenario of investigating the preferences of a group of target customers where the environment is dynamic and the customers are variable, the decision maker may be interested in finding how the group's preferences change over time (Chen and Cheng 2010). In such a case, the preference information of the customers is collected from different stages. In order to represent the information from multiple stages, Liao et al. (2014) introduced the concept of hesitant fuzzy variable (HFV).

Definition 5.5 (*Liao et al.* 2014). Let t be a time instant and X be a given set. Then $h_A(x(t))$ is called a HFV at the time t, where $h_A(x(t))$ is a set of some possible values in $[0, 1]$ at the time t, denoting the possible membership degrees of the element $x \in X$ to the given set $A \subseteq X$ at the time t.

For a hesitant fuzzy variable $h_A(x(t))$ (we herein and thereafter denote it as h_x^t for short), if $t = (t_1, t_2, \ldots, t_p)^T$, then $h_x^{t_1}, h_x^{t_2}, \ldots, h_x^{t_p}$ indicate p HFEs of the element $x \in X$ collected from p different stages. That is to say, the values of a HFV are HFEs. For simplicity, let V be the set of all HFVs for short.

It should be noted that the HFV, the HFS and the HFE are three different concepts, but they are closely related. The relationships of them are as follows: the HFS is a set of HFEs and the HFE is the element of a HFS, whereas the HFE is the value of a HFV. The HFV is somehow like the random variable in probability theory. The value of a HFV varies over time. The significant characteristic of the HFV is that it involves the dimension of time. The reason for introducing the concept of HFV is to represent the changes of people's evaluating values on the given alternatives over criteria regarding to different time instants in the process of MSMCDM. MSMCDM is common in our daily life. For example, in the websites such as Youtube, Yahoo and Blogger, different customers are invited to provide the ratings of the most popular videos, movies and songs over time. In this example, the preference h_{ij}^t of the criterion C_j (for instance, the plot of a movie) is a HFV, while the preference value, such as $h_{ij}^t = \{0.3, 0.4\}$ is a HFE which denotes the preference for the alternative A_i at the stage t.

It is worthy pointing out that if omitting the parameter t, the HFV can be mathematically taken as the HFE. That is to say, all the operational laws and properties on HFEs also hold for HFVs. For the simplicity of future presentation, below we just set out some prominent properties for HFVs.

Property 5.1 (Liao et al. 2014). Let h_x^t, $h_x^{t_1}$, $h_x^{t_2}$ and $h_x^{t_k}$ be HFVs defined as above. Then

(1) $\left(h_x^t\right)^\lambda = \cup_{\gamma \in h_x^t}\{\gamma^\lambda\}, \lambda > 0.$

(2) $\lambda h_x^t = \cup_{\gamma \in h_x^t}\{1 - (1-\gamma)^\lambda\}, \lambda > 0.$

(3) $h_x^{t_1} \oplus h_x^{t_1} = \cup_{\gamma(t_1) \in h_x^{t_1}, \gamma(t_2) \in h_x^{t_1}}\{\gamma(t_1) + \gamma(t_2) - \gamma(t_1)\gamma(t_2)\}.$

(4) $h_x^{t_1} \ominus h_x^{t_2} = \cup_{\gamma(t_1) \in h_x^{t_1}, \gamma(t_2) \in h_x^{t_2}}\{\xi\}$, where

$$\xi = \begin{cases} \frac{\gamma(t_1) - \gamma(t_2)}{1 - \gamma(t_2)}, & \text{if } \gamma(t_1) \geq \gamma(t_2) \text{ and } \gamma(t_2) \neq 1 \\ 0, & \text{otherwise} \end{cases}$$

(5) $h_x^{t_1} \otimes h_x^{t_2} = \cup_{\gamma(t_1) \in h_x^{t_1}, \gamma(t_2) \in h_x^{t_2}}\{\gamma(t_1)\gamma(t_2)\}.$

(6) $h_x^{t_1} \oslash h_x^{t_2} = \cup_{\gamma(t_1) \in h_x^{t_1}, \gamma(t_2) \in h_x^{t_2}}\{\xi\}$, where

$$\xi = \begin{cases} \frac{\gamma(t_1)}{\gamma(t_2)}, & \text{if } \gamma(t_1) \leq \gamma(t_2) \text{ and } \gamma(t_2) \neq 0 \\ 1, & \text{otherwise} \end{cases}.$$

(7) $\overset{p}{\underset{k=1}{\oplus}} h_x^{t_k} = \cup_{\gamma(t_k) \in h_x^{t_k}}\{1 - \prod_{k=1}^{p}(1 - \gamma(t_k))\}.$

(8) $\overset{p}{\underset{k=1}{\otimes}} h_x^{t_k} = \cup_{\gamma(t_k) \in h_x^{t_k}}\{\prod_{k=1}^{p} \gamma(t_k)\}.$

Property 5.2 (Liao et al. 2014). Let h_x^t, $h_x^{t_1}$, $h_x^{t_2}$ and $h_x^{t_3}$ be four HFVs, and λ, λ_1 and λ_2 be three positive real numbers. Then $\left(h_x^t\right)^\lambda$, λh_x^t, $h_x^{t_1} \oplus h_x^{t_1}$ and $h_x^{t_1} \otimes h_x^{t_2}$ are HFVs, and

(1) $h_x^{t_1} \oplus h_x^{t_1} = h_x^{t_2} \oplus h_x^{t_1}.$

(2) $h_x^{t_1} \otimes h_x^{t_2} = h_x^{t_2} \otimes h_x^{t_1}.$

(3) $[h_x^{t_1} \oplus h_x^{t_2}] \oplus h_x^{t_3} = h_x^{t_1} \oplus [h_x^{t_2} \oplus h_x^{t_3}].$

(4) $[h_x^{t_1} \otimes h_x^{t_2}] \otimes h_x^{t_3} = h_x^{t_1} \otimes [h_x^{t_2} \otimes h_x^{t_3}].$

(5) $\lambda(h_x^{t_1} \oplus h_x^{t_2}) = \lambda h_x^{t_1} \oplus \lambda h_x^{t_2}.$

(6) $\left(h_x^{t_1} \otimes h_x^{t_2}\right)^\lambda = \left(h_x^{t_1}\right)^\lambda \otimes \left(h_x^{t_2}\right)^\lambda.$

(7) $\lambda_1 h_x^t \oplus \lambda_2 h_x^t = (\lambda_1 + \lambda_2) h_x^t.$

(8) $\left(h_x^t\right)^{\lambda_1} \otimes \left(h_x^t\right)^{\lambda_2} = \left(h_x^t\right)^{\lambda_1 + \lambda_2}.$

Chapter 3 introduced a series of aggregation operators for HFEs. But all these operators can only be used to deal with time-independent hesitant fuzzy arguments. If the time t is taken into account, the aggregation operators and their associated weights should not be static. Since the values of a HFV are HFEs, those aggregation operators in time independent hesitant fuzzy circumstances can be easily extended into the multiple stages hesitant fuzzy environment. We hereby just extend those time-independent hesitant fuzzy aggregation operators into to the multiple stages circumstance.

Definition 5.6 (*Liao et al.* 2014). Let $h_x^{t_1}$, $h_x^{t_2}, \ldots, h_x^{t_p}$ be a collection of HFVs collected at p different stages t_k $(k = 1, 2, \ldots, p)$, and $\lambda(t) = (\lambda(t_1), \lambda(t_2), \ldots, \lambda(t_p))^T$ be the weight vector of the stages t_k $(k = 1, 2, \ldots, p)$, where $\lambda(t_k)$ indicates the importance degree of $h_x^{t_k}$ satisfying $\lambda(t_k) \in [0, 1], k = 1, 2, \ldots, p$, and $\sum_{k=1}^{p} \lambda(t_k) = 1$. Then we call

$$\text{DHFWA}_{\lambda(t)}\left(h_x^{t_1}, h_x^{t_2}, \ldots, h_x^{t_p}\right) = \overset{p}{\underset{k=1}{\oplus}}\left(\lambda(t_k)h_x^{t_k}\right)$$

$$= \bigcup_{\gamma(t_1)\in h_x^{t_1},\,\gamma(t_2)\in h_x^{t_2},\,\ldots,\gamma(t_p)\in h_x^{t_p}}\left\{1 - \prod_{k=1}^{p}(1 - \gamma(t_k))^{\lambda(t_k)}\right\}$$

$$(5.44)$$

a dynamic hesitant fuzzy weighted averaging (DHFWA) operator. Especially, if $\lambda(t) = (1/p, 1/p, \ldots, 1/p)^T$, then the DHFWA operator reduces to the dynamic hesitant fuzzy averaging (DHFA) operator:

$$\text{DHFA}\left(h_x^{t_1}, h_x^{t_2}, \ldots, h_x^{t_p}\right) = \overset{p}{\underset{k=1}{\oplus}}\left(\frac{1}{p}h(t_k)\right)$$

$$= \bigcup_{\gamma(t_1)\in h_x^{t_1},\,\gamma(t_2)\in h_x^{t_2},\,\ldots,\gamma(t_p)\in h_x^{t_p}}\left\{1 - \prod_{k=1}^{p}(1 - \gamma(t_k))^{1/p}\right\}$$

$$(5.45)$$

Definition 5.7 (*Liao et al.* 2014). Let $h_x^{t_1}, h_x^{t_2}, \ldots, h_x^{t_p}$ be a collection of HFVs collected at p different stages $t_k\ (k = 1, 2, \ldots, p)$, and $\lambda(t) = (\lambda(t_1), \lambda(t_2), \ldots, \lambda(t_p))^T$ be the weight vector of the stages $t_k\ (k = 1, 2, \ldots, p)$, where $\lambda(t_k)$ indicates the importance degree of $h_x^{t_k}$ satisfying $\lambda(t_k) \in [0, 1]$, $k = 1, 2, \ldots, p$, and $\sum_{k=1}^{p}\lambda(t_k) = 1$. Then we call

$$\text{DHFWG}_{\lambda(t)}\left(h_x^{t_1}, h_x^{t_2}, \ldots, h_x^{t_p}\right) = \overset{p}{\underset{k=1}{\oplus}}\left(\left(h_x^{t_k}\right)^{\lambda(t_k)}\right)$$

$$= \bigcup_{\gamma(t_1)\in h_x^{t_1},\,\gamma(t_2)\in h_x^{t_2},\,\ldots,\gamma(t_p)\in h_x^{t_p}}\left\{\prod_{k=1}^{p}\gamma(t_k)^{\lambda(t_k)}\right\} \quad (5.46)$$

a dynamic hesitant fuzzy weighted geometric (DHFWG) operator. Especially, if $\lambda(t) = (1/p, 1/p, \ldots, 1/p)^T$, then the DHFWG operator reduces to the dynamic hesitant fuzzy geometric (DHFG) operator:

$$\text{DHFG}\left(h_x^{t_1}, h_x^{t_2}, \ldots, h_x^{t_p}\right) = \overset{p}{\underset{k=1}{\otimes}}\left(\left(h_x^{t_k}\right)^{1/p}\right)$$

$$= \bigcup_{\gamma(t_1)\in h_x^{t_1},\,\gamma(t_2)\in h_x^{t_2},\,\ldots,\gamma(t_p)\in h_x^{t_p}}\left\{\prod_{k=1}^{p}\gamma(t_k)^{1/p}\right\} \quad (5.47)$$

Definition 5.8 (*Liao et al.* 2014). Let $h_x^{t_1}, h_x^{t_2}, \ldots, h_x^{t_p}$ be a collection of HFVs collected at p different stages $t_k\ (k = 1, 2, \ldots, p)$, and $\lambda(t) = (\lambda(t_1), \lambda(t_2), \ldots, \lambda(t_p))^T$ be the weight vector of the stages $t_k\ (k = 1, 2, \ldots, p)$, where $\lambda(t_k)$ indicates the importance degree of $h_x^{t_k}$ satisfying $\lambda(t_k) \in [0, 1]$, $k = 1, 2, \ldots, p$, and $\sum_{k=1}^{p}\lambda(t_k) = 1$. Suppose that $h_x^{t_{\sigma(1)}}, h_x^{t_{\sigma(2)}}, \ldots, h_x^{t_{\sigma(p)}}$ is the permutation of $h_x^{t_1}, h_x^{t_2}, \ldots, h_x^{t_p}$ where $h_x^{t_{\sigma(k)}}$ is the kth largest of $h_x^{t_1}, h_x^{t_2}, \ldots, h_x^{t_p}$. Then

(1) A dynamic hesitant fuzzy ordered weighted averaging (DHFOWA) operator is a mapping DHFOWA: $V^n \rightarrow V$, where

$$\text{DHFOWA}\left(h_x^{t_1}, h_x^{t_2}, \ldots, h_x^{t_p}\right) = \overset{p}{\underset{k=1}{\oplus}} \left(\lambda(t_k) h_x^{t_{\sigma(k)}}\right)$$

$$= \underset{\gamma(t_{\sigma(1)}) \in h_x^{t_{\sigma(1)}}, \gamma(t_{\sigma(2)}) \in h_x^{t_{\sigma(2)}}, \ldots, \gamma(t_{\sigma(p)}) \in h_x^{t_{\sigma(p)}}}{\cup} \left\{1 - \prod_{k=1}^{p} (1 - \gamma(t_{\sigma(k)}))^{\lambda(t_k)}\right\}$$

$$(5.48)$$

(2) A dynamic hesitant fuzzy ordered weighted geometric (DHFOWG) operator is a mapping DHFOWG: $V^n \rightarrow V$, where

$$\text{DHFOWG}\left(h_x^{t_1}, h_x^{t_2}, \ldots, h_x^{t_p}\right) = \overset{p}{\underset{k=1}{\otimes}} \left(h_x^{t_{\sigma(k)}}\right)^{\lambda(t_k)}$$

$$= \underset{\gamma(t_{\sigma(1)}) \in h_x^{t_{\sigma(1)}}, \gamma(t_{\sigma(2)}) \in h_x^{t_{\sigma(2)}}, \ldots, \gamma(t_{\sigma(p)}) \in h_x^{t_{\sigma(p)}}}{\cup} \left\{\prod_{k=1}^{p} \left(\gamma(t_{\sigma(k)})\right)^{\lambda(t_k)}\right\}$$

$$(5.49)$$

In the case where $\lambda(t) = (1/p, 1/p, \ldots, 1/p)^T$, the DHFOWA operator reduces to the DHFA operator, and the DHFOWG operator reduces to the DHFG operator.

Definition 5.9 (*Liao et al.* 2014). Let $h_x^{t_1}, h_x^{t_2}, \ldots, h_x^{t_p}$ be a collection of HFVs collected at p different stages $t_k (k = 1, 2, \ldots, p)$, and $\lambda(t) = (\lambda(t_1), \lambda(t_2), \ldots, \lambda(t_p))^T$ be the weight vector of the stages t_k $(k = 1, 2, \ldots, p)$, where $\lambda(t_k)$ indicates the importance degree of $h_x^{t_k}$ satisfying $\lambda(t_k) \in [0, 1]$, $k = 1, 2, \ldots, p$, and $\sum_{k=1}^{p} \lambda(t_k) = 1$. p is the balancing coefficient which plays a role of balance. Then we define the following aggregation operators, which are all based on the mapping $V^n \rightarrow V$ with an aggregation-associated vector $w(t) = (w(t_1), w(t_2), \ldots, w(t_p))$ such that $w(t_k) \in [0, 1]$, $k = 1, 2, \ldots, p$, and $\sum_{k=1}^{p} w(t_k) = 1$:

(1) The dynamic hesitant fuzzy hybrid averaging (DHFHA) operator:

$$\text{DHFHA}\left(h_x^{t_1}, h_x^{t_2}, \ldots, h_x^{t_p}\right) = \overset{p}{\underset{k=1}{\oplus}} \left(w(t_k)\dot{h}_{\sigma(k)}\right)$$

$$= \underset{\dot{\gamma}_{\sigma(1)} \in \dot{h}_{\sigma(1)}, \dot{\gamma}_{\sigma(2)} \in \dot{h}_{\sigma(2)}, \ldots, \dot{\gamma}_{\sigma(p)} \in \dot{h}_{\sigma(p)}}{\cup} \left\{1 - \prod_{k=1}^{p} (1 - \dot{\gamma}_{\sigma(k)})^{w(t_k)}\right\}$$

$$(5.50)$$

where $\dot{h}_{\sigma(k)}$ is the kth largest of $\dot{h} = p\lambda(t_k)h_x^{t_k}$ $(k = 1, 2, \ldots, p)$.

(2) The dynamic hesitant fuzzy hybrid geometric (DHFHG) operator:

$$\text{DHFHG}\left(h_x^{t_1}, h_x^{t_2}, \ldots, h_x^{t_p}\right) = \overset{p}{\underset{k=1}{\otimes}} \left(\ddot{h}_{\sigma(k)}^{w(t_k)}\right)$$

$$= \underset{\ddot{\gamma}_{\sigma(1)} \in \ddot{h}_{\sigma(1)}, \ddot{\gamma}_{\sigma(2)} \in \ddot{h}_{\sigma(2)}, \ldots, \ddot{\gamma}_{\sigma(p)} \in \ddot{h}_{\sigma(p)}}{\cup} \left\{\prod_{k=1}^{p} \ddot{\gamma}_{\sigma(k)}^{w(t_k)}\right\}$$

$$(5.51)$$

where $\ddot{h}_{\sigma(k)}$ is the kth largest of $\ddot{h} = \left(h_x^{t_k}\right)^{p\lambda(t_k)} (k = 1, 2, \ldots, p)$.

Especially, if $\lambda(t) = (1/p, 1/p, \ldots, 1/p)^T$, then the DHFHA operator reduces to the DHFOWA operator, and the DHFHG operator reduces to the DHFOWG operator.

It is noted that the DHFWA and DHFWG operators only weight the hesitant fuzzy arguments, but ignore the importance of the order positions, while the DHFOWA and DHFOWG operators only weight the order positions, but ignore the importance degrees of the arguments. The DHFHA and DHFHG weight all the given arguments and their order positions.

5.3.2   Methods to Determine the Weight Vector of Multiple Stages

As we have pointed out previously, the crucial process of tackling the MSMCDM problem is to determine the weight vector $\lambda(t) = (\lambda(t_1), \lambda(t_2), \ldots, \lambda(t_p))^T$ for the values of a HFV h_x^t over the stages t_k $(k = 1, 2, \ldots, p)$. After assigning the weights to different stages, it is easy to aggregate the multiple stages hesitant fuzzy information by the dynamic hesitant fuzzy aggregation operators. The following process of ranking and selecting the alternatives can be conducted by the comparison scheme, which is very easy. Thus, in this subsection, we shall pay attention to develop some novel methods to derive the multiple stages weight vector.

(1) **The Improved Maximum Entropy (IME) Method**

Let us first analyze the minimum variance model in Xu (2008). By introducing the measure of "orness" (Yager 1988) associated with the time series weight vector $\lambda(t) = (\lambda(t_1), \lambda(t_2), \ldots, \lambda(t_p))^T$, Xu (2008) improved the minimum variance model (Fullér and Majlender 2003) to derive the time series weights $\lambda(t_k)$ $(k = 1, 2, \ldots, p)$.

Model 5.22

$$\min D^2(\lambda(t)) = \sum_{k=1}^{p} \frac{1}{p} (\lambda(t_k) - E(\lambda(t_k)))^2 = \frac{1}{p} \sum_{k=1}^{p} (\lambda(t_k))^2 - \frac{1}{p^2}$$

$$\text{s.t.} \, orness(\lambda(t)) = \frac{1}{p-1} \sum_{k=1}^{p} (p - k)\lambda(t_k) = \alpha, \; 0 \le \alpha \le 1$$

$$\lambda(t_k) \ge 0, \; k = 1, 2, \ldots, p$$

$$\sum_{k=1}^{p} \lambda(t_k) = 1$$

where $E(\lambda(t_k)) = \frac{1}{p}\sum_{k=1}^{p}\lambda(t_k)$ stands for the arithmetic mean of the weights $\lambda(t_k)$ $(k = 1, 2, \ldots, p)$, and α $(0 \leq \alpha \leq 1)$ is the given level of orness, which depicts the degree to which the aggregation is like an "or" operation.

Solving the above constrained mathematical programming problem, the time series weights $\lambda(t_k)$ $(k = 1, 2, \ldots, p)$ can be obtained in terms of

$$\lambda(t_k) = \frac{(6p - 12k + 6)\alpha - 2p + 6k - 2}{p(p+1)}, \quad k = 1, 2, \ldots, p \qquad (5.52)$$

under the condition:

$$\frac{p-2}{3p-3} \leq \alpha \leq \frac{2p-1}{3p-3} \qquad (5.53)$$

It should be pointed out that the time series weights derived from Model 5.22 under the given orness level α are monotonic increasing with $\frac{p-2}{3p-3} \leq \alpha < \frac{1}{2}$ and are monotonic decreasing with $\frac{1}{2} < \alpha \leq \frac{2p-1}{3p-3}$. Particularly, if $\alpha = 1/2$, then all the derived time series weights are equal. However, the time series weights which were determined by Xu (2008)'s method, to some extent, cannot reflect the real situation in the temporal environment, especially if we use it to aggregate the multiple stages hesitant fuzzy information with the purpose of selecting the best alternative. In the multiple stages decision making problem, the data may be updated with the time going on. In other words, new readings are obtained constantly. Hence, we shall not treat all the observations over different stages as the same. More weights should be assigned to the latest data, which means we ought to have preference to the weight vector as $\lambda(t_i) > \lambda(t_j)$, for $i > j$. However, as presented in Xu (2008)'s method, when $\frac{1}{2} < \alpha \leq \frac{2p-1}{3p-3}$, $\lambda(t_k)$ is monotonic decreasing and when $\alpha = 1/2$, all the derived time series weights are equal, which both contradict the real temporal situation. Actually, Model 5.22 is invalid or cannot reflect the underlying information exactly in the dynamic circumstance due to the fact that our purpose of determining the weight vector is not to forecast but to select the best alternative. Especially, if we take all the $\lambda(t_k) = 1/p$, $k = 1, 2, \ldots, p$, it satisfies the minimum variance, but the dynamic information may not reach its maximum utilization. Hence, Xu (2008)'s method is not appropriate to solve the MSMCDM problem.

It is noted that we shall prefer to the fresh data. Some old data may be out of date and we would not use it any more. To measure the average age of the data, Yager (2008) introduced the definition of $\overline{AGE}$.

Definition 5.10 (*Yager* 2008). Let t_p be the current time, then the age of the piece of data x_k is $AGE(t_k) = p - k$. Using this, we get the average age of the data as:

$$\overline{AGE} = \frac{\sum_{k=1}^{p}(\lambda(t_k)AGE(k))}{\sum_{k=1}^{p}\lambda(t_k)} = \sum_{k=1}^{p}(\lambda(t_k)(p - k)) = p - \sum_{k=1}^{p}(k\lambda(t_k)) \qquad (5.54)$$

where $\lambda(t_k)$ $(k = 1, 2, \ldots, p)$ is the weight vector of the stages t_k $(k = 1, 2, \ldots, p)$ such that $\lambda(t_k) \in [0, 1]$, $k = 1, 2, \ldots, p$, and $\sum_{k=1}^{p} \lambda(t_k) = 1$.

In order to obtain the weight vector to aggregate the past observations for prediction, Yager (2008) viewed the process as a time series smoothing problem and the OWA aggregation problem simultaneously, and then proposed some methods to determine the weight vector. Nevertheless, all these weight determining methods also take the minimum variance as an objective function and simultaneously minimize the $\overline{AGE}$ of the data. As we have pointed out that the aim of our study is to find the most desirable alternatives from a discrete set of feasible alternatives with respect to a finite set of criteria but not to predict, the minimum variance cannot be taken as an objective function any more. In addition, in Yager (2008), the variance is obtained under the assumption that the observations are not correlative. However, since we are investigating the same alternatives over different periods, the observations may be with some autocorrelations. From the above analysis, we can see that both Xu (2008)'s and Yager (2008)'s methods are not appropriate to generate the multiple stages weight vector. Thus, we need to find other ways to determine the weights for the MSMCDM problem within the context of hesitant fuzzy information.

The concept of entropy was introduced into the OWA operator to depict the completeness of utilizing the original information in the aggregated value (Yager 1988). It is suitable to derive the multiple stages weight vector (Nasibova and Nasibov 2010) and can be mathematically shown as $E(\lambda(t)) = -\sum_{k=1}^{p} (\lambda(t_k) \ln \lambda(t_k))$, where $\lambda(t) = (\lambda(t_1), \lambda(t_2), \ldots, \lambda(t_p))^T$ is the time series weight vector. Based on that, O'Hagan (1988) developed a model to generate the weights, which has maximum entropy under a given level of orness, shown as follows:

Model 5.23

$$\max E(\lambda(t)) = -\sum_{k=1}^{p} (\lambda(t_k) \ln \lambda(t_k))$$

$$\text{s.t. } orness(\lambda(t)) = \frac{1}{p-1} \sum_{k=1}^{p} ((p - k)\lambda(t_k)) = \alpha, \ \ 0 \le \alpha \le 1$$

$$\lambda(t_k) \ge 0, \ \ k = 1, 2, \ldots, p$$

$$\sum_{k=1}^{p} \lambda(t_k) = 1$$

Using the Lagrange multiplier to solve this model, Filev and Yager (1995) obtained an analytic form for the weight vector and described some of its properties. Later, Fuller and Majlender (2000) solved this constrained optimization problem analytically as well by using the Lagrange multiplier and derived a polynomial equation to determine the optimal weight vector, which is quite different from that of Filev and Yager (1995). Furthermore, based on the concept of parametric entropy (Rényi 1961), Majlender (2005) extended the maximal entropy model. All these entropy models used in the above literature do not consider the dynamic nature of the data, which leads to some limitations in practical application. Hence, in the following, we shall improve the maximum entropy model to make it more objective to the time series circumstance.

In the MSMCDM problem, there is no doubt that we shall not want the observations in adjacent stages changed significantly. Hence, if the deviations of observations in adjacent stages change largely, we should certainly assign small weights to them. In other words, the entropy of the deviations between the adjacent observations multiplied the corresponding weights should achieve its maximum value. Inspired by this, Liao et al. (2014) introduced the definition of the improved entropy.

Definition 5.11 (*Liao et al.* 2014). The improved entropy of the multiple stages hesitant fuzzy information $h_x^{t_k}(k = 1, 2, \ldots, p)$ with the time series weight vector $\lambda(t) = (\lambda(t_1), \lambda(t_2), \ldots, \lambda(t_p))^T$, where $\lambda(t_k) \in [0, 1]$, $k = 1, 2, \ldots, p$, and $\sum_{k=1}^{p} \lambda(t_k) = 1$ can be formulated as:

$$IE_p(\lambda(t)) = -\sum_{k=2}^{p} \left(\lambda(t_k) s(\Delta h_x^{t_k}) \ln \left(\lambda(t_k) s(\Delta h_x^{t_k}) \right) \right) \tag{5.55}$$

where $s(\Delta h_x^{t_k}) = s(h_x^{t_k}) - s(h_x^{t_{k-1}})$, $k = 2, \ldots, p$, and $s(h(t_k))$ is the score function of $h_x^{t_k}$.

It is appropriate to take $\lambda(t_1) = \lambda(t_2)$ because the preferences to the initial observations would be smaller and smaller with the increase of t. In order to get a consistent formulation, we can take $s(\Delta h(t_1)) = s(\Delta h(t_2))$ as a definition. Then, the improved entropy of the multiple stages hesitant fuzzy information can be expressed in terms of the following mathematical form:

$$IE_p(\lambda(t)) = -\sum_{k=1}^{p} \left(\lambda(t_k) s(\Delta h_x^{t_k}) \ln \left(\lambda(t_k) s(\Delta h_x^{t_k}) \right) \right) \tag{5.56}$$

where $s(\Delta h_x^{t_k}) = s(h_x^{t_k}) - s(h_x^{t_{k-1}})$, $k = 1, \ldots, p$, and $s(\Delta h_x^{t_1}) = s(\Delta h_x^{t_2})$.

The value of the improved entropy depends not only on the weight vector but also on the deviation between the adjacent evaluation values. Hence, the improved entropy can represent the actual dynamic situation more objectively.

As preferring to fresh data or youthful data in the process of MSMCDM, in order to derive the reasonable result, we should maximize the improved entropy of the hesitant fuzzy variables and minimize the average age of the data simultaneously in the process of determining the multiple stages weight vector. Hence, Model 5.24 can be obtained:

Model 5.24

$$\max IE_p(\lambda(t)) = - \sum_{k=1}^{p} \left[\lambda(t_k)s(\Delta h_x^{t_k}) \ln\left(\lambda(t_k)s(\Delta h_x^{t_k}) \right) \right]$$

$$\min \overline{AGE} = \sum_{k=1}^{p} (\lambda(t_k)(p-k)) = p - \sum_{k=1}^{p} (k\lambda(t_k))$$

$$\text{s.t. } \sum_{k=1}^{p} \lambda(t_k) = 1, \ \lambda(t_k) \geq 0, \ k = 1, 2, \ldots, p$$

Since both of the two objectives have the same degree, we can transfer the above multiple objective programming model into the single objective programming model as:

Model 5.25

$$\min T_p(\lambda(t)) = \sum_{k=1}^{p} \left[\lambda(t_k)s(\Delta h_x^{t_k}) \ln\left(\lambda(t_k)s(\Delta h_x^{t_k}) \right) \right] + p - \sum_{k=1}^{p} (k\lambda(t_k))$$

$$\text{s.t. } \sum_{k=1}^{p} \lambda(t_k) = 1, \ \lambda(t_k) \geq 0, \ k = 1, 2, \ldots, p$$

To solve this model, the following Lagrange function is constructed:

$$L(\lambda(t), \lambda) = \sum_{k=1}^{p} \left(\lambda(t_k)s(\Delta h_x^{t_k}) \ln\left(\lambda(t_k)s(\Delta h_x^{t_k}) \right) \right) + p - \sum_{k=1}^{p} (k\lambda(t_k)) + \lambda \left(\sum_{k=1}^{p} \lambda(t_k) - 1 \right)$$

(5.57)

Differentiating Eq. (5.57) with respect to $\lambda(t_k)$ $(k = 1, 2, \ldots, p)$ and λ, and setting these partial derivatives equal to zero, the following equations can be obtained easily:

$$\frac{\partial L}{\partial \lambda(t_k)} = s(\Delta h_x^{t_k}) \ln\left(\lambda(t_k)s(\Delta h_x^{t_k}) \right) + s\left(\Delta h_x^{t_k} \right) - k + \lambda = 0, \ k = 1, 2, \ldots, p \quad (5.58)$$

$$\frac{\partial L}{\partial \lambda} = \sum_{k=1}^{p} \lambda(t_k) - 1 = 0 \quad (5.59)$$

Solving Eq. (5.58), we get

$$\lambda(t_k) = \frac{1}{s\left(\Delta h_x^{t_k}\right)} e^{\frac{k-\lambda}{s\left(\Delta h_x^{t_k}\right)} - 1}, \ k = 1, 2, \ldots, p \tag{5.60}$$

Combining Eqs. (5.59) and (5.60), the following normal equation regarding to λ can be derived:

$$\sum_{k=1}^{p} \frac{1}{s\left(\Delta h_x^{t_k}\right)} e^{\frac{k-\lambda}{s\left(\Delta h_x^{t_k}\right)} - 1} = 1 \tag{5.61}$$

Since λ is the only unknown parameter in the normal equation, we can solve it to get the value of λ. Thus, the weight vector can be calculated according to Eq. (5.60).

In the following, let us compare our solution of the improved maximum entropy (IME) method with the result of Filev and Yager (1995). Filev and Yager (1995) gave the solution of Model 5.23 like this:

$$\lambda_j = \frac{e^{\beta \frac{p-j}{p-1}}}{\sum_{k=1}^{p} e^{\beta \frac{p-k}{p-1}}} \tag{5.62}$$

where $\beta \in (-\infty, +\infty)$ is a parameter dependent upon the value of α. Specifically, they pointed out that $\beta = (n-1)\ln(g)$, where g is a positive solution of the equation:

$$\sum_{j=1}^{p} \left((p-j)/(p-1) - \alpha\right) g^{p-j} = 0 \tag{5.63}$$

Comparing Eq. (5.60) with Eq. (5.62), we can see that both of them have one parameter which can be calculated by Eqs. (5.61) and (5.63), respectively. The difference between them is that the weight vector determined by Eq. (5.60) contains the variety of the adjacent periods whereas the later one determined by Eq. (5.62) does not consider the decision information at multiple stages. Hence, the improved maximum entropy method can obtain a more objective weight vector than the original model.

Although our IME method has many advantages over Xu (2008)'s method, Yager (2008)'s method, O'Hagan (1988)'s method and Filev and Yager (1995)'s method in determining the weights under dynamic environment, it also has a flaw. It is very hard to obtain the value of the parameter λ by solving the normal equation especially when k is very large. This requires a numerical method to obtain the roots of a polynomial equation, which is the same as Filev and Yager (1995) have pointed out. Fortunately, Darroch and Ratcliff (1972) solved this problem by using the GIS (Generalized Iterative scaling) method. Berger et al. (1996) also proposed an IIS (improved iterative scaling) algorithm, which is elegant in calculating the parameter λ. In addition, Yager (2009) developed a new way to get the ordinary maximum entropy weight vector based on the weight-generating function. In the following, we do not want to give some algorithms to show how to derive the solution of the parameter λ, but to develop another easier model to determine the weight vector.

(2) **The Minimum Average Deviation (MAD) Method**

In the above subsection, we have indicated that the deviation of the observations in adjacent stages should not be changed largely. Once the observations vary largely at certain stage, a small weight should be assigned to that stage. Inspired by Definitions 5.10 and 5.11, Liao et al. (2014) introduced a definition to measure the average deviation of the hesitant fuzzy observations over different stages.

Definition 5.12 (*Liao et al.* 2014). The average deviation (AD) of the multiple stages hesitant fuzzy variables $h_x^{t_k}$ $(k = 1, 2, \ldots, p)$ with the time series weight vector $\lambda(t) = (\lambda(t_1), \lambda(t_2), \ldots, \lambda(t_p))^T$ where $\lambda(t_k) \in [0, 1]$, $k = 1, 2, \ldots, p$, and $\sum_{k=1}^{p} \lambda(t_k) = 1$, can be formulated as:

$$AD = \frac{\sum_{k=1}^{p} \left(\lambda(t_k) s(\Delta h_x^{t_k}) \right)}{\sum_{k=1}^{p} \lambda(t_k)} = \sum_{k=1}^{p} \left(\lambda(t_k) s(\Delta h_x^{t_k}) \right) \tag{5.64}$$

where $s(\Delta h_x^{t_k}) = s(h_x^{t_k}) - s(h_x^{t_{k-1}})(k = 2, \ldots, p)$, denoting the deviation between the adjacent periods, $s(h_x^{t_k})$ is the score function of $h_x^{t_k}$. For $k = 1$, let $s(\Delta h_x^{t_1}) = s(\Delta h_x^{t_2})$ and $\lambda(t_1) = \lambda(t_2)$.

Motivated by Xu (2008), Yager (2008) and the IME method, the following model is constructed:

Model 5.26

$$\min AD = \sum_{k=1}^{p} \left(\lambda(t_k) s(\Delta h_x^{t_k}) \right)^2$$

$$\min \overline{AGE} = \sum_{k=1}^{p} \left(\lambda(t_k)(p - k) \right) = p - \sum_{k=1}^{p} (k\lambda(t_k))$$

$$\text{s.t.} \sum_{k=1}^{p} \lambda(t_k) = 1, \ \lambda(t_k) \geq 0, \ k = 1, 2, \ldots, p$$

The main idea of Model 5.26 is to minimize the average deviation of the hesitant fuzzy variables and minimize the average age of the data. This model is named as the minimum average deviation (MAD) method. Since both of the two objectives have the same degree, we can transfer the above multiple objective programming model into the single objective programming model.

Model 5.27

$$\min \sum_{k=1}^{p} \left(\lambda(t_k)s(\Delta h_x^{t_k})\right)^2 + p - \sum_{k=1}^{p} (k\lambda(t_k))$$

$$\text{s.t.} \sum_{k=1}^{p} \lambda(t_k) = 1, \ \lambda(t_k) \geq 0, \ \ k = 1,2,\ldots,p$$

To solve this single programming model, the following Lagrange function is introduced:

$$L(\lambda(t),\lambda) = \sum_{k=1}^{p} \left(\lambda(t_k)s(\Delta h_x^{t_k})\right)^2 + p - \sum_{k=1}^{p} (k\lambda(t_k)) + \lambda(\sum_{k=1}^{p} \lambda(t_k) - 1) \quad (5.65)$$

Differentiating Eq. (5.65) with respect to $\lambda(t_k)$ $(k = 1,2,\ldots,p)$ and λ, and setting these partial derivatives equal to zero, the following equations can be obtained easily:

$$\frac{\partial L}{\partial \lambda(t_k)} = 2s^2(\Delta h_x^{t_k})\lambda(t_k) - k + \lambda = 0, \ k = 1,2,\ldots,p \quad (5.66)$$

$$\frac{\partial L}{\partial \lambda} = \sum_{k=1}^{p} \lambda(t_k) - 1 = 0 \quad (5.67)$$

Solving Eq. (5.66), it follows

$$\lambda(t_k) = \frac{k - \lambda}{2s^2(\Delta h_x^{t_k})}, \ k = 1,2,\ldots,p \quad (5.68)$$

Combining Eqs. (5.67) and (5.68), the following normal equation regarding to λ can be derived:

$$\sum_{k=1}^{p} \frac{k - \lambda}{2s^2(\Delta h_x^{t_k})} = 1 \quad (5.69)$$

There is only one parameter in this equation, and we always can solve it easily.

Comparing Eq. (5.61) with Eq. (5.69), we can see that although both of these two normal equations have only one parameter, solving Eq. (5.69) is far easier than solving Eq. (5.61). Thus, it has more effectiveness in practical applications.

5.3.3 Approach to Hesitant Fuzzy Multiple Stages Multiple Criteria Decision Making and Its Application in Ecosystem Management

A hesitant fuzzy MSMCDM problem can be described as follows: let $A = \{A_1, A_2, \ldots, A_m\}$ be a finite set of alternatives, $C = \{C_1, C_2, \ldots, C_n\}$ be the set of criteria, and $t_k(k = 1, 2, \ldots, p)$ be p different periods, whose weight vector is $\lambda(t) = (\lambda(t_1), \lambda(t_2), \ldots, \lambda(t_p))^T$, where $\lambda(t_k) \geq 0$, $k = 1, 2, \ldots, p$, $\sum_{k=1}^{p} \lambda(t_k) = 1$, and $\omega = (\omega_1, \omega_2, \ldots, \omega_n)^T$ be the weight vector of the criteria, where $\omega_j \geq 0, j = 1, 2, \ldots, n$, and $\sum_{j=1}^{n} \omega_j = 1$. Let $H(t_k) = (h_{ij}^{t_k})_{m \times n}$ $(i = 1, 2, \ldots, m; \quad j = 1, 2, \ldots, n; k = 1, 2, \ldots, p)$ be the hesitant fuzzy decision matrices over p different stages, where $h_{ij}^{t_k}$ denotes the HFE of the alternative A_i on the criterion C_j at the stage t_k. Based on the proposed weight determining method, we can develop an algorithm to solve the hesitant fuzzy MSMCDM problem.

Algorithm 5.6

Step 1. Use the HFWA operator:

$$h_i^{t_k} = \mathrm{HFWA}\,(h_{i1}^{t_k}, h_{i2}^{t_k}, \ldots, h_{in}^{t_k}) = \mathop{\oplus}_{j=1}^{n} \left(\omega_j h_{ij}^{t_k}\right), \ i = 1, 2, \ldots, m; \tag{5.70}$$
$$k = 1, 2, \ldots, p$$

to calculate the overall values of the alternatives $A_i(i = 1, 2, \ldots, m)$ at the stages $t_k(k = 1, 2, \ldots, p)$. Then go to the next step.

Step 2. Construct a model via Model 5.25 or Model 5.27 and derive the weight vector $\lambda(t)$ of different stages for the alternatives $A_i(i = 1, 2, \ldots, m)$ by using the IME method or the MAD method. Then go to the next step.

Step 3. Utilize the DHFWA operator:

$$h_i = \mathrm{DHFWA}_{\lambda(t)}(h_i^{t_1}, h_i^{t_2}, \ldots, h_i^{t_p}) = \mathop{\oplus}_{k=1}^{p} \left(\lambda(t_k)h_i^{t_k}\right), \ i = 1, 2, \ldots, m \tag{5.71}$$

to aggregate the overall values $h_i^{t_k}$ $(k = 1, 2, \ldots, p)$ collected from p different periods, and get the overall value h_i of the alternative $A_i(i = 1, 2, \ldots, m)$, then go to the next step.

Step 4. Rank the alternatives $A_i(i = 1, 2, \ldots, m)$ according to the overall values $h_i i = (1, 2, \ldots, m)$ by using the comparison law given in Sect. 1.1.3, then go the next step.

Step 5. Pick out the best alternative(s) and the procedure ends.

In what follows, let us present a numerical example concerning the selection of suitable plan for rangeland area to validate our proposed models and illustrate how to implement the approach in hesitant fuzzy MSMCDM:

Example 5.5 (Liao et al. 2014) The rangeland is a complex ecosystem which provides many ecological, social and economic services, including food, water supply, wildlife diversity, recreational facilities, animal husbandry, climate regulation, erosion control as well as ethical and social services. The rangeland has attracted a large number of social groups' attentions. Basically, there are six social groups involved: ranchers, citizens, NGOs, environmental managers, watershed managers, range managers and nomad management departments (Zendehedl et al. 2009). Different groups have different interests over the rangelands. For example, the ranchers aim to increase animal grazing rate to extend their profits, whereas other social groups, such as local citizens and environmental agencies, would like to minimize the ranchers' activities to support biodiversity. In order to establish a sustainable policy, four alternative plans have been formulated:

A_1 (Livestock control): Reduce the livestock by 40 % in the area, and introduce new legislation to facilitate grazing license transaction;

A_2 (Rangeland rehabilitation): Introduce hand planting, seedling and a grazing system (no change in the number of animals);

A_3 (Watershed management): Harvest water through contour furrow, gabion, bio-mechanical treatment, and reduce the livestock by 20 % in the area;

A_4 (Environmental preservation): Change the area into a national park without any ranchers, and implement a number of plans for ecotourism and wildlife diversity.

Suppose that the six social groups cannot persuade each other to reach a consensus plan. In order to select an appropriate policy to ensure that those services will be available for generations to come, the government decided to test each of the plans for three years in four rangelands under similar conditions, and then choose the best one to implement in the future. Every year the government evaluated the four rangelands A_i $(i = 1, 2, 3, 4)$ over three different kinds of criteria $C_j (j = 1, 2, 3)$ and got some evaluation values. The weights of the three criteria were established as $\omega = (0.3, 0.3, 0.4)^T$. A set of rules were employed to help evaluate these four rangelands over the three criteria (shown as Table 5.7).

Since many of the rules are qualitative, it is suitable for the decision maker to use fuzzy set to express their assessments. Meanwhile, the traditional fuzzy set cannot express more than one rule simultaneously, but HFS is suitable to express such information. For instance, if we want to measure the ecological criterion, there are three rules, so it is hard to represent the evaluation values of the ecological criteria with the ordinary fuzzy numbers, but it can be done using HFEs. In other words, it is adequate to take the criteria of these four plans as HFVs. Once all the 3-year

Table 5.7 The rules for different criteria

Criteria	Rules		
Ecological criterion C_1	Climate regulation	Soil conservation	Species diversity
Social criterion C_2	Cultural criteria	Social education	Recreation
Economic criterion C_3	Part-time job	Water supply	Cost of plan

Table 5.8 The evaluation values of the four plans in 1st year ($t = 1$)

	C_1	C_2	C_3
A_1	{0.6, 0.5, 0.6}	{1, 0.6, 0.4}	{0.4, 0.1, 0.3}
A_2	{0, 0.7, 0.6}	{0.1, 0.3, 0.5}	{0.6, 0.05, 0.6}
A_3	{0, 0.4, 0.6}	{0.1, 0.5, 0.3}	{0.5, 0.2, 0.3}
A_4	{0.6, 0.6, 0.6}	{0, 0.7, 0.8}	{1, 0.15, 0.9}

Table 5.9 The evaluation values of the four plans in 2nd year ($t = 2$)

	C_1	C_2	C_3
A_1	{0.7, 0.65, 0.65}	{1, 0.7, 0.4}	{0.5, 0.2, 0.4}
A_2	{0, 0.75, 0.65}	{0.15, 0.4, 0.6}	{0.7, 0.1, 0.7}
A_3	{0, 0.45, 0.7}	{0.15, 0.6, 0.4}	{0.6, 0.4, 0.4}
A_4	{0.65, 0.65, 0.8}	{0, 0.8, 0.9}	{1, 0.3, 0.95}

Table 5.10 The evaluation values of the four plans in 3rd year ($t = 3$)

	C_1	C_2	C_3
A_1	{0.8, 0.65, 0.7}	{1, 0.8, 0.6}	{0.6, 0.3, 0.5}
A_2	{0, 0.8, 0.7}	{0.2, 0.5, 0.7}	{0.8, 0.15, 0.8}
A_3	{0, 0.5, 1}	{0.2, 0.7, 0.5}	{0.7, 0.6, 0.5}
A_4	{0.7, 0.7, 1}	{0, 0.9, 1}	{1, 0.45, 1}

evaluation values were determined, three hesitant fuzzy matrices were obtained, shown as Tables 5.8, 5.9 and 5.10.

In this example, the evaluation values of the four plans A_i ($i = 1, 2, 3, 4$) over the criteria $C_j (j = 1, 2, 3)$ are collected from three years. It can be seen as a MSMCDM problem. In addition, as presented above, it is appropriate to take the criteria of these four plans as HFEs. Thus, this is a hesitant fuzzy MSMCDM problem. Hence, we can use Algorithm 5.6 to solve the problem and then obtain the appropriate policy.

Step 1 Firstly, the three hesitant fuzzy matrices at different stages can be rewritten as:

$$
H(t_1) = \begin{bmatrix}
\{0.5, 0.6\} & \{0.4, 0.6, 1\} & \{0.1, 0.3, 0.4\} \\
\{0, 0.6, 0.7\} & \{0.1, 0.3, 0.5\} & \{0.05, 0.6\} \\
\{0, 0.4, 0.6\} & \{0.1, 0.3, 0.5\} & \{0.2, 0.3, 0.5\} \\
\{0.6\} & \{0, 0.7, 0.8\} & \{0.15, 0.9, 1\}
\end{bmatrix}
$$

$$
H(t_2) = \begin{bmatrix}
\{0.65, 0.7\} & \{0.4, 0.7, 1\} & \{0.2, 0.4, 0.5\} \\
\{0, 0.65, 0.75\} & \{0.15, 0.4, 0.6\} & \{0.1, 0.7\} \\
\{0, 0.45, 0.7\} & \{0.15, 0.4, 0.6\} & \{0.4, 0.6\} \\
\{0.65, 0.8\} & \{0, 0.8, 0.9\} & \{0.3, 0.95, 1\}
\end{bmatrix}
$$

$$H(t_3) = \begin{bmatrix} \{0.65, 0.7, 0.8\} & \{0.6, 0.8, 1\} & \{0.3, 0.5, 0.6\} \\ \{0, 0.7, 0.8\} & \{0.2, 0.5, 0.7\} & \{0.15, 0.8\} \\ \{0, 0.5, 1\} & \{0.2, 0.5, 0.7\} & \{0.5, 0.6, 0.7\} \\ \{0.7, 1\} & \{0, 0.9, 1\} & \{0.45, 1\} \end{bmatrix}$$

Then, via Eq. (5.70), we can aggregate the HFEs by the HFWA operator. The results are set out as follows:

$h_1(t_1) = \{0.3319, 0.3752, 0.3958, 0.4084, 0.4319, 0.4349, 0.4467, 0.4650, 0.4687, 0.4970,$
$0.4996, 0.5296, 1\}$

$h_2(t_1) = \{0.0508, 0.1197, 0.2042, 0.2789, 0.3284, 0.3313, 0.3385, 0.3772, 0.3866, 0.3955,$
$0.4370, 0.4455, 0.4898, 0.5269, 0.5320, 0.5660, 0.5723, 0.6077\}$

$h_3(t_1) = \{0.1138, 0.1599, 0.1782, 0.2209, 0.2398, 0.2571, 0.2657, 0.2793, 0.2950, 0.2957,$
$0.3190, 0.3268, 0.3316, 0.3618, 0.3627, 0.3700, 0.3757, 0.3844, 0.3958, 0.4082,$
$0.4158, 0.4357, 0.4422, 0.4650, 0.4719, 0.4827, 0.5324\}$

$h_4(t_1) = \{0.2882, 0.5040, 0.5608, 0.6976, 0.7893, 0.8134, 1\}$
$h_1(t_2) = \{0.4273, 0.4532, 0.4896, 0.5126, 0.5255, 0.5349, 0.5469, 0.5559, 0.5854, 0.6041,$
$0.6146, 0.6320, 1\}$

$h_2(t_2) = \{0.0869, 0.1775, 0.2717, 0.3336, 0.3976, 0.3997, 0.4116, 0.4573, 0.4685, 0.4700,$
$0.5195, 0.5307, 0.5706, 0.6118, 0.6132, 0.6503, 0.6575, 0.6904\}$

$h_3(t_2) = \{0.2236, 0.3006, 0.3398, 0.3511, 0.3807, 0.4053, 0.4155, 0.4482, 0.4590, 0.4734,$
$0.4824, 0.5030, 0.5126, 0.5400, 0.5599, 0.5685, 0.5856, 0.6331\}$

$h_4(t_2) = \{0.3672, 0.4650, 0.6095, 0.6699, 0.6829, 0.7319, 0.7798, 0.8138, 0.8641, 0.8851,$
$0.8896, 0.9067, 1\}$

$h_1(t_3) = \{0.5193, 0.5410, 0.5798, 0.5936, 0.5988, 0.6095, 0.6157, 0.6272, 0.6331, 0.6448,$
$0.6587, 0.6699, 0.6741, 0.6751, 0.6879, 0.7020, 0.7115, 0.7361, 1\}$

$h_2(t_3) = \{0.1236, 0.2389, 0.3470, 0.3893, 0.4592, 0.4696, 0.5087, 0.5304, 0.5450, 0.5733,$
$0.5971, 0.6339, 0.6576, 0.6969, 0.7027, 0.7367, 0.7449, 0.7741\}$

$h_3(t_3) = \{0.2912, 0.3517, 0.3844, 0.4222, 0.4243, 0.4370, 0.4719, 0.4734, 0.4982, 0.5000,$
$\qquad\quad 0.5170, 0.5307, 0.5427, 0.5695, 0.5710, 0.5924, 0.6077, 0.6503, 1\}$
$h_4(t_3) = \{0.4514, 0.7250, 1\}$

Step 2. Use the MAD method to calculate the weight vector $\lambda(t)$ for different stages. By Eq. (1.17), we can calculate:

$$s(h_1(t_1)) = 0.4834,\ s(h_2(t_1)) = 0.3882,\ s(h_3(t_1)) = 0.3403,\ s(h_4(t_1)) = 0.6648$$
$$s(h_1(t_2)) = 0.5755,\ s(h_2(t_2)) = 0.4621,\ s(h_3(t_2)) = 0.4546,\ s(h_4(t_2)) = 0.7435$$
$$s(h_1(t_3)) = 0.6567,\ s(h_2(t_3)) = 0.5405,\ s(h_3(t_3)) = 0.5177,\ s(h_4(t_3)) = 0.7255$$

Thus, we have

$$s(\triangle h_1(t_1)) = s(\triangle h_1(t_2)) = 0.0923,\ s(\triangle h_1(t_3)) = 0.0810$$
$$s(\triangle h_2(t_1)) = s(\triangle h_2(t_2)) = 0.0739,\ s(\triangle h_2(t_3)) = 0.0784$$
$$s(\triangle h_3(t_1)) = s(\triangle h_3(t_2)) = 0.1143,\ s(\triangle h_1(t_3)) = 0.0631$$
$$s(\triangle h_4(t_1)) = s(\triangle h_4(t_2)) = 0.0787,\ s(\triangle h_4(t_3)) = 0.0180$$

Using Eq. (5.69), we can obtain the values of the parameter λ. Then the associated dynamic weight vectors for different alternatives can be calculated according to Eq. (5.68), which are $\lambda_1 = (0.0953, 0.2860, 0.6186)^T$, $\lambda_2 = (0.1184, 0.3553, 0.5262)^T$, $\lambda_3 = (0.0490, 0.1470, 0.8040)^T$, and $\lambda_4 = (0.0101, 0.0301, 0.9598)^T$.

Step 3. Utilize the DHFWA operator to aggregate the overall values $h_i(t_k)(k = 1, 2, 3)$, and get the overall values $h_i(i = 1, 2, 3, 4)$ for the plans $A_i\ (i = 1, 2, 3, 4)$.

Step 4. The score function values of $h_i\ (i = 1, 2, 3, 4)$ are $s(h_1) = 0.6207$, $s(h_2) = 0.4973$, $s(h_3) = 0.5013$, and $s(h_4) = 0.7255$. Since $s(h_4) > s(h_1) > s(h_3) > s(h_2)$, the ranking of these four plans is $A_4 \succ A_1 \succ A_3 \succ A_2$, where "$\succ$" denotes "prior to". That is to say, changing the area into a National Park without any ranchers and implementing a number of plans for eco-tourism and wildlife diversity is the most appropriate plan for such a rangeland.

In Example 5.5, the most important criterion is the Economic criterion C_3. Let us look back to the judgment values in Tables 5.8, 5.9 and 5.10. We can find that in each year, the HFE of the fourth alternative is higher than those of the other three alternatives. In addition, as to the criteria C_1 and C_2, the HFEs of the fourth alternative are also slightly higher than the others. Thus, it is intuitive that the fourth alternative A_4 is the best one. That is to say, the result derived from our approach is consistent with our intuition.

References

Berger AL, Pietra SAD, Pietra VJD (1996) A maximum entropy approach to natural language processing. Comput Linguist 22(1):39–71

Chen YL, Cheng LC (2010) An approach to group ranking decision in dynamic environment. Decis Support Syst 48(4):622–634

Chen Y, Li B (2011) Dynamic multi-attribute decision making model based on triangular intuitionistic fuzzy number. Scientia Iranica B 18(2):268–274

Chen N, Xu ZS, Xia MM (2013) Interval-valued hesitant preference relations and their applications to group decision making. Knowl-Based Syst 37:528–540

Darroch JN, Ratcliff D (1972) Generalized iterative scaling for Log-Linear models. Ann Math Stat 43(5):1470–1480

Filev D, Yager RR (1995) Analytic properties of maximum entropy OWA operators. Inf Sci 85(1–3):11–27

French S, Hartley R, Thomas LC, White DJ (1983) Multi-objective decision making. Academic Press, New York

Fullér R, Majlender P (2000) An analytic approach for obtaining maximal entropy OWA operator weights. Fuzzy Sets Syst 124(1):53–57

Fullér R, Majlender P (2003) On obtaining minimal variability OWA operator weights. Fuzzy Sets Syst 136(2):203–215

Liao HC, Xu ZS (2014a) Priorities of intuitionistic fuzzy preference relation based on multiplicative consistency. IEEE Trans Fuzzy Syst 22(6):1669–1681

Liao HC, Xu ZS (2014b) Satisfaction degree based interactive decision making method under hesitant fuzzy environment with incomplete weights. Int J Uncertainty Fuzziness Knowl-Based Syst 22(4):553–572

Liao HC, Xu ZS, Xu JP (2014) An approach to hesitant fuzzy multi-stage multi-criterion decision making. Kybernetes 43(9/10):1447–1468

Majlender P (2005) OWA operators with maximal Rényi entropy. Fuzzy Sets Syst 155(3):340–360

Nasibova RA, Nasibov EN (2010) Linear aggregation with weighted ranking. Autom Control Comput Sci 44(2):96–102

O'Hagan M (1988) Aggregating template or rule antecedents in real-time expert systems with fuzzy set logic. Proceedings 22nd annual IEEE asilomar conference on signals, systems and computers. IEEE & Maple Press, Pacific Grove, CA, pp 681–689

Perez IJ, Cabrerizo FJ, Herrera-Viedma E (2011) A mobile group decision making model for heterogeneous information and changeable decision contexts. Int J Uncertainty Fuzziness Knowl-Based Syst 19(33):33–52

Rényi A (1961) On measures of entropy and information. In: Proceedings 4th Berkeley symposium on mathematical statistics and probability. University of California Press, California, pp. 547–561

Terlaky T (1996) Interior point methods in mathematics programming. Kluwer Press, Boston, MA

Xia MM, Xu ZS, Chen N (2013) Some hesitant fuzzy aggregation operators with their application in group decision making. Group Decis Negot 22(2):259–279

Xu ZS (2004) Method based on expected values for fuzzy multiple attribute decision making problems with preference information on alternatives. Syst Eng-Theory Pract 119(1):109–113

Xu ZS (2008) On multi-period multi-attribute decision making. Knowl-Based Syst 21(2):164–171

Xu ZS (2009) Multi-period multi-attribute group decision making under linguistic assessments. Int J Gen Syst 38(8):823–850

Xu ZS, Xia MM (2011) Distance and similarity measures for hesitant fuzzy sets. Inf Sci 181:2128–2138

Xu ZS, Yager RR (2008) Dynamic intuitionistic fuzzy multi-attribute decision making. Int J Approximate Reasoning 48(1):246–262

Xu ZS, Zhang XL (2013) Hesitant fuzzy multi-attribute decision making based on TOPSIS with incomplete weight information. Knowl-Based Syst 52:53–64

Yager RR (1988) On ordered weighted averaging aggregation operators in multi-criteria decision making. IEEE Trans Syst Man Cybern 18(1):183–190

Yager RR (2008) Time series smoothing and OWA aggregation. IEEE Trans Fuzzy Syst 16 (4):994–1007

Yager RR (2009) Weighted maximum entropy OWA aggregation with applications to decision making under risk. IEEE Trans Syst Man Cybern Part A Syst Hum 39(3):555–564

Zendehedl K, Rademaker M, De Baets B, Van Huylenbroeck G (2009) Improving tractability of group decision making on environmental problems through the use of social intensities of preferences. Environ Model Softw 24(12):1457–1466

Zhao H, Xu ZS, Wang H, Liu SS (2016) Hesitant fuzzy multi-attribute decision making based on the minimum deviation method. Soft Comput doi:10.1007/s00500-015-2020-y

Zimmermann HJ, Zysno P (1980) Latent connectives in human decision making. Fuzzy Sets Syst 4:37–51

Chapter 6
Decision Making with Hesitant Fuzzy Preference Relation

In the process of decision making, the decision maker may be more rational and suitable to express his/her preferences by comparing each pair of objects and constructs a preference relation. The preference relation, as the most common and paramount representation of information, has attracted great attention from scholars and has been widely applied, especially in multiple criteria decision making. Up to now, many different types of preference relations have been proposed, such as the fuzzy preference relation (Tanino 1984), the multiplicative preference relation (Saaty 1980), the linguistic preference relation (Herrera and Herrera-Viedma 2000; Xu 2006), the intuitionistic fuzzy preference relation (Xu 2007; Xu and Liao 2014), the intuitionistic multiplicative preference relation (Xia et al. 2013), and the interval-valued intuitionistic preference relation (Xu and Ygaer 2009). However, most of the existing preference relations do not consider the hesitant fuzzy information which allows the decision makers to provide all the possible values when comparing two alternatives (or criteria). To solve this drawback, Liao et al. (2014b) introduced the definition of hesitant fuzzy preference relation (HFPR) and investigated its distinctive properties. This chapter explores the properties of the HFPR and introduces the concepts of multiplicative consistency, perfect multiplicative consistency and acceptable multiplicative consistency for a HFPR, based on which, two algorithms are given to improve the inconsistency degree of a HFPR.

As the consistency index of a HFPR determines the accuracy and reliability, in order to improve the accuracy in checking the multiplicative consistency of a HFPR, we then provide a method to determine the values of the consistency index for the HFPRs with different orders. We point out the weaknesses of the existing method in checking the multiplicative consistency of a HFPR. As there is no any theoretical evidence to support the given consistency threshold, we investigate the density function of the consistency index of a HFPR in the second part of this chapter and introduce an algorithm to determine the value of the multiplicative consistency index of a HFPR. Based on some simulations, a value table of critical values of the multiplicative consistency index of a HFPR is determined, whose elements vary with respect to the order of the HFPR and the measure used on

© Springer Nature Singapore Pte Ltd. 2017   221
H. Liao and Z. Xu, *Hesitant Fuzzy Decision Making Methodologies and Applications*, Uncertainty and Operations Research,
DOI 10.1007/978-981-10-3265-3_6

distance calculations. Finally, we study the consensus reaching process of group decision making based on the HFPRs. Several illustrative examples are given to demonstrate the practicality of the algorithms. In the following parts of this chapter, we apply the HFPR to group decision making problems and investigate its interval-valued forms.

6.1 Hesitant Fuzzy Preference Relation and Its Multiplicative Consistency

In the process of decision making, in order to avoid the influence of the limited ability of human thinking and obtain the best ranking result, people prefer to take pairwise comparison of one alternative over another and construct a preference relation. Therefore, the preference relations turn out to be the most common representation formats for expressing the decision makes' preferences. Let $A = \{A_1, A_2, \ldots, A_n\}$ be a set of alternatives, then $R = (r_{ij})_{n \times n}$ is called a fuzzy preference relation on $A \times A$ with the condition that $r_{ij} \geq 0$, $r_{ij} + r_{ji} = 1$, $i, j = 1, 2, \ldots, n$, where r_{ij} denotes the degree to which the alternative A_i is prior to the alternative A_j. The values in a fuzzy preference relation are certain values between 0 and 1. However, when people establish the preference degree of one object over another, they may have a set of possible values but not one single value due to the complexity of the decision making problem, the lack of knowledge about the problem domain, and so on. In such cases, it is very suitable and reasonable to represent the preference information by HFEs, which permit the membership degree of an element to a set represented by several possible values. Thus, Liao et al. (2014b) defined the HFPR:

Definition 6.1 (*Liao et al.* 2014b). Let $A = \{A_1, A_2, \ldots, A_n\}$ be a fixed set, a HFPR H on A is presented by a matrix $H = (h_{ij})_{n \times n} \subset A \times A$ with h_{ij} being a HFE indicating all the possible degrees to which A_i is preferred to A_j. Moreover, h_{ij} should satisfy the following conditions:

$$h_{ij}^{\sigma(t)} + h_{ji}^{\sigma(l_{h_{ji}} - t + 1)} = 1, h_{ii} = \{0.5\}, l_{h_{ij}} = l_{h_{ji}}, i, j = 1, 2, \ldots, n \qquad (6.1)$$

where $h_{ij}^{\sigma(t)}$ is the tth smallest value in h_{ij}, $t = 1, 2, \ldots, l_{h_{ij}}$.

With Definition 6.1, we can easily derive the following results:

Theorem 6.1 (Liao et al. 2014b). *The transpose $H^T = (h_{ij}^T)_{n \times n}$ of the HFPR $H = (h_{ij})_{n \times n}$ is also a HFPR, where $h_{ij}^T = h_{ji}$, $i, j = 1, 2, \ldots, n$.*

Theorem 6.2 (Liao et al. 2014b). *Let $H = (h_{ij})_{n \times n}$ be a HFPR, then, if we remove the ith row and the ith column, then the remaining matrix $\overline{H} = (h_{ij})_{(n-1) \times (n-1)}$ is also a HFPR.*

When the decision maker evaluates the alternatives, he/she may provide inconsistent preference values, and consequently constructs the inconsistent preference relation due to the complexity of the considered problem or other reasons. The investigation on consistency of a preference relation generally involves two phases: (1) judge whether the preference relation considered is consistent or not; (2) adjust or repair the inconsistent preference relation until it is with acceptable consistency. As for the first phase, the concept of consistency was traditionally defined in terms of transitivity, such as weak transitivity, max-max transitivity, max-min transitivity, restricted max-min transitivity, restricted max-max transitivity, additive transitivity, and multiplicative transitivity. Based on the above transitivity properties, some methods to measure the consistency of a preference relation can be developed. Saaty (1980) firstly derived a consistency ratio for multiplicative preference relation and developed the concept of perfect consistency and acceptable consistency. He further pointed out that the multiplicative preference relation is of acceptable consistency if its consistency ratio is less than 0.1. However, the more common situation in practice is the preference relation possessing unacceptable consistency, which may mislead the ranking results. Therefore, we need to repair the inconsistent preference relation, which is the target of the second phase.

The HFPR $H = (h_{ij})_{n \times n}$ should satisfy the following transitivity properties:

(1) If $h_{ik} \oplus h_{kj} \geq h_{ij}$, for all $i, j, k = 1, 2, \ldots, n$, then we say H satisfies the triangle condition.
(2) If $h_{ik} \geq \{0.5\}$, $h_{kj} \geq \{0.5\}$, then $h_{ij} \geq \{0.5\}$, for all $i, j, k = 1, 2, \ldots, n$, then we say H satisfies the weak transitivity property.
(3) If $h_{ij} \geq \min\{h_{ik}, h_{kj}\}$, for all $i, j, k = 1, 2, \ldots, n$, then we say H satisfies max-min transitivity property.
(4) If $h_{ij} \geq \max\{h_{ik}, h_{kj}\}$, for all $i, j, k = 1, 2, \ldots, n$, then we say H satisfies max-max transitivity property.
(5) If $h_{ik} \geq \{0.5\}$, $h_{kj} \geq \{0.5\}$, then $h_{ij} \geq \min(h_{ik}, h_{kj})$, for all $i, j, k = 1, 2, \ldots, n$, then we say H satisfies the restricted max-min transitivity property.
(6) If $h_{ik} \geq \{0.5\}$, $h_{kj} \geq \{0.5\}$, then $h_{ij} \geq \max(h_{ik}, h_{kj})$, for all $i, j, k = 1, 2, \ldots, n$, then we say H satisfies the restricted max-max transitivity property.

The weak transitivity is the usual and basic property which can be interpreted as follows: If the alternative A_i is preferred to A_k, and A_k is preferred to A_j, then A_i should be preferred to A_j. If the person who is logic and consistent does not want to draw inconsistent conclusions, he/she should first ensure that the HFPR satisfies the weak transitivity. However, the weak transitivity is the minimum requirement condition to make sure that the HFPR is consistent. There are another two conditions named additive transitivity and multiplicative transitivity which are more restrictive than weak transitivity and can imply reciprocity.

Associated with the study of transitivity properties, Herrera-Viedma et al. (2004) proposed the additive transitivity property of fuzzy preference relation as a new characterization of the consistency property. Gong et al. (2010) investigated the property of additive consistent intuitionistic fuzzy preference relation. Ma et al. (2006) presented two methods derived from graph theory to judge whether a fuzzy preference relation has weak transitivity or not, and then via a synthesis matrix which reflects the relationship between the fuzzy preference relation with additive consistency and the original one given by the decision maker, an algorithm was developed to repair the inconsistent fuzzy preference relation. Many scholars have applied the additive transitivity property of fuzzy preference relations to practice. But as the additive transitivity property of a fuzzy preference relation $P = (p_{ij})_{n \times n}$ is represented as $p_{ij} = p_{ik} + p_{kj} - 0.5$, $i, j, k = 1, 2, \ldots, n$, where p_{ij}, p_{ik} and p_{kj} are the preference information on the alternatives A_i, A_j and A_k, if we take $p_{ik} = 0.8$ and $p_{kj} = 0.9$ as an example, then, $p_{ij} = 1.2 > 1$, which does not belong to the unit closed interval [0,1] and thus is unreasonable. To solve this problem, based on the multiplicative consistency of the fuzzy preference relation $P = (p_{ij})_{n \times n}$, which is represented as $p_{ij} p_{jk} p_{ki} = p_{ik} p_{kj} p_{ji}$, $i, j, k = 1, 2, \ldots, n$, where p_{ij}, p_{ik} and p_{kj} are the preference values given by the decision maker, Chiclana et al. (2009a) developed a method to construct the consistent fuzzy preference relation from a set of $n - 1$ preference values. Xia and Xu (2011a) proposed a new method which can get the complete consistent fuzzy preference relation quickly without any transformation based on the multiplicative consistency of a fuzzy preference relation. Xia and Xu (2011b) also developed some methods to get the perfect multiplicative consistent interval reciprocal relation from the inconsistent one and estimate the missing values from an incomplete interval reciprocal relation. Liao et al. (2014a) investigated the multiplicative consistency of interval-valued intuitionistic fuzzy preference relations. In the following, we utilize the multiplicative consistency to investigate the consistency of HFPR.

The additive transitivity can be generalized to accommodate the HFPR in terms of $(h_{ik} - \{0.5\}) \oplus (h_{kj} - \{0.5\}) = (h_{ij} - \{0.5\})$, for all $i, j, k = 1, 2, \ldots, n$. The multiplicative transitivity is an important property of the fuzzy preference relation $P = (p_{ij})_{n \times n}$, which was firstly introduced by Tanino (1984) and shown as:

$$\frac{p_{ji}}{p_{ij}} \cdot \frac{p_{kj}}{p_{jk}} = \frac{p_{ki}}{p_{ik}} \tag{6.2}$$

where p_{ij} denotes a ratio of preference intensity for the alternative A_i to that for A_j. In other words, A_i is p_{ij} times as good as A_j, and $p_{ij} \in [0, 1]$, for all $i, j = 1, 2, \ldots, n$.

Even though both additive transitivity and multiplicative transitivity can be used to measure the consistency, the additive consistency may produce the unreasonable results as discussed above. Thus, we shall take the multiplicative transitivity to verify the consistency of a HFPR. The condition of multiplicative transitivity can be rewritten as follows:

$$p_{ij} p_{jk} p_{ki} = p_{ik} p_{kj} p_{ji} \tag{6.3}$$

In the case where $(p_{ik}, p_{kj}) \notin \{(0, 1), (1, 0)\}$, Eq. (6.3) is equivalent to (Chiclana et al. 2009b):

$$p_{ij} = \frac{p_{ik}p_{kj}}{p_{ik}p_{kj} + (1 - p_{ik})(1 - p_{kj})} \tag{6.4}$$

and if $(p_{ik}, p_{kj}) \in \{(0, 1), (1, 0)\}$, we stipulate $p_{ij} = 0$.

Inspired by Eq. (6.4), Liao et al. (2014b) defined the concept of multiplicative consistent HFPR:

Definition 6.2 (*Liao et al.* 2014b). Let $H = (h_{ij})_{n \times n}$ be a HFPR on a fixed set $A = \{A_1, A_2, \ldots, A_n\}$, then $H = (h_{ij})_{n \times n}$ is multiplicative consistent if

$$h_{ij}^{\sigma(t)} = \begin{cases} 0, & (h_{ik}, h_{kj}) \in \{(\{0\}, \{1\}), (\{1\}, \{0\})\} \\ \frac{h_{ik}^{\sigma(t)} h_{kj}^{\sigma(t)}}{h_{ik}^{\sigma(t)} h_{kj}^{\sigma(t)} + (1 - h_{ik}^{\sigma(t)})(1 - h_{kj}^{\sigma(t)})}, & otherwise \end{cases}, \quad i \le k \le j \tag{6.5}$$

where $h_{ik}^{\sigma(t)}$ and $h_{kj}^{\sigma(t)}$ are the tth smallest values in h_{ik} and h_{kj}, respectively.

For the convenience of checking whether a HFPR is consistent or not, based on Definition 6.2, we can introduce the corresponding perfect multiplicative consistent HFPR for any HFPR:

Definition 6.3 (*Liao et al.* 2014b). Let $H = (h_{ij})_{n \times n}$ be a HFPR on a fixed set $A = \{A_1, A_2, \ldots, A_n\}$, then we call $\overline{H} = (\bar{h}_{ij})_{n \times n}$ a prefect multiplicative consistent HFPR, where

$$\bar{h}_{ij}^{\sigma(t)} = \begin{cases} \frac{1}{j - i - 1} \sum_{k=i+1}^{j-1} \frac{h_{ik}^{\sigma(t)} h_{kj}^{\sigma(t)}}{h_{ik}^{\sigma(t)} h_{kj}^{\sigma(t)} + (1 - h_{ik}^{\sigma(t)})(1 - h_{kj}^{\sigma(t)})}, & i + 1 < j \\ h_{ij}^{\sigma(t)}, & i + 1 = j \\ \{0.5\}, & i = j \\ 1 - \bar{h}_{ji}^{\sigma(t)}, & i > j \end{cases} \tag{6.6}$$

and $\bar{h}_{ij}^{\sigma(t)}$, $h_{ik}^{\sigma(t)}$ and $h_{kj}^{\sigma(t)}$ are the tth smallest values in $\bar{h}_{ij}$, h_{ik} and h_{kj}, respectively, $t = 1, 2, \ldots, l$, $l = \max\{l_{h_{ik}}, l_{h_{kj}}\}$.

Definition 6.4 (*Liao et al.* 2014b). Let $H = (h_{ij})_{n \times n}$ be a HFPR on a fixed set $A = \{A_1, A_2, \ldots, A_n\}$, then we call $H = (h_{ij})_{n \times n}$ an acceptable multiplicative consistent HFPR, if

$$d(H, \overline{H}) < \tau \tag{6.7}$$

where τ is the consistency threshold. Without loss of generality, we usually let $\tau = 0.1$ in practice (the values of τ will be discussed in-depth in Sect. 6.2). $d(H, \overline{H})$ is the distance measure between H and $\overline{H}$ which can be calculated by using the hesitant normalized Hamming distance measure:

$$d_{hnh}(H,\overline{H}) = \frac{1}{(n-1)(n-2)}\sum_{i=1}^{n}\sum_{j=1}^{n}\left[\frac{1}{l_{h_{ij}}}\sum_{t=1}^{l_{h_{ij}}}\left|h_{ij}^{\sigma(t)}-\overline{h}_{ij}^{\sigma(t)}\right|\right] \tag{6.8}$$

or the hesitant normalized Euclidean distance measure:

$$d_{hne}(H,\overline{H}) = \frac{1}{(n-1)(n-2)}\sum_{i=1}^{n}\sum_{j=1}^{n}\left[\frac{1}{l_{h_{ij}}}\sum_{t=1}^{l_{h_{ij}}}\left|h_{ij}^{\sigma(t)}-\overline{h}_{ij}^{\sigma(t)}\right|^{2}\right]^{\frac{1}{2}} \tag{6.9}$$

The hesitant normalized Hamming distance and the hesitant normalized Euclidean distance are drawn on the well-known Hamming distance and the Euclidean distance. The difference of them takes place in the power of the absolute distance $\left|h_{ij}^{\sigma(t)}-\overline{h}_{ij}^{\sigma(t)}\right|$.

Theorem 6.3 (Liao et al. 2014b). *Any HFPR $H = (h_{ij})_{2\times 2}$ is multiplicative consistent.*

Proof Suppose that $h_{12} = \{h_{12}^{\sigma(1)}, h_{12}^{\sigma(2)}, \ldots, h_{12}^{\sigma(n)}\}$, then, $h_{21} = \{1 - h_{12}^{\sigma(n)}, 1 - h_{12}^{\sigma(n-1)}, \ldots, 1 - h_{12}^{\sigma(1)}\}$. Thus,

$$\frac{0.5h_{12}^{\sigma(1)}}{0.5h_{12}^{\sigma(1)} + 0.5(1 - h_{12}^{\sigma(1)})} = \frac{0.5h_{12}^{\sigma(1)}}{0.5} = h_{12}^{\sigma(1)}$$

Similarly,

$$\frac{0.5h_{12}^{\sigma(2)}}{0.5h_{12}^{\sigma(2)} + 0.5(1 - h_{12}^{\sigma(2)})} = \frac{0.5h_{12}^{\sigma(2)}}{0.5} = h_{12}^{\sigma(2)}$$

$$\vdots$$

$$\frac{0.5h_{12}^{\sigma(n)}}{0.5h_{12}^{\sigma(n)} + 0.5(1 - h_{12}^{\sigma(n)})} = \frac{0.5h_{12}^{\sigma(n)}}{0.5} = h_{12}^{\sigma(n)}$$

which satisfies Eq. (6.5), additionally, when $h_{12} = \{0\}$, Eq. (6.5) also holds. Thus, $H = (h_{ij})_{2\times 2}$ is multiplicative consistent, which completes the proof of Theorem 6.3.   □

Theorem 6.3 reveals that any two-order HFPR is multiplicative consistent. However, when the order of a HFPR is greater than two, the HFPR constructed by an expert is sometimes unacceptably consistent, which means $d(H,\overline{H}) > \tau$. Thus, we need to adjust the elements in the HFPR to improve its consistency degree till it accomplishes the required consistency. Below we propose an iterative algorithm to repair the consistency degree of a HFPR.

Algorithm 6.1

Step 1. Suppose that p is the number of iterations, δ is the step size, $0 \leq \xi = p\delta \leq 1$ and τ is the consistency threshold. Let $p = 1$, and construct the prefect multiplicative consistent HFPR $\overline{H} = (\overline{h}_{ij})_{n \times n}$ from $H^{(p)} = (h_{ij}^{(p)})_{n \times n}$ by Eq. (6.6).

Step 2. Calculate the deviation $d(H^{(p)}, \overline{H})$ between $\overline{H}$ and $H^{(p)}$ by

$$d_{hnh}(H^{(p)}, \overline{H}) = \frac{1}{(n-1)(n-2)} \sum_{i=1}^{n} \sum_{j=1}^{n} \left[\frac{1}{l_{h_{ij}}} \sum_{t=1}^{l_{h_{ij}}} \left| h_{ij}^{(p)\sigma(t)} - \overline{h}_{ij}^{\sigma(t)} \right| \right] \quad (6.10)$$

or

$$d_{hne}(H^{(p)}, \overline{H}) = \frac{1}{(n-1)(n-2)} \sum_{i=1}^{n} \sum_{j=1}^{n} \left(\frac{1}{l_{h_{ij}}} \sum_{t=1}^{l_{h_{ij}}} \left| h_{ij}^{(p)\sigma(t)} - \overline{h}_{ij}^{\sigma(t)} \right|^2 \right)^{1/2} \quad (6.11)$$

where $h_{ij}^{(p)\sigma(t)}$ and $\overline{h}_{ij}^{\sigma(t)}$ are the tth smallest values in $h_{ij}^{(p)}$ and $\overline{h}_{ij}$, respectively. If $d(H^{(p)}, \overline{H}) < \tau$, then output $H^{(p)}$; Otherwise, go to the next step.

Step 3. Repair the multiplicative inconsistent HFPR $H^{(p)}$ to $\widehat{H}^{(p)} = (h_{ij}^{(p)})_{n \times n}$ by using the following equations:

$$\widehat{h}_{ij}^{(p)\sigma(t)} = \frac{(h_{ij}^{(p)\sigma(t)})^{1-\xi}(\overline{h}_{ij}^{\sigma(t)})^{\xi}}{(h_{ij}^{(p)\sigma(t)})^{1-\xi}(\overline{h}_{ij}^{\sigma(t)})^{\xi} + (1 - h_{ij}^{(p)\sigma(t)})^{1-\xi}(1 - \overline{h}_{ij}^{\sigma(t)})^{\xi}},$$

$$i, j = 1, 2, \ldots, n$$

where $\widehat{h}_{ij}^{(p)\sigma(t)}$, $h_{ij}^{(p)\sigma(t)}$ and $\overline{h}_{ij}^{\sigma(t)}$ are the tth smallest values in $\widehat{h}_{ij}^{(p)}$, $h_{ij}^{(p)}$ and $\overline{h}_{ij}$, respectively. Let $H^{(p+1)} = \widehat{H}^{(p)}$ and $p = p + 1$, then go to Step 2.

In Algorithm 6.1, the parameter δ is an iteration step size. Such a step size controls the speed of convergence of this algorithm. That is to say, by setting different values of δ, the acceptable multiplicative consistent HFPR can be obtained in different time. One special case is to set the step size $\delta = 1/N$, where N is the maximum number of iteration. Then after $p(=N)$ times of iteration, $\xi = p\delta = N \cdot (1/N) = 1$, thus, $\widehat{H}^{(p)} = \overline{H}$, which is of perfect multiplicative consistency. In other words, the above algorithm is convergent.

It should also be noted that in Algorithm 6.1, the repaired HFPR shares the same perfect multiplicative consistent HFPR with the original HFPR H. This can be formulated into the following theorem:

Theorem 6.4 (Liu et al. 2016). *For a HFPR $H = \left(h_{ij}\right)_{n \times n}$ on a fixed set $A = \{A_1, A_2, \ldots, A_n\}$, let $\overline{H} = \left(\overline{h}_{ij}\right)_{n \times n}$ be the corresponding perfect multiplicative consistent HFPR of H, and $\widehat{H} = \left(\widehat{h}_{ij}\right)_{n \times n}$ be the repaired HFPR of H through Algorithm 6.1 within limited times of iteration. Then the corresponding perfect multiplicative consistent HFPR of $\widehat{H}$ is $\overline{H}$ as well.*

Proof We first define $\Omega: \{\text{HFPR}\} \mapsto \{\text{HFPR}\}$, $\Omega(H) = \overline{H}$, where $H = \left(h_{ij}\right)_{n \times n}$ is any fixed HFPR, and $\overline{H} = \left(\overline{h}_{ij}\right)_{n \times n}$ is the corresponding perfect multiplicative consistent HFPR of H.

Then we define $\Pi: \{\text{HFPR}\} \mapsto \{\text{HFPR}\}$, $\Pi\left(\widehat{H}^{(k)}\right) = \widehat{H}^{(k+1)}$, where $k \in N^+$, and $\widehat{H}^{(k)} = \left(\widehat{h}_{ij}^{(k)}\right)_{n \times n}$ is the repaired HFPR of H by Algorithm 6.1 for k iterations, and define φ: $\varphi_\xi(x, y) = \frac{x^{1-\xi}y^\xi}{x^{1-\xi}y^\xi + (1-x)^{1-\xi}(1-y)^\xi}$, where x, y, $\xi \in (0, 1)$. Now we illustrate that for any HFPR H:

$$\Pi \circ \Omega(H) = \Omega \circ \Pi(H) \tag{6.13}$$

Owing to the fact that $\Omega(H) = \overline{H}$ is a perfect multiplicative consistent HFPR, $\Pi \circ \Omega(H) = \Omega(H) = \overline{H}$. While $\Pi(H) = \widehat{H} = \left(\widehat{h}_{ij}\right)_{n \times n}$. For $i = 1, 2, \ldots, n$, $\overline{h}_{ii} = 0.5$, and for $i = 1, 2, \ldots, n-1$ and $t = 1, 2, \ldots, l_{h_{i(i+1)}}$, $\overline{h}_{i(i+1)} = h_{i(i+1)}$. Hence, for $i = 1, 2, \ldots, n$, $\widehat{h}_{ii} = 0.5$, and for $i = 1, 2, \ldots, n-1$ and $t = 1, 2, \ldots, l_{h_{i(i+1)}}$, $\widehat{h}_{i(i+1)}^{\sigma(t)} = \varphi\left(h_{i(i+1)}^{\sigma(t)}, \overline{h}_{i(i+1)}^{\sigma(t)}\right) = h_{i(i+1)}^{\sigma(t)}$. Thereby in $H' = \Omega \circ \Pi(H)$, $h_{i(i+1)}'^{\sigma(t)} = h_{i(i+1)}^{\sigma(t)}$, $i = 1, 2, \ldots, n-1$. Notice the point that the transformation Ω keeps the elements of diagonal and secondary diagonal remain, and the else elements determined by the ones of secondary diagonal, so $\Omega \circ \Pi(H) = \overline{H}$. Therefore, the assumption $\Pi \circ \Omega(H) = \Omega \circ \Pi(H)$ is true. For the repaired HFPR $\widehat{H}$, $\exists P_0 \in \mathbb{N}^+$, where $\mathbb{N}^+$ is the set of all natural numbers, such that $\widehat{H} = H^{(P_0)}$, where $H^{(P_0)} = \left(h_{ij}^{(P_0)}\right)_{n \times n}$ is the operated HFPR of H by Algorithm 6.1 for P_0 times. Now we use mathematical induction to further prove the proposition:

(1) When $P_0 = 1$, $\widehat{H} = H^{(1)} = H$. Therefore, it holds that H and $\widehat{H}$ share the same perfect multiplicative consistent HFPR $\overline{H}$.

(2) For $k \in \mathbb{N}^+$, $1 < k < P_0$, suppose that $\overline{H}^{(k)} = \overline{H}$, where $\overline{H}^{(k)} = \left(\overline{h}_{ij}^{(k)} \right)_{n \times n}$ is the corresponding perfect multiplicative consistent HFPR of $H^{(k)} = \left(h_{ij}^{(k)} \right)_{n \times n}$, then

$$\Omega \left(H^{(k+1)} \right) = \Omega \circ \Pi \left(H^{(k)} \right) = \Pi \circ \Omega \left(H^{(k)} \right) = \Pi \left(\overline{H}^{(k)} \right) = \Pi(\overline{H}) = \overline{H} \quad (6.14)$$

To sum up, the repaired HFPR shares the same perfect multiplicative consistent HFPR with the original HFPR H. This completes the proof of Theorem 6.2.$\Box$

Example 6.1 (Liao et al. 2014b). Suppose that a decision maker provides his/her preference information over a set of alternatives A_1, A_2, A_3, A_4 in HFEs and constructs the following HFPR:

$$H = \begin{pmatrix} \{0.5\} & \{0.1, 0.4\} & \{0.1, 0.2\} & \{0.4, 0.5, 0.6\} \\ \{0.6, 0.9\} & \{0.5\} & \{0.3, 0.8\} & \{0.3, 0.6\} \\ \{0.8, 0.9\} & \{0.2, 0.7\} & \{0.5\} & \{0.2, 0.7\} \\ \{0.4, 0.5, 0.6\} & \{0.4, 0.7\} & \{0.3, 0.8\} & \{0.5\} \end{pmatrix}$$

Firstly, let $p = 1$ and $H^{(1)} = H$, then we construct the prefect multiplicative HFPR $\overline{H} = (\overline{h}_{ij})_{n \times n}$ from $H^{(1)}$ by Eq. (6.6). Taking $\overline{h}_{14}$ as an example, we have

$$\overline{h}_{14}^{\sigma(1)} = \frac{1}{2} \left(\frac{h_{12}^{\sigma(1)} h_{24}^{\sigma(1)}}{h_{12}^{\sigma(1)} h_{24}^{\sigma(1)} + (1 - h_{12}^{\sigma(1)})(1 - h_{24}^{\sigma(1)})} + \frac{h_{13}^{\sigma(1)} h_{34}^{\sigma(1)}}{h_{13}^{\sigma(1)} h_{34}^{\sigma(1)} + (1 - h_{13}^{\sigma(1)})(1 - h_{34}^{\sigma(1)})} \right)$$

$$= \frac{1}{2} \left(\frac{0.1 \times 0.3}{0.1 \times 0.3 + (1 - 0.1)(1 - 0.3)} + \frac{0.1 \times 0.2}{0.1 \times 0.2 + (1 - 0.1)(1 - 0.2)} \right) = 0.036$$

$$\overline{h}_{14}^{\sigma(2)} = \frac{1}{2} \left(\frac{h_{12}^{\sigma(2)} h_{24}^{\sigma(2)}}{h_{12}^{\sigma(2)} h_{24}^{\sigma(2)} + (1 - h_{12}^{\sigma(2)})(1 - h_{24}^{\sigma(2)})} + \frac{h_{13}^{\sigma(2)} h_{34}^{\sigma(2)}}{h_{13}^{\sigma(2)} h_{34}^{\sigma(2)} + (1 - h_{13}^{\sigma(2)})(1 - h_{34}^{\sigma(2)})} \right)$$

$$= \frac{1}{2} \left(\frac{0.4 \times 0.6}{0.4 \times 0.6 + (1 - 0.4)(1 - 0.6)} + \frac{0.2 \times 0.7}{0.2 \times 0.7 + (1 - 0.2)(1 - 0.7)} \right) = 0.434$$

In the similar way, we can obtain

$$\overline{H} = \begin{pmatrix} \{0.5\} & \{0.1, 0.4\} & \{0.046, 0.727\} & \{0.036, 0.434\} \\ \{0.6, 0.9\} & \{0.5\} & \{0.3, 0.8\} & \{0.097, 0.903\} \\ \{0.273, 0.954\} & \{0.2, 0.7\} & \{0.5\} & \{0.2, 0.7\} \\ \{0.566, 0.964\} & \{0.097, 0.903\} & \{0.3, 0.8\} & \{0.5\} \end{pmatrix}$$

Then, we use Eq. (6.10) to calculate the hesitant normalized Hamming distance between $H^{(1)}$ and $\overline{H}$:

$$d_{hnh}(H^{(1)}, \overline{H}) = \frac{1}{6} \sum_{i=1}^{4} \sum_{j=1}^{4} \left[\frac{1}{l_{x_{ij}}} \sum_{t=1}^{l_{x_{ij}}} \left| h_{H^{(1)}}^{\sigma(t)}(x_{ij}) - h_{\overline{H}}^{\sigma(t)}(x_{ij}) \right| \right]$$

$$= \frac{1}{6} \left[\frac{1}{2}(|0.1 - 0.046| + |0.2 - 0.727|) + \frac{1}{3}(|0.4 - 0.036| + |0.5 - 0.434| + |0.6 - 0.5|) \right.$$

$$+ \frac{1}{2}(|0.3 - 0.097| + |0.6 - 0.903|) + \frac{1}{2}(|0.8 - 0.273| + |0.9 - 0.954|) + \frac{1}{3}(|0.4 - 0.5|$$

$$+ |0.5 - 0.566| + |0.6 - 0.964|) + \frac{1}{2}(|0.4 - 0.097| + |0.7 - 0.903|)] = 0.2401$$

Without loss of generality, let $\tau = 0.1$, then $d_{hnh}(H^{(1)}, \overline{H}) = 0.2401 > \tau$, which means that $H^{(1)}$ is not a multiplicative consistent HFPR. Therefore, it is needed to repair the multiplicative inconsistent HFPR $H^{(1)}$ according to $\widehat{H}^{(1)}$ by Eq. (6.12). We hereby let $\xi = 0.8$, then

$$\widehat{H}^{(1)} = \begin{pmatrix} \{0.5\} & \{0.1, 0.4\} & \{0.054, 0.624\} & \{0.062, 0.447, 0.52\} \\ \{0.6, 0.9\} & \{0.5\} & \{0.3, 0.8\} & \{0.124, 0.866\} \\ \{0.376, 0.946\} & \{0.2, 0.7\} & \{0.5\} & \{0.2, 0.7\} \\ \{0.48, 0.553, 0.938\} & \{0.134, 0.876\} & \{0.3, 0.8\} & \{0.5\} \end{pmatrix}$$

Let $H^{(2)} = \widehat{H}^{(1)}$ and $p = 2$, then the hesitant normalized Hamming distance between $H^{(2)}$ and $\overline{H}$ can be calculated, i.e., $d_{hnh}(H^{(2)}, \overline{H}) = 0.039 < 0.1$. Since the hesitant normalized Hamming distance is less than the given consistency threshold, we can draw a conclusion that $H^{(2)}$ is the repaired multiplicative consistent HFPR of H.

In Example 6.1, we can also use Eq. (6.11) to calculate the hesitant normalized Euclidean distance instead of the hesitant normalized Hamming distance, and both of them can get the same result.

Beside Algorithm 6.1, the most directive method to repair the inconsistent HFPR is returning the multiplicative inconsistent HFPR to the decision maker to reconsider and construct a new HFPR according to his/her new comparisons until it has acceptable consistency. This algorithm can be described in details as follows:

Algorithm 6.2

Step 1. Same as that in Algorithm 6.1.

Step 2. Same as that in Algorithm 6.1.

Step 3. Return the multiplicative inconsistent HFPR $H^{(p)}$ to the decision maker to reconsider and construct a new HFPR $H^{(p+1)}$ according to the new judgments. Let $p = p + 1$, then go to Step 2.

Example 6.2 (Liao et al. 2014b). Suppose the analyst does not repair the multiplicative inconsistent HFPR $H^{(1)}$ in Example 6.1 by Eq. (6.12), but returns it to the decision maker to reconsider their opinions with reference to the prefect multiplicative HFPR $\overline{H}$. After re-evaluation, the decision maker provides a new HFPR $H^{(1)}$ as:

$$
H^{(2)} = \begin{pmatrix}
\{0.5\} & \{0.1, 0.4\} & \{0.1, 0.7\} & \{0.1, 0.4\} \\
\{0.6, 0.9\} & \{0.5\} & \{0.3, 0.8\} & \{0.1, 0.9\} \\
\{0.3, 0.9\} & \{0.2, 0.7\} & \{0.5\} & \{0.2, 0.7\} \\
\{0.6, 0.9\} & \{0.1, 0.9\} & \{0.3, 0.8\} & \{0.5\}
\end{pmatrix}
$$

Afterwards, we use Eq. (6.6) to construct the prefect multiplicative HFPR $\overline{H}'$ from $H^{(2)}$. It is also the same as $\overline{H}$ in Example 6.1. Then, we use Eq. (6.10) to calculate the hesitant normalized Hamming distance between $H^{(2)}$ and $\overline{H}'$:

$$
\begin{aligned}
d_{hnh}(H^{(2)}, \overline{H}') &= \frac{1}{6} \sum_{i=1}^{4} \sum_{j=1}^{4} \left[\frac{1}{l_{x_{ij}}} \sum_{t=1}^{l_{x_{ij}}} \left| h_{H^{(2)}}^{\sigma(t)}(x_{ij}) - h_{\overline{H}'}^{\sigma(t)}(x_{ij}) \right| \right] \\
&= \frac{1}{6} [\frac{1}{2} (|0.1 - 0.046| + |0.7 - 0.727|) + \frac{1}{2} (|0.1 - 0.036| + |0.4 - 0.434|) \\
&\quad + \frac{1}{2} (|0.1 - 0.097| + |0.9 - 0.903|) + \frac{1}{2} (|0.3 - 0.273| + |0.9 - 0.954|) \\
&\quad + \frac{1}{2} (|0.6 - 0.566| + |0.9 - 0.964|) + \frac{1}{2} (|0.1 - 0.097| + |0.9 - 0.903|)] = 0.0308
\end{aligned}
$$

Since the hesitant normalized Hamming distance is less than the consistency threshold, i.e., $d_{hnh}(H^{(2)}, \overline{H}') = 0.0308 < 0.1$, we can draw a conclusion that $H^{(2)}$ is the multiplicative consistent HFPR of H.

Both of Algorithms 6.1 and 6.2 can transfer the multiplicative inconsistent HFPR to the acceptable consistent HFPR. But in practice, we usually use Algorithm 6.1 because the latter one wastes a lot of time and resources.

6.2 The Multiplicative Consistency Index of Hesitant Fuzzy Preference Relation

6.2.1 The Necessity to Derive the Consistency Index Value for Hesitant Fuzzy Preference Relation

It is noted that there are some weaknesses in Algorithm 6.1. In view of Definition 6.4, H is regarded as an acceptable multiplicative consistent HFPR if $d(H, \overline{H}) < \tau$. Practically, Algorithm 6.1 suggested that $\tau = 0.1$ can be taken as a consistency threshold without providing any theoretical reasons. Absolutely we can promise the consistency checking of a HFPR to be more effective if we limit $d(H, \overline{H})$ to a much smaller scale. Although in most occasions it works in checking the consistency of a HFPR, setting consistency index to be 0.1 is lack of theoretical foundation and sometimes it may be unreasonable. In addition, if we just limit $d(H, \overline{H})$ to a small value τ, taking $\tau = 0.1$ as an example, we have no idea what a statistical degree of the consistency threshold 0.1 can provide. The following practical example concerning project evaluation (adapted from Ngwenyama and Bryson (1999)) illustrates the drawbacks of setting the consistency threshold to be 0.1.

Example 6.3 (Liu et al. 2016). Consider a complicated decision making problem that the information management steering committee of Midwest American Manufacturing Corp. (MAMC) needs to prioritize a sort of information technology improvement projects, which are proposed by area managers, for development and implementation. In order to rank these projects from high to low potential contribution to the firm's strategic goal of gaining competitive advantage in the industry, the committee divided this problem into a multiple criteria decision making problem and three main factors $C = \{C_1, C_2, C_3\}$ = {productivity, differentiation, management} are constructed as criteria in the process of assessing these projects, where

(1) the productivity factor assesses the potential of a proposed project to increase the effectiveness and efficiency of the firm's manufacturing and service operations;
(2) the differentiation factor assesses the potential of a proposed project to fundamentally differentiate the firm's products and services from its competitors, and to make them more desirable to its customers;
(3) the management factor assesses the potential of a proposed project to assist management in improving their planning, controlling and decision making activities.

In order to rank these projects, we should first determine the weights of these factors. To do so, the pairwise comparisons over these three factors are conducted. Different individuals in the committee may provide different assessments over these factors and we can use HFEs to maintain all the possible preference values. To simplify the presentation, suppose that the committee provides their preference information over the three factors by a HFPR:

$$H = \begin{pmatrix} \{0.5\} & \{0.53, 0.54\} & \{0.46, 0.47\} \\ \{0.46, 0.47\} & \{0.5\} & \{0.52, 0.53\} \\ \{0.53, 0.54\} & \{0.47, 0.48\} & \{0.5\} \end{pmatrix}$$

By Definition 6.3, the corresponding perfect multiplicative consistent HFPR of H is determined as:

$$\overline{H} = \begin{pmatrix} \{0.5\} & \{0.53, 0.54\} & \{0.55, 0.57\} \\ \{0.46, 0.47\} & \{0.5\} & \{0.52, 0.53\} \\ \{0.43, 0.45\} & \{0.47, 0.48\} & \{0.5\} \end{pmatrix}$$

The distance between H and $\overline{H}$ can be calculated by Eq. (6.8), and then we obtain

$$d_{hnh}(H,\overline{H}) = \frac{1}{(3-1)\times(3-2)}\left(\frac{1}{2}\times(|0.46-0.55|+|0.47-0.57|)\right.$$

$$\left.+\frac{1}{2}\times(|0.53-0.43|+|0.54-0.45|)\right) = 0.095 < 0.1$$

Therefore, according to Definition 6.4, H is supposed to be a multiplicative consistent HFPR.

Considering three HFEs of H: $h_{12} = \{0.53, 0.54\}$, $h_{23} = \{0.52, 0.53\}$ and $h_{13} = \{0.46, 0.47\}$, by Eq. (1.17), we have $s(h_{12}) = 0.535$, which indicates that $A_1 \succ A_2$, i.e., A_1 is preferred to A_2; $s(h_{23}) = 0.525$, which indicates that $A_2 \succ A_3$. To guarantee H to be consistent enough, at least we expect that $A_1 \succ A_3$. As $\overline{h}_{13} = \{0.55, 0.57\}$, whose every single preference is greater than 0.5, we get the score $s(\overline{h}_{13}) = 0.56$. It makes our expectation reasonable. However, $s(h_{13}) = 0.465$ fails what we expect. Combing the distance $d_{hnh}(H,\overline{H}) = 0.095$ provided from the perspective of distribution of $d_{hnh}(H,\overline{H})$, it is not persuadable enough to accept H as a multiplicative consistent HFPR.

Though the method for consistency checking process on HFPR provided in Algorithm 6.1 is efficient in most time, from the discussion above, we can see that it still presents some weaknesses. One disadvantage is that in some cases a HFPR may be accepted as a multiplicative consistent HFPR, but it is somehow unreasonable or not persuadable enough. Another problem is that by simply taking 0.1 or any other constant as a consistent threshold, we have no idea what a percentile of consistency degree the value of consistency index equals to. For example, if we set 0.1 as the consistency index of three-order HFPR, actually we can see in the later of this section that the three-order HFPR which satisfies the condition that $d_{hnh}(H,\overline{H}) < 0.1$ takes 45.4 % of the whole three-order HFPRs. It is so big a percentile that can barely be reliable.

From the discussion above, it is definitely essential to do some more research on the consistency index of a HFPR and to provide more reasonable consistency index values for a HFPR. In the following, we will study the HFPR's consistency index from the statistical point of view and then present a novel method to derive the critical value of consistency index varying the orders of HFPR from 3 to 14. This novel method allows the decision makers to decide the consistency level on which a HFPR is accepted or rejected in particular cases.

6.2.2 Density Function of the Consistency Index of a Hesitant Fuzzy Preference Relation

For a HFPR $H = (h_{ij})_{n\times n}$ on a fixed set $A = \{A_1, A_2, \ldots, A_n\}$, Liao et al. (2014b) suggested to use $d(H,\overline{H})$ as a consistency checking index, where $\overline{H} = (\overline{h}_{ij})_{n\times n}$ is the corresponding perfect multiplicative consistent HFPR of H. Motivated by this idea, naturally, if we can find out the distribution of such consistency checking index values, we can value the consistency checking index from the perspective of statistics. To do so, first of all, we need to determine the density function of $d(H,\overline{H})$. For convenience, we give the definition of error coefficient.

Definition 6.5 (*Liu et al.* 2016). Let $H = \left(h_{ij}\right)_{n\times n}$ be a HFPR on a fixed set $A = \{A_1, A_2, \ldots, A_n\}$, and $\overline{H} = \left(\overline{h}_{ij}\right)_{n\times n}$ be the corresponding perfect multiplicative consistent HFPR, then $d_{ij} = \left|h_{ij} - \overline{h}_{ij}\right|$ is called an error coefficient.

There are two commonly used distance measures for HFPRs: the hesitant normalized Hamming distance presented as Eq. (6.8) and the hesitant normalized Euclidean distance presented as Eq. (6.9). If we calculate the distance between H and $\overline{H}$ by Eq. (6.8), then

$$d_{ij} = \left|h_{ij} - \overline{h}_{ij}\right| = \frac{1}{l_{h_{ij}}} \sum_{t=1}^{l_{h_{ij}}} \left|h_{ij}^{\sigma(t)} - \overline{h}_{ij}^{\sigma(t)}\right| \tag{6.15}$$

and if we use Eq. (6.9) to calculate the distance, then

$$d_{ij} = \left|h_{ij} - \overline{h}_{ij}\right| = \left[\frac{1}{l_{h_{ij}}} \sum_{t=1}^{l_{h_{ij}}} \left|h_{ij}^{\sigma(t)} - \overline{h}_{ij}^{\sigma(t)}\right|^2\right]^{1/2} \tag{6.16}$$

Technically, for the calculations of both hesitant normalized Hamming distance and hesitant normalized Euclidean distance, time consuming is a big problem. From the expression of hesitant normalized Hamming distance:

$$d_{hnh}(H,\overline{H}) = \frac{1}{(n-1)(n-2)} \sum_{i=1}^{n} \sum_{j=1}^{n} \left[\frac{1}{l_{h_{ij}}} \sum_{t=1}^{l_{h_{ij}}} \left|h_{ij}^{\sigma(t)} - \overline{h}_{ij}^{\sigma(t)}\right|\right] \tag{6.17}$$

we can see that the expression is linear and the computational complexity is n, which is a small scale. For the hesitant normalized Euclidean distance, which is

$$d_{hne}(H,\overline{H}) = \frac{1}{(n-1)(n-2)} \sum_{i=1}^{n} \sum_{j=1}^{n} \left[\frac{1}{l_{h_{ij}}} \sum_{t=1}^{l_{h_{ij}}} \left|h_{ij}^{\sigma(t)} - \overline{h}_{ij}^{\sigma(t)}\right|^2\right]^{1/2} \tag{6.18}$$

the computational complexity is n^3. So theoretically, the time consuming concern is more or less a small one. In a decision making process in practice, the number of elements from the alternatives set is usually limited to a small scale, which the computer can handle easily.

Combing Eqs. (6.15) and (6.17), we have

$$d_{hnh}(H,\overline{H}) = \frac{1}{(n-1)(n-2)}\sum_{i=1}^{n}\sum_{j=1}^{n}\left|h_{ij}-\overline{h}_{ij}\right| = \frac{1}{(n-1)(n-2)}\sum_{i=1}^{n}\sum_{j=1}^{n}d_{ij}$$

(6.19)

Combing Eqs. (6.16) and (6.18), we have

$$d_{hne}(H,\overline{H}) = \frac{1}{(n-1)(n-2)}\sum_{i=1}^{n}\sum_{j=1}^{n}\left|h_{ij}-\overline{h}_{ij}\right| = \frac{1}{(n-1)(n-2)}\sum_{i=1}^{n}\sum_{j=1}^{n}d_{ij}$$

(6.20)

For the simplicity of presentation, we unify $d_{hnh}(H,\overline{H})$ and $d_{hne}(H,\overline{H})$ by the same expression $d(H,\overline{H})$ when there is unnecessary to distinguish them. Then it follows

$$d(H,\overline{H}) = \frac{1}{(n-1)(n-2)}\sum_{i=1}^{n}\sum_{j=1}^{n}d_{ij}$$

(6.21)

where $d(H,\overline{H})$ can be regarded as the summation of numbers of independent identically distributed random variables multiplied by a constant.

Although a perfect multiplicative consistent HFPR is hard to achieve in practice, the fact is that people's subjective judgment is tending to consistency, which arises two notes:

Firstly, when a HFPR provided by a group of experts is returned to them for repairing repeatedly, the repaired HFPR is tending to a perfect multiplicative consistent one. Secondly, a lot of HFPRs on the same issue is tending to the consistent one on distribution.

Based on these two notes, d_{ij} is tending to 0 at low side of values. It is obvious that $d_{ij} \in [0, 1]$. Let

$$d'_{ij} = \frac{1}{1-d_{ij}}$$

(6.22)

and $d'_{ij} \in [1, +\infty)$. Therefore, without loss of generality, it is appropriate to assume that d'_{ij} satisfies a half-normal distribution, whose density function is:

$$f(x) = \begin{cases} \frac{2}{\sigma\sqrt{2\pi}}e^{-\frac{(x-1)^2}{2\sigma^2}}, & x \geq 1 \\ 0, & x < 1 \end{cases}$$

(6.23)

The density function of d_{ij} can be derived from Eqs. (6.22) and (6.23):

$$g(x) = \begin{cases} \dfrac{\sqrt{2}}{\sqrt{\pi}\sigma(1-x)^2} e^{-\frac{x^2}{2\sigma^2(x-1)^2}}, & x \in [0, 1] \\ 0, & else \end{cases} \tag{6.24}$$

To estimate the threshold of the consistency index of $d(H, \overline{H})$, we can integrate the density function of $d(H, \overline{H})$ from 0 to the threshold, which is the definite value we are trying to achieve. To present the exact expression of the density function of $d(H, \overline{H})$, we have to see the fact that $d(H, \overline{H})$ can be regarded as the summation of n^2 numbers of $\frac{d_{ij}}{(n-1)(n-2)}$ (see Eq. (6.21)). From Eq. (6.24), we can get the density function of $\frac{d_{ij}}{(n-1)(n-2)}$ as:

$$g_1(x) = \begin{cases} \dfrac{\sqrt{2}(n-1)(n-2)}{\sqrt{\pi}\sigma(1-(n-1)(n-2)x)^2} e^{-\frac{(n-1)^2(n-2)^2 x^2}{2\sigma^2(1-(n-1)(n-2)x)^2}}, & x \in [0, \frac{1}{(n-1)(n-2)}] \\ 0, & else \end{cases} \tag{6.25}$$

Hence, the density function of $d(H, \overline{H})$ is $(n-1)(n-1)$-dimensional convolution of $g_1(x)$, which is shown as:

$$g_2(x) = \underbrace{g_1(x) * g_1(x) * \cdots * g_1(x)}_{(n-1)(n-2)} \tag{6.26}$$

Up to now, theoretically we can achieve the exact expression of the density function of $d(H, \overline{H})$, but it should be very complicated. In fact, there is no need for us to determine the exact expression of the density function of $d(H, \overline{H})$. In the following, let us pay our attention to the expectation and variance of $d(H, \overline{H})$. For simplicity, suppose that the density function of $d(H, \overline{H})$ is given in form of

$$g_2(x) = \begin{cases} m(x), & x \in [0, 1] \\ 0, & else \end{cases} \tag{6.27}$$

According to Eqs. (6.24)–(6.27), the following results can be easily derived:

Theorem 6.5 (Liu et al. 2016). *The expectation of $d(H, \overline{H})$ can be presented as follows:*

$$E\big(d(H, \overline{H})\big) = \int_{-\infty}^{+\infty} t g_2(t) dt = \int_0^1 t m(t) dt \tag{6.28}$$

$$E(d(H,\overline{H})) = \frac{1}{(n-1)(n-2)} \sum_{i=1}^{n} \sum_{j=1}^{n} E(d_{ij}(x))$$

$$= \frac{n^2}{(n-1)(n-2)} \int_0^1 \frac{\sqrt{2}t}{\sqrt{\pi}\sigma(1-t)^2} e^{-\frac{t^2}{2\sigma^2(1-t)^2}} dt \qquad (6.29)$$

Proof According to the definition of expectation, $E(d(H,\overline{H})) = \int_{-\infty}^{+\infty} tg_2(t)dt$, and with the density function of $d(H,\overline{H})$ given by Eq. (6.27), we have $\int_{-\infty}^{+\infty} tg_2(t)dt = \int_0^1 tm(t)dt$. Owing to Eq. (6.21), it follows

$$E(d(H,\overline{H})) = E\left(\frac{1}{(n-1)(n-2)} \sum_{i=1}^{n} \sum_{j=1}^{n} d_{ij}(x)\right)$$

$$= \frac{1}{(n-1)(n-2)} \sum_{i=1}^{n} \sum_{j=1}^{n} E(d_{ij}(x))$$

Notice that the density function of d_{ij} is given by Eq. (6.24). Then,

$$E(d_{ij}) = \int_0^1 \frac{\sqrt{2}t}{\sqrt{\pi}\sigma(1-t)^2} e^{-\frac{t^2}{2\sigma^2(1-t)^2}} dt.$$

Thus,

$$\frac{1}{(n-1)(n-2)} \sum_{i=1}^{n} \sum_{j=1}^{n} E(d_{ij}) = \frac{n^2}{(n-1)(n-2)} \int_0^1 \frac{\sqrt{2}t}{\sqrt{\pi}\sigma(1-t)^2} e^{-\frac{t^2}{2\sigma^2(1-t)^2}} dt$$

This completes the proof of the theorem. □

Theorem 6.6 (Liu et al. 2016). *The variance $D(d(H,\overline{H}))$ can be obtained as:*

$$D(d(H,\overline{H})) = \int_{-\infty}^{+\infty} (t - E(d(H,\overline{H})))^2 g_2(t)dt = \int_0^1 t^2 m(t)dt - E^2(d(H,\overline{H})) \qquad (6.30)$$

$$D(d(H,\overline{H})) = \frac{1}{(n-1)^2(n-2)^2} \sum_{i=1}^{n} \sum_{j=1}^{n} D(d_{ij})$$

$$= \frac{n^2}{(n-1)^2(n-2)^2} \left(\int_0^1 \frac{\sqrt{2}t^2}{\sqrt{\pi}\sigma(1-t)^2} e^{-\frac{t^2}{2\sigma^2(1-t)^2}} dt - E^2(d_{ij})\right) \qquad (6.31)$$

Proof According to the definition of variance, it follows

$$D\big(d\big(H,\overline{H}\big)\big) = \int_{-\infty}^{+\infty} \big(t - E\big(d\big(H,\overline{H}\big)\big)\big)^2 g_2(t)dt$$

Furthermore, with the result of Theorem 6.3, we have

$$\int_{-\infty}^{+\infty} \big(t - E\big(d\big(H,\overline{H}\big)\big)\big)^2 g_2(t)dt$$

$$= \int_{-\infty}^{+\infty} t^2 g_2(t)dt - 2\int_{-\infty}^{+\infty} tE\big(d\big(H,\overline{H}\big)\big)g_2(t)dt + \int_{-\infty}^{+\infty} E\big(d\big(H,\overline{H}\big)\big)^2 g_2(t)dt$$

$$= \int_0^1 t^2 m(t)dt - 2E\big(d\big(H,\overline{H}\big)\big)^2 \int_{-\infty}^{+\infty} tg_2(t)dt + E\big(d\big(H,\overline{H}\big)\big)^2 \int_{-\infty}^{+\infty} g_2(t)dt$$

$$= \int_0^1 t^2 m(t)dt - 2E\big(d\big(H,\overline{H}\big)\big)^2 + E\big(d\big(H,\overline{H}\big)\big)^2$$

$$= \int_0^1 t^2 m(t)dt - E\big(d\big(H,\overline{H}\big)\big)^2$$

Owing to Eq. (6.21), it yields

$$D\big(d\big(H,\overline{H}\big)\big) = D\left(\frac{1}{(n-1)(n-2)} \sum_{i=1}^{n} \sum_{j=1}^{n} d_{ij}\right) = \frac{1}{(n-1)^2(n-2)^2} \sum_{i=1}^{n} \sum_{j=1}^{n} D(d_{ij})$$

Similarly, we have

$$D\big(d_{ij}(x)\big) = \int_0^1 \frac{\sqrt{2}t^2}{\sqrt{\pi}\sigma(1-t)^2} e^{-\frac{t^2}{2\sigma^2(1-t)^2}} dt - E^2\big(d_{ij}\big)$$

Hence,

$$\frac{1}{(n-1)^2(n-2)^2} \sum_{i=1}^{n} \sum_{j=1}^{n} D(d_{ij})$$

$$= \frac{n^2}{(n-1)^2(n-2)^2} \left(\int_0^1 \frac{\sqrt{2}t^2}{\sqrt{\pi}\sigma(1-t)^2} e^{-\frac{t^2}{2\sigma^2(1-t)^2}} dt - E^2\big(d_{ij}\big)\right)$$

This completes the proof of the theorem.   □

6.2.3   Methods to Determine the Value of Consistency Index of a Hesitant Fuzzy Preference Relation

The definition of consistency level of a HFPR is introduced as follows:

Definition 6.6 (*Liu et al.* 2016). Let $H = (h_{ij})_{n \times n}$ be a HFPR on a fixed set $A = \{A_1, A_2, \ldots, A_n\}$ and $\overline{H} = (\overline{h}_{ij})_{n \times n}$ be the corresponding perfect multiplicative consistent HFPR, assume that the density function of $d(H, \overline{H})$ is $g(x)$, a percentage ρ is called consistency level, which is given by the decision makers and meanwhile it satisfies that $\int_0^\tau g(t)dt = \rho$, where τ is the consistency index of HFPRs.

   Note: It is noted that the consistency level ρ and the consistency index τ present the consistency degree from different perspectives. Consistency level is a percentage and it is given by decision makers to show the consistency degree from the perspective of statistics, which makes the consistency degree more intuitive. While consistency index is the measure of $d(H, \overline{H})$ and it is the definite value that we try to achieve. When the consistency level ρ is given, theoretically we integrate the density function of $d(H, \overline{H})$ from 0 up to a number until the integrations is ρ, then the number is the value of consistency index τ.

   Technically, there are three methods which can be used to achieve the critical value of consistency index. The first one is to integrate the density function of $d(H, \overline{H})$ from 0 to some value until the integration reaches the consistency level. The second one is to make use of the Chebyshev inequality to estimate the critical value of consistency. The last one is to calculate the critical value by generating a great numbers of points satisfying the distribution of $d(H, \overline{H})$.

(1) Integrating the density function

   Since the density function of $d(H, \overline{H})$ has been given by Eqs. (6.25) and (6.26), a direct idea is to integrate $g_2(x)$ from 0 to the critical value of consistency index until reaching the consistency level. However, as the expression of the density function of $d(H, \overline{H})$ is very complicated, this method is very hard to be implemented especially when the order n is large. In addition, the difficulty of convolution operation and the further large calculation cost of integration also suggest that it is not a good idea to integrate Eq. (6.26) directly in practice.

(2) Estimating the consistency index by Chebyshev inequality

   Instead of integrating the density function of $d(H, \overline{H})$, this method makes use of the mean value and the mean square deviation of $d(H, \overline{H})$ to estimate the consistency index.

Suppose that $h(x)$ is a transformation of $g_2(x)$, where

$$h(x) = \begin{cases} \frac{1}{2}g_2(x), & x \in [0,1] \\ \frac{1}{2}g_2(2-x), & x \in [1,2] \\ 0, & else \end{cases} \tag{6.32}$$

Then, the following theorem is obtained:

Theorem 6.7 (Liu et al. 2016). *There exists a probability space $(\Theta, \mathbb{F}, P)$, where P, the probability measure, has total mass 1, and a random variable $\xi(\omega)$ whose density function is $h(x)$.*

Proof Assume that $\Theta = [0,2]$, $\mathbb{F}$ contains every Borel subset of Θ and P is the Lebesgue measure of line. Defining $\theta(\omega) = 2\omega$, then $\theta(\omega)$ is a random variable from $(\Theta, \mathbb{F}, P)$. Hence, for any $x \in [0,2]$, $P(\theta(\omega) < x) = P(\omega \in [0,x)) = \frac{1}{2}x$ and $\theta(\omega)$ has a uniform distribution on $[0,2]$. In particular, the corresponding distribution function is given by $H(x) = \int_{-\infty}^{x} h(t)dt$. Since $h(t)$ is non-negative, then $H(x)$ is monotonic increasing. Furthermore, we have $H(x) = 0$ when $x \in (-\infty, 0]$ and $H(x) = 1$ when $x \in [2, +\infty)$. Therefore, we can define $H^{-1}(y) = \inf\{x : H(x) > y\}$ as the invers function of $H(x)$ and similarly it is monotonic, so it is Borel function. Letting $\xi(\omega) = H^{-1}(\theta(\omega))$, we get the random variable $\xi(\omega)$ from $(\Theta, \mathbb{F}, P)$ with the density function $h(x)$. This completes the proof of Theorem 6.5.

Assume that ξ is a random variable with the density function $h(x)$, then its mean value $E(\xi)$ and the mean square deviation $D(\xi)$ are given as:

$$E(\xi) = \int_{-\infty}^{+\infty} th(t)dt = 1$$

$$D(\xi) = \int_{-\infty}^{+\infty} (t - E(\xi))^2 h(t)dt = \int_{-\infty}^{+\infty} t^2 h(t)dt - 1$$

Note that the density function $h(x)$ of ξ is symmetric at $x = 1$. In view of the Chebyshev inequality: $P(|\xi - E(\xi)| \geq k\sqrt{D(\xi)}) \leq \frac{1}{k^2}$, the following result can be inferred:

$$P(0 \leq \xi \leq 1 - D(\xi)) \leq \frac{1}{2k^2} \tag{6.33}$$

Equation (6.33) can be transformed into

$$\int_{0}^{1-k\sqrt{D(\xi)}} h(t)dt \leq \frac{1}{2k^2} \tag{6.34}$$

Therefore,

$$\int_0^{1-k\sqrt{D(\xi)}} 2h(t)dt = \int_0^{1-k\sqrt{D(\xi)}} g_2(t)dt \leq \frac{1}{k^2} \tag{6.35}$$

So the following conclusion can be inferred:

$$P\left(0 \leq d\left(H, \overline{H}\right) \leq 1 - kD(\xi)\right) \leq \frac{1}{k^2} \tag{6.36}$$

Hence, we can use $1 - kD(\xi)$ to estimate the consistency index, and the consistency level $\rho = \frac{1}{k^2}$. This is absolutely a brief method to estimate the consistency index. However, the drawback is that Eq. (6.36) is an inequality instead of an equality, and simply using $1 - kD(\xi)$ to estimate the consistency causes the error existing in Chebyshev inequality.

(3) Generating points to fit the distribution of $d\left(H, \overline{H}\right)$

If the density function of $d\left(H, \overline{H}\right)$ is a randomly given density function, it will be difficult to find such points that fit it. It is noted that the density function of $d\left(H, \overline{H}\right)$ is based on the assumption that $d'_{ij}(x)$ satisfies the half-normal distribution and Eqs. (6.21)–(6.24) show the exact process to achieve it. As a result, a large number of discrete points fitting the density function of $d\left(H, \overline{H}\right)$ can be generated. Supposing that N points are generated fitting the density of $d\left(H, \overline{H}\right)$, to calculate the corresponding consistency index, we can just value it by $\lfloor \rho N \rfloor$, where ρ is the consistency level and $\lfloor \rho N \rfloor$ rounds ρN down. Actually it can be proven that ignoring the error brought by the assumption to infer the density function of $d\left(H, \overline{H}\right)$, this method can ensure the calculation error as small as required by increasing the number of the points.

Comparing the three methods discussed above and concerning the difficulty and complexity in practice as well as the calculation errors, we choose the third method to achieve the value of consistency index.

Before generating the point range obeying the density function of $d\left(H, \overline{H}\right)$, we first give an algorithm to obtain the table of σ depending on the order of HFPRs:

Algorithm 6.3

Step 1. Generate the large numbers of HFPRs $H_1, H_2, \ldots, H_N$ randomly with the fixed order n and calculate the corresponding perfect multiplicative consistent HFPR $\overline{H}_1, \overline{H}_2, \ldots, \overline{H}_N$ by Eq. (6.6), then go to the next step.

Step 2. Calculate the error coefficient $d_{11}^i, d_{12}^i, \ldots, d_{1n}^i, d_{21}^i, \ldots, d_{2n}^i, \ldots, d_{nn}^i$, $i = 1, 2, \ldots, N$, by Eq. (6.15) or Eq. (6.16), then go to the next step.

Step 3. Transform $d_{11}^i, d_{12}^i, \ldots, d_{1n}^i, d_{21}^i, \ldots, d_{2n}^i, \ldots, d_{nn}^i$, $i = 1, 2, \ldots, N$, from the interval $[0, 1]$ to the interval $[1, +\infty)$ by Eq. (6.22), then go to the next step.

Step 4. For any d'_{ij}, if $(i+j)$ is an odd number, then we update d'_{ij} by using $d'_{ij} = 2 - d'_{ij}$, else if $(i+j)$ is an even number, then d'_{ij} remains, and go to the next step.

Step 5. Calculate the standard deviation σ^i of $d'^{i}_{11}, d'^{i}_{12}, \ldots, d'^{i}_{1n}, d'^{i}_{21}, \ldots, d'^{i}_{2n}, \ldots, d'^{i}_{nn}$, $i = 1, 2, \ldots, N$, and let $\sigma^n = \sum_{i=1}^{N} \sigma^i / N$, where n is the order of HFPRs $H_1, H_2, \ldots, H_N$, then go to the next step.

Step 6. Range n from 3 to 14 and derive $\sigma^3, \sigma^4, \ldots, \sigma^{15}$ for the HFPR of different order respectively. This ends the algorithm.

To verify the above algorithm, 240,000 HFPRs are generated by Fortran Software Package, among which 120,000 are operated by the hesitant normalized Hamming distance and the other 120,000 by the hesitant normalized Euclidean distance, with 10,000 matrices for each order respectively, from 3 to 14, whose HFEs are generated randomly using a uniform distribution. By applying the algorithm above to the generated HFPRs, a table of σ varying the order of HFPRs from 3 to 14 are established, which can be seen in Table 6.1.

In this part, we can give the point ranges to fit the density function of $d(H, \overline{H})$ varying the order from 3 to 14, and with these point ranges obtaining the density distribution of $d(H, \overline{H})$ and the consistency level provided by the decision maker, we can finally calculate the critical value of the consistency index τ. Actually, Eqs. (6.21)–(6.24) have told the method to gain the point ranges and the only unknown σ has been valued as listed in Table 6.1. Firstly, we can generate the point ranges obeying the normal distribution with σ valued in Table 6.1. Then, applying Eq. (6.22) to every single point from point ranges above and implement Eq. (6.21) to each point range derived, we can finally generate the point ranges which enjoy the distribution of $d(H, \overline{H})$.

Without loss of generality, here we set the consistency level $\rho = 20\%$. So the critical value of consistency index τ satisfies that $\int_0^\tau g_2(x)dx = 0.2$ where $g_2(x)$ is the density function of $d(H, \overline{H})$. As a result, we take H as an acceptable multiplicative consistent HFPR when $d(H, \overline{H})$ belongs to the top 20 % distribution of the corresponding-order- HFPRs scaling by the increase of $d(H, \overline{H})$ from 0 to 1, which is equivalent to $d(H, \overline{H}) < \tau$.

Now with the method introduced above, the consistency index values τ_i ($i = 3, 4, \ldots, 14$) respectively using Eqs. (6.15) and (6.16) with the order of H varying

Table 6.1 The standard deviation σ depending on the number of alternatives

Order n	3	4	5	6	7	8
σ_{hnh}	0.184	0.223	0.260	0.298	0.310	0.370
σ_{hne}	0.188	0.298	0.332	0.352	0.396	0.402
Order n	9	10	11	12	13	14
σ_{hnh}	0.375	0.378	0.384	0.427	0.460	0.474
σ_{hne}	0.408	0.420	0.422	0.434	0.452	0.462

from 3 to 14 can be derived, which are listed in Table 6.2. All the calculations are accomplished by Matlab software package.

From Table 6.2, we can get the following conclusions:

Firstly, the critical value of consistency index value appears to increase monotonically with the order of H increasing from 3 to 14 and the fixed consistency level $\rho = 20\%$.

Secondly, the average occasion is that the critical value of consistency index of different orders by the hesitant normalized Euclidean distance is bigger than the corresponding one by the hesitant normalized Hamming distance, which can be explained by the mean value inequality combined with the definition of two distance definitions.

Thirdly, the critical value of the consistency index in Table 6.2 is based on the assumption that the consistency level is 20 %. In fact, this chapter is not only aimed to get the critical value of consistency index, but also to provide a different method to calculate it. So the decision makers can resize the consistency level greater or smaller depending on specific situations.

Last but not the least, we can see that most critical values of consistency index are less than 0.1, except when the order values are 13 and 14, which indicates that setting the critical value of consistency as 0.1 is not appropriate in most occasions, especially when the order is not so large in practice. Actually when the order $n = 3$, to reach the original consistency index value 0.1, the consistency level should reach 45.4 %, which is such a large percentage that it can hardly absolutely to guarantee the consistency of a HFPR whose order is 3. Therefore, in some certain occasions, the original critical consistency index valued as 0.1 is not restrictive enough. To see this more clearly, we use a similar method to achieve the table of the consistency level when the consistency index of different-order-HFPR values 0.1 as usual, which can be seen in Table 6.3.

Comparing the assumptions in this paper that the consistency level $\rho = 20.0\%$ with the consistency level in Table 6.3, we can see that in most occasions the consistency level when setting the consistency index $\tau = 0.1$ is much bigger than $\rho = 20.0\%$. This reflects that the result of consistency checking by simply setting $\tau = 0.1$ is usually not strict enough in the perspective of statistics.

Let us go back to Example 6.3. $d_{hnh}(H, \overline{H}) = 0.095 > \tau_{hnh}(3) = 0.045$ indicates that the HFPR is not a multiplicative consistent HFPR. Actually, in our method, H will not be accepted as a multiplicative consistent HFPR until the decision maker sets the consistency level up to 43.1 %.

Table 6.2 The new derived critical values of the consistency index τ

Order n	3	4	5	6	7	8
τ_{hnh}	0.045	0.054	0.062	0.070	0.072	0.086
τ_{hne}	0.046	0.070	0.077	0.082	0.091	0.093
Order n	9	10	11	12	13	14
τ_{hnh}	0.087	0.088	0.089	0.097	0.104	0.107
τ_{hne}	0.094	0.096	0.097	0.099	0.102	0.105

Table 6.3 The consistency levels when the consistency index value is 0.1

Order n	3	4	5	6	7	8
ρ_{hnh}	45.4 %	38.2 %	33.0 %	29.2 %	28 %	23.6 %
ρ_{hne}	44.6 %	29.0 %	26.2 %	24.8 %	22.2 %	21.8 %
Order n	9	10	11	12	13	14
ρ_{hnh}	23.4 %	23.2 %	22.8 %	20.6 %	19.0 %	18.6 %
ρ_{hne}	21.4 %	20.8 %	20.7 %	20.2 %	19.4 %	19.0 %

Now we give two examples to see exactly how the new critical value table of consistency index works.

Example 6.4 (Liu et al. 2016). Assume that a decision maker gives his/her preference information over a collection of alternatives A_1, A_2, A_3, A_4, and A_5 in HFEs and construct the following HFPR:

$$H = \begin{pmatrix}
\{0.5\} & \{0.6, 0.7\} & \{0.6, 0.8\} & \{0.3, 0.9\} & \{0.4, 0.8\} \\
\{0.3, 0.4\} & \{0.5\} & \{0.4, 0.7\} & \{0.2, 0.7\} & \{0.1, 0.8\} \\
\{0.2, 0.4\} & \{0.3, 0.6\} & \{0.5\} & \{0.2, 0.6\} & \{0.2, 0.6\} \\
\{0.1, 0.7\} & \{0.3, 0.8\} & \{0.4, 0.8\} & \{0.5\} & \{0.3, 0.4\} \\
\{0.2, 0.6\} & \{0.2, 0.9\} & \{0.4, 0.8\} & \{0.6, 0.7\} & \{0.5\}
\end{pmatrix}$$

The corresponding perfect multiplicative consistent HFPR of H can be obtained by Eq. (6.6), and it follows

$$\overline{H} = \begin{pmatrix}
\{0.5\} & \{0.6, 0.7\} & \{0.5, 0.845\} & \{0.2, 0.891\} & \{0.097, 0.845\} \\
\{0.3, 0.4\} & \{0.5\} & \{0.4, 0.7\} & \{0.143, 0.778\} & \{0.067, 0.7\} \\
\{0.155, 0.5\} & \{0.3, 0.6\} & \{0.5\} & \{0.2, 0.6\} & \{0.097, 0.5\} \\
\{0.109, 0.8\} & \{0.222, 0.857\} & \{0.4, 0.8\} & \{0.5\} & \{0.3, 0.4\} \\
\{0.155, 0.903\} & \{0.3, 0.933\} & \{0.5, 0.903\} & \{0.6, 0.7\} & \{0.5\}
\end{pmatrix}$$

Then, the hesitant normalized Hamming distance between H and $\overline{H}$ can be calculated as:

$$d_{hnh}(H, \overline{H}) = \frac{1}{12} \sum_{i=1}^{5} \sum_{j=1}^{5} \left[\frac{1}{l_{x_{ij}}} \sum_{t=1}^{l_{x_{ij}}} \left| h_{ij}^{\sigma(t)}(x) - \overline{h}_{ij}^{\sigma(t)} \right| \right] = 0.0894$$

According to Algorithm 6.1, $d_{hnh}(H, \overline{H}) = 0.0894 < 0.1$, which means that H is a multiplicative consistent HFPR. While according to Table 6.2, the consistency index value is 0.062 with the order 5 and the hesitant normalized Hamming distance measure. Then $d_{hnh}(H, \overline{H}) = 0.0894 > 0.062$ indicates that H is not a multiplicative consistent HFPR. Utilizing the similar process to approach the critical value of consistency index, we can calculate that it has to take as much as top 29.4 %

HFPRs of all over the 5-order ones in the increasing order of $d_{hnh}(H,\overline{H})$ from 0 to 1.

To repair the multiplicative inconsistent HFPR H above, an effective choice is to return it to the decision makers repeatedly so that they can adjust the preference information until the HFPR H is multiplicative consistent. While in certain occasions, the preference information cannot be returned to the original decision makers, we can apply Algorithm 6.1 on H to improve its consistency level.

In Algorithm 6.1, the correction factor $\xi = p\delta$ actually shows the original information retention ratio. When $\xi = 0$, $\widehat{h}_{ij}^{(p)\sigma(t)} = h_{ij}^{(p)\sigma(t)}$, which indicates that it totally saves the original information, and the original information retention ratio is 100 %. Similarly, when $\xi = 1$, $\widehat{h}_{ij}^{(p)\sigma(t)} = \overline{h}_{ij}^{(p)\sigma(t)}$, which means that the original information is totally replaced by the derived information from the corresponding perfect multiplicative consistent HFPR $\overline{H}$. So the closer the correction factor ξ approaching 0, the more the original information saved. Hereby, we let $\xi = 0.3$, and the repaired HFPR is given by Algorithm 6.1 as:

$$\widehat{H} = \begin{pmatrix} \{0.5\} & \{0.6,0.7\} & \{0.571,0.814\} & \{0.267,0.897\} & \{0.278,0.814\} \\ \{0.3,0.4\} & \{0.5\} & \{0.4,0.7\} & \{0.181,0.72\} & \{0.088,0.772\} \\ \{0.186,0.429\} & \{0.3,0.6\} & \{0.5\} & \{0.2,0.6\} & \{0.162,0.57\} \\ \{0.103,0.733\} & \{0.28,0.819\} & \{0.4,0.8\} & \{0.5\} & \{0.3,0.4\} \\ \{0.186,0.722\} & \{0.228,0.912\} & \{0.43,0.838\} & \{0.6,0.7\} & \{0.5\} \end{pmatrix}$$

Furthermore, $d_{hnh}(\widehat{H},\overline{H}) = 0.059 < 0.062$, and now the repaired HFPR $\widehat{H}$ from the HFPR H is multiplicative consistent.

Example 6.5 (Liu et al. 2016). Suppose that the decision maker gives the preference information in HFE and constructs a HFPR as:

$$H = \begin{pmatrix} \{0.5\} & \{0.3,0.7\} & \{0.5,0.7\} & \{0.2,0.6\} \\ \{0.3,0.7\} & \{0.5\} & \{0.5,0.7\} & \{0.4,0.8\} \\ \{0.3,0.5\} & \{0.3,0.5\} & \{0.5\} & \{0.6,0.7\} \\ \{0.4,0.8\} & \{0.2,0.6\} & \{0.3,0.4\} & \{0.5\} \end{pmatrix}$$

Then, the corresponding perfect multiplicative consistent HFPR is calculated as:

$$\overline{H} = \begin{pmatrix} \{0.5\} & \{0.3,0.7\} & \{0.3,0.845\} & \{0.391,0.927\} \\ \{0.3,0.7\} & \{0.5\} & \{0.5,0.7\} & \{0.6,0.845\} \\ \{0.155,0.7\} & \{0.3,0.5\} & \{0.5\} & \{0.6,0.7\} \\ \{0.073,0.609\} & \{0.155,0.4\} & \{0.3,0.4\} & \{0.5\} \end{pmatrix}$$

Instead of the hesitant normalized Hamming distance measure, here we measure the distance between H and $\overline{H}$ by the hesitant normalized Euclidean distance and we have $d_{hne}(H,\overline{H}) = 0.203$. From Table 6.2, we can see that the critical value of the consistency index with the order 4 and the hesitant normalized Euclidean distance is 0.070, which illustrates H inconsistent. Using Algorithm 6.1 to repair H, and let $\xi = 0.7$, then we can get the repaired HFPR:

$$\widehat{H} = \begin{pmatrix} \{0.5\} & \{0.3,0.7\} & \{0.3559,0.8087\} & \{0.3261,0.87\} \\ \{0.3,0.7\} & \{0.5\} & \{0.5,0.7\} & \{0.5405,0.8324\} \\ \{0.1913,0.6441\} & \{0.3,0.5\} & \{0.5\} & \{0.6,0.7\} \\ \{0.13,0.6739\} & \{0.1676,0.4595\} & \{0.3,0.4\} & \{0.5\} \end{pmatrix}$$

The hesitant normalized Euclidean distance between $\widehat{H}$ and $\overline{H}$ can be calculated as $d_{hne}(H,\overline{H}) = 0.051 < 0.070$. Thus, the repaired HFPR is multiplicative consistent.

6.3   Approaches to Group Decision Making with Hesitant Fuzzy Preference Relations

In our daily life, in order to choose the most desirable and reasonable solution(s) for a decision making problem, people prefer to form a commitment or organization constructed by several decision makers coming from different aspects instead of single decision maker for the sake of avoiding the limited knowledge, personal background, private emotion, and so on (Liao et al. 2015). As mentioned above, people may provide the preference information by pairwise comparisons and thus construct the preference relations. If people in the commitment or organization express their preference values in HFEs, then some HFPRs can be constructed.

The group decision making problem in hesitant fuzzy circumstance can be described as follows. Suppose that $A = \{A_1, A_2, \ldots, A_n\}$ is a discrete set of alternatives; $d_k (k = 1, 2, \ldots, m)$ are the decision organizations (each of which contains a collection of decision makers), and $\omega = (\omega_1, \omega_2, \ldots, \omega_m)^T$ is the weight vector of the decision organizations with $\sum_{k=1}^{m} \omega_k = 1$, $\omega_k \in [0, 1]$, $k = 1, 2, \ldots, m$. The decision organization d_k provides all the possible preference values for each pair of alternatives, and constructs a HFPR $H^{(k)} = (h_{ij}^{(k)})_{n \times n}$.

Definition 6.7 (*Liao et al.* 2014b). Let $H_k = (h_{ij}^{(k)})_{n \times n} (k = 1, 2, \ldots, m)$ be a collection of m HFPRs on a fixed set $A = \{A_1, A_2, \ldots, A_n\}$, and $\omega = (\omega_1, \omega_2, \ldots, \omega_m)^T$ be the weight vector of $H_k (k = 1, 2, \ldots, m)$, where $\sum_{k=1}^{m} \omega_k = 1$ and $0 \leq \omega_k \leq 1$. Then we call $H = (h_{ij})_{n \times n}$ a collective HFPR of H_k $(k = 1, 2, \ldots, m)$, where h_{ij} is obtained by the AHFWA or AHFWG operator, i.e.,

$$h_{ij} = \overset{m}{\underset{k=1}{\oplus}} \left(\omega_k h_{ij}^{(k)} \right) = \left\{ 1 - \prod_{k=1}^{m} (1 - h_{ij}^{(k)\sigma(t)})^{\omega_k} | t = 1, 2, \ldots, l \right\}, \quad i, j = 1, 2, \ldots, n$$

$$(6.37)$$

and

$$h_{ij} = \overset{m}{\underset{k=1}{\otimes}} (h_{ij}^{(k)})^{\omega_k} = \left\{ \prod_{k=1}^{m} (h_{ij}^{(k)\sigma(t)})^{\omega_k} | t = 1, 2, \ldots, l \right\}, \ i, j = 1, 2, \ldots, n \quad (6.38)$$

with $h_{ij}^{(k)\sigma(t)}$ being the tth smallest value in $h_{ij}^{(k)}$.

Different experts (or decision makers) can have different preferences and then it is needed to propose some consensus reaching methods. There are many researches on consensus of preference relations (Xu 2005; Fan et al. 2006; Herrera-Viedma et al. 2007; Alonso et al. 2009; Xu and Cai 2011; Liao and Xu 2014a, 2015; Xu and Liao 2015). However, no work has done on HFPRs. In the following, based on the multiplicative consistency, we also develop two consensus reaching processes for HFPRs in group decision making.

6.3.1 Automatic Consensus Reaching Process

As to the other preference relations, many scholars have proposed some automatic consensus improving procedures (Liao and Xu 2014b). Based on two soft consensus criteria—a consensus measure and a proximity measure, Tapia García et al. (2012) presented a consensus model for group decision making problems with interval fuzzy preference relations. They also designed an automatic feedback mechanism to help the decision makers in consensus reaching process. Cabrerizo et al. (2009) developed a consensus model for group decision making problems with unbalanced fuzzy linguistic information based on the above two soft consensus criteria. In a multigranular fuzzy linguistic context, Mata et al. (2009) proposed an adaptive consensus support model for group decision making problems, which increases the convergence toward the consensus and reduces the number of rounds to reach it. These works are all automatic which transform the decision makers' opinions themselves, without the decision makers' interactivity. Below we propose an automatic consensus reaching process for HFPRs.

Algorithm 6.4

Step 1. Let $(H^{(k)})^{(p)} = ((h_{ij}^{(k)})_{n \times n})^{(p)} (k = 1, 2, \ldots, m)$ and $p = 1$. We construct the prefect multiplicative consistent HFPRs $(\bar{H}^{(k)})^{(p)} = ((\bar{h}_{ij}^{(k)})_{n \times n})^{(p)}$ from $(H^{(k)})^{(p)} = ((h_{ij}^{(k)})_{n \times n})^{(p)}$ by Algorithm 6.1 (or Algorithm 6.2), then go to the next step.

Step 2. Aggregate all the individual prefect multiplicative consistent HFPRs $(\bar{H}^{(k)})^{(p)} = ((\bar{h}_{ij}^{(k)})_{n \times n})^{(p)}$ into a collective HFPR $(\bar{H})^{(p)} = ((\bar{h}_{ij})_{n \times n})^{(p)}$ by the AHFWA or AHFWG operator, where

$$\bar{h}_{ij} = \overset{m}{\underset{k=1}{\oplus}} \left(w_k \bar{h}_{ij}^{(k)} \right) = \left\{ 1 - \prod_{k=1}^{m} \left(1 - \bar{h}_{ij}^{(k)\sigma(t)} \right)^{w_k} \Big| t = 1, 2, \ldots, l \right\}, \ i, j = 1, 2, \ldots, n$$

$$(6.39)$$

and

$$\bar{h}_{ij} = \overset{m}{\underset{k=1}{\otimes}} (\bar{h}_{ij}^{(k)})^{w_k} = \left\{ \prod_{k=1}^{m} (\bar{h}_{ij}^{(k)\sigma(t)})^{w_k} \Big| t = 1, 2, \ldots, l \right\}, \ i, j = 1, 2, \ldots, n$$

$$(6.40)$$

with $\bar{h}_{ij}^{(k)\sigma(t)}$ being the tth smallest value in $\bar{h}_{ij}^{(k)}$, then go to the next step.

Step 3. Calculate the deviation between each individual HFPR $(\overline{H}^{(k)})^{(p)} = ((\bar{h}_{ij}^{(k)})_{n \times n})^{(p)}$ and the collective HFPR $\overline{H}^{(p)} = ((\bar{h}_{ij})_{n \times n})^{(p)}$, i.e.,

$$d_{hnh}((\overline{H}^{(k)})^{(p)}, \overline{H}^{(p)}) = \frac{1}{(n-1)(n-2)} \sum_{i=1}^{n} \sum_{j=1}^{n} \left[\frac{1}{l_{\bar{h}_{ij}}} \sum_{t=1}^{l_{\bar{h}_{ij}}} \left| \bar{h}_{ij}^{(k)(p)\sigma(t)} - \bar{h}_{ij}^{(p)\sigma(t)} \right| \right]$$

$$(6.41)$$

or

$$d_{hne}((\overline{H}^{(k)})^{(p)}, \overline{H}^{(p)}) = \frac{1}{(n-1)(n-2)} \sum_{i=1}^{n} \sum_{j=1}^{n} \left(\frac{1}{l_{\bar{h}_{ij}}} \sum_{t=1}^{l_{\bar{h}_{ij}}} \left| h_{ij}^{(k)(p)\sigma(t)} - \bar{h}_{ij}^{(p)\sigma(t)} \right|^2 \right)^{1/2}$$

$$(6.42)$$

If $d((\overline{H}^{(k)})^{(p)}, \overline{H}^{(p)}) \leq \tau^*$, for all $k = 1, 2, \ldots, m$, where τ^* is the consensus threshold, then go to Step 5; Otherwise, go to the next step.

Step 4. Let $(\overline{H}^{(k)})^{(p+1)} = ((\bar{h}_{ij}^{(k)})_{n \times n})^{(p+1)}$, where

$$\bar{h}_{ij}^{(k)(p+1)\sigma(t)} = \frac{(\bar{h}_{ij}^{(k)(p)\sigma(t)})^{1-\xi}(\bar{h}_{ij}^{(p)\sigma(t)})^{\xi}}{(h_{ij}^{(k)(p)\sigma(t)})^{1-\xi}(\bar{h}_{ij}^{(p)\sigma(t)})^{\xi} + (1 - h(p_{ij}^{(k)(p)\sigma(t)})^{1-\xi}(1 - \bar{h}_{ij}^{(p)\sigma(t)})^{\xi}}, i, j = 1, 2, \ldots, n$$

$$(6.43)$$

$\bar{h}_{ij}^{(k)(p+1)\sigma(t)}$, $\bar{h}_{ij}^{(k)(p)\sigma(t)}$ and $\bar{h}_{ij}^{(p)\sigma(t)}$ are the tth smallest values in $\bar{h}_{ij}^{(k)(p+1)}$, $\bar{h}_{ij}^{(k)(p)}$ and $\bar{h}_{ij}^{(p)}$, respectively. Let $p = p + 1$, then go to Step 2.

Step 5. Let $H = \overline{H}^{(p)}$, and employ the AHFA or AHFG operator to fuse all the hesitant preference values $h_{ij}(j = 1, 2, \ldots, n)$ corresponding to the object A_i into the overall hesitant preference value h_i, i.e.,

$$h_i = \text{AHFA}(h_{i1}, h_{i2}, \ldots, h_{in}) = \overset{n}{\underset{j=1}{\oplus}} \left(\frac{1}{n} h_{ij} \right)$$
$$= \left\{ 1 - \prod_{j=1}^{n} (1 - h_{ij}^{\sigma(t)})^{1/n} \middle| t = 1, 2, \ldots, l \right\} \qquad (6.44)$$

or

$$h_i = \text{AHFG}(h_{i1}, h_{i2}, \ldots, h_{in}) = \overset{n}{\underset{j=1}{\otimes}} (h_{ij})^{1/n}$$
$$= \left\{ \prod_{j=1}^{n} (h_{ij}^{\sigma(t)})^{1/n} \middle| t = 1, 2, \ldots, l \right\} \qquad (6.45)$$

where $h_{ij}^{\sigma(t)}$ is the tth smallest values in h_{ij}, then go to the next step.

Step 6. Rank all the objects corresponding to the methods given in Sect. 1.1.3, and then go to the next step.

Step 7. End.

This consensus reaching process can be interpreted like this. Firstly, we construct the prefect multiplicative consistent HFPRs for the individual HFPRs given by the different decision organizations. Then, a collective HFPR can be obtained by aggregating the constructed prefect multiplicative consistent HFPRs. Afterwards, we can easily calculate the distance between each individual HFPR and the collective HFPR respectively. If the distance is greater than the given consensus level, we need to improve it; otherwise, it is acceptable. To improve the individual HFPR, we fuse it with the collective HFPR by using Eq. (6.43), and then get some new individual HFPRs. We can iterate until all the individual HFPRs are acceptable.

We now consider a group decision making problem that concerns the evaluation and ranking of the main factors of electronic learning to illustrate our procedure.

Example 6.6 (Liao et al. 2014b). The electronic learning (e-learning) not only can provide expediency for learners to study courses and professional knowledge without the constraint of time and space especially in an asynchronous distance e-learning system, but also may save internal training cost for some enterprises organizations in a long-term strategy. Meanwhile, it also can be used as an alternative self-training for assisting or improving the traditional classroom teaching. The e-learning becomes more and more popular along with the advancement of information technology and has played an important role in teaching and learning not only in different levels of schools but also in various commercial or industrial companies. Many schools and businesses invest manpower and money in e-learning to enhance their hardware facilities and software contents. Thus, it is meaningful and urgent to determinate which is the most important among the main factors which influence the e-learning effectiveness. Based on the research of Tzeng et al. (2007), there are four key factors (or criteria) to evaluate the effectiveness of an e-learning system, which are

C_1: the synchronous learning;
C_2: the e-learning material;
C_3: the quality of web learning platform;
C_4: the self-learning.

In order to rank the above four factors, a committee comprising three decision makers $d_l(l = 1, 2, 3)$ (whose weight vector is $\omega = (0.3, 0.4, 0.3)^T$) is founded. After comparing pairs of the factors (or criteria) $C_i(i = 1, 2, 3, 4)$, the decision makers $d_l(l = 1, 2, 3)$ give their preferences using HFEs, and then obtain the HFPRs as follows:

$$
H_1 = \begin{pmatrix}
\{0.5\} & \{0.2, 0.3, 0.4\} & \{0.4, 0.5, 0.6\} & \{0.3, 0.7\} \\
\{0.6, 0.7, 0.8\} & \{0.5\} & \{0.5, 0.6\} & \{0.3, 0.4\} \\
\{0.4, 0.5, 0.6\} & \{0.4, 0.5\} & \{0.5\} & \{0.4, 0.5\} \\
\{0.3, 0.7\} & \{0.6, 0.7\} & \{0.5, 0.6\} & \{0.5\}
\end{pmatrix}
$$

$$
H_2 = \begin{pmatrix}
\{0.5\} & \{0.3, 0.4\} & \{0.5, 0.6, 0.7\} & \{0.3, 0.4, 0.6\} \\
\{0.6, 0.7\} & \{0.5\} & \{0.4, 0.7\} & \{0.4, 0.6\} \\
\{0.3, 0.4, 0.5\} & \{0.3, 0.6\} & \{0.5\} & \{0.6, 0.7\} \\
\{0.4, 0.6, 0.7\} & \{0.4, 0.6\} & \{0.3, 0.4\} & \{0.5\}
\end{pmatrix}
$$

$$
H_3 = \begin{pmatrix}
\{0.5\} & \{0.2, 0.4\} & \{0.4, 0.7\} & \{0.3, 0.6, 0.7\} \\
\{0.6, 0.8\} & \{0.5\} & \{0.5, 0.7\} & \{0.3, 0.6\} \\
\{0.3, 0.6\} & \{0.3, 0.5\} & \{0.5\} & \{0.4, 0.6\} \\
\{0.3, 0.4, 0.7\} & \{0.4, 0.7\} & \{0.4, 0.6\} & \{0.5\}
\end{pmatrix}
$$

To solve this problem, the following steps are given according to Algorithm 6.4.

Step 1: Let $(H^{(k)})^{(p)} = H_k$ and $p = 1$. We first construct the prefect multiplicative consistent HFPRs $(\overline{H}^{(k)})^{(1)} = ((\bar{h}_{ij}^{(k)})_{n \times n})^{(1)}$ $(k = 1, 2, 3)$ from $(H^{(k)})^{(1)} = ((h_{ij}^{(k)})_{n \times n})^{(1)}$ $(k = 1, 2, 3)$ by Algorithm 6.1, respectively.

$$
\left(\overline{H}^{(1)}\right)^{(1)} = \begin{pmatrix}
\{0.5\} & \{0.2, 0.3, 0.4\} & \{0.2, 0.3, 0.5\} & \{0.202, 0.361, 0.5\} \\
\{0.6, 0.7, 0.8\} & \{0.5\} & \{0.5, 0.6\} & \{0.4, 0.6\} \\
\{0.5, 0.7, 0.8\} & \{0.4, 0.5\} & \{0.5\} & \{0.4, 0.5\} \\
\{0.5, 0.639, 0.798\} & \{0.4, 0.6\} & \{0.5, 0.6\} & \{0.5\}
\end{pmatrix}
$$

$$
\left(\overline{H}^{(2)}\right)^{(1)} = \begin{pmatrix}
\{0.5\} & \{0.3, 0.4\} & \{0.222, 0.609\} & \{0.361, 0.596, 0.672\} \\
\{0.6, 0.7\} & \{0.5\} & \{0.4, 0.7\} & \{0.5, 0.845\} \\
\{0.391, 0.778\} & \{0.3, 0.6\} & \{0.5\} & \{0.6, 0.7\} \\
\{0.328, 0.404, 0.639\} & \{0.155, 0.5\} & \{0.3, 0.4\} & \{0.5\}
\end{pmatrix}
$$

$$
\left(\overline{H}^{(3)}\right)^{(1)} = \begin{pmatrix}
\{0.5\} & \{0.2, 0.4\} & \{0.2, 0.609\} & \{0.202, 0.639\} \\
\{0.6, 0.8\} & \{0.5\} & \{0.5, 0.7\} & \{0.4, 0.778\} \\
\{0.391, 0.8\} & \{0.3, 0.5\} & \{0.5\} & \{0.4, 0.6\} \\
\{0.361, 0.798\} & \{0.222, 0.6\} & \{0.4, 0.6\} & \{0.5\}
\end{pmatrix}
$$

Step 2. Fuse the individual prefect multiplicative consistent HFPRs $(\overline{H}^{(k)})^{(1)} = ((\bar{h}_{ij}^{(k)})_{n\times n})^{(1)}$ into a collective prefect HFPR $(\overline{H})^{(1)} = ((\bar{h}_{ij})_{n\times n})^{(1)}$ by the AHFWA or AHFWG operator. We hereby take the AHFWA operator, i.e., Eq. (6.39), as an example, and then we obtain

$$
\overline{H}^{(1)} = \begin{pmatrix}
\{0.5\} & \{0.242, 0.372, 0.472\} & \{0.209, 0.447, 0.579\} & \{0.27, 0.506, 0.617\} \\
\{0.532, 0.633, 0.765\} & \{0.5\} & \{0.462, 0.673\} & \{0.442, 0.771\} \\
\{0.426, 0.571, 0.792\} & \{0.332, 0.543\} & \{0.5\} & \{0.49, 0.619\} \\
\{0.394, 0.514, 0.745\} & \{0.256, 0.563\} & \{0.396, 0.53\} & \{0.5\}
\end{pmatrix}
$$

Step 3 Calculate the deviation between each individual prefect multiplicative consistent HFPR $(\overline{H}^{(k)})^{(1)} = ((\bar{h}_{ij}^{(k)})_{n\times n})^{(1)}$ and the collective HFPR $\overline{H}^{(1)} = ((\bar{h}_{ij})_{n\times n})^{(1)}$. In this example, we use Eq. (6.41), i.e., the hesitant normalized Hamming distance, as a representation, and then

$$
d_{hnh}((\overline{H}^{(1)})^{(1)}, \overline{H}^{(1)}) = 0.162, \ d_{hnh}((\overline{H}^{(2)})^{(1)}, \overline{H}^{(1)}) = 0.128,
$$
$$
d_{hnh}((\overline{H}^{(3)})^{(1)}, \overline{H}^{(1)}) = 0.07
$$

Without loss of generality, we let the consensus level $\tau^* = 0.1$. We can see that both $d_{hnh}((\overline{H}^{(1)})^{(1)}, \overline{H}^{(1)})$ and $d_{hnh}((\overline{H}^{(2)})^{(1)}, \overline{H}^{(1)})$ are bigger than 0.1, and thus, we need to improve these individual prefect multiplicative consistent HFPRs.

Step 4. Let $\xi = 0.7$, and by Eq. (6.43), we can construct respectively the new individual HFPRs $(\overline{H}^{(k)})^{(2)} = ((\bar{h}_{ij}^{(k)})_{n\times n})^{(2)} (k = 1, 2, 3)$ as follows:

$$
(\overline{H}^{(1)})^{(2)} = \begin{pmatrix}
\{0.5\} & \{0.229, 0.35, 0.45\} & \{0.206, 0.401, 0.556\} & \{0.248, 0.461, 0.583\} \\
\{0.553, 0.654, 0.78\} & \{0.5\} & \{0.473, 0.652\} & \{0.429, 0.725\} \\
\{0.445, 0.612, 0.794\} & \{0.352, 0.53\} & \{0.5\} & \{0.463, 0.584\} \\
\{0.425, 0.552, 0.762\} & \{0.296, 0.574\} & \{0.427, 0.551\} & \{0.5\}
\end{pmatrix}
$$

$$
(\overline{H}^{(2)})^{(2)} = \begin{pmatrix}
\{0.5\} & \{0.259, 0.38, 0.48\} & \{0.213, 0.463, 0.588\} & \{0.296, 0.533, 0.634\} \\
\{0.522, 0.623, 0.747\} & \{0.5\} & \{0.443, 0.681\} & \{0.459, 0.796\} \\
\{0.412, 0.55, 0.788\} & \{0.322, 0.56\} & \{0.5\} & \{0.523, 0.644\} \\
\{0.374, 0.481, 0.715\} & \{0.222, 0.544\} & \{0.366, 0.491\} & \{0.5\}
\end{pmatrix}
$$

$$
(\overline{H}^{(3)})^{(2)} = \begin{pmatrix}
\{0.5\} & \{0.229, 0.38, 0.48\} & \{0.206, 0.463, 0.588\} & \{0.248, 0.504, 0.624\} \\
\{0.522, 0.623, 0.776\} & \{0.5\} & \{0.473, 0.681\} & \{0.429, 0.773\} \\
\{0.412, 0.55, 0.794\} & \{0.322, 0.53\} & \{0.5\} & \{0.463, 0.613\} \\
\{0.384, 0.51, 0.762\} & \{0.246, 0.574\} & \{0.397, 0.551\} & \{0.5\}
\end{pmatrix}
$$

Let $p = 2$, then go back to Step 2. We fuse the individual HFPRs $(\overline{H}^{(k)})^{(2)} (k = 1, 2, 3)$ into a collective HFPR $(\overline{H})^{(2)} = ((\bar{h}_{ij})_{n\times n})^{(2)}$ by the AHFWA operator:

$$\overline{H}^{(2)} = \begin{pmatrix} \{0.5\} & \{0.241, 0.371, 0.471\} & \{0.209, 0.445, 0.579\} & \{0.268, 0.504, 0.616\} \\ \{0.532, 0.633, 0.766\} & \{0.5\} & \{0.461, 0.673\} & \{0.442, 0.77\} \\ \{0.422, 0.57, 0.792\} & \{0.331, 0.542\} & \{0.5\} & \{0.488, 0.618\} \\ \{0.393, 0.512, 0.744\} & \{0.252, 0.562\} & \{0.394, 0.528\} & \{0.5\} \end{pmatrix}$$

Thus, the hesitant normalized Hamming distance between each individual prefect multiplicative consistent HFPR $(\overline{H}^{(k)})^{(2)}$ and the collective HFPR $\overline{H}^{(2)}$ can be calculated as:

$$d_{hnh}((\overline{H}^{(1)})^{(2)}, \overline{H}^{(2)}) = 0.048, \quad d_{hnh}((\overline{H}^{(2)})^{(2)}, \overline{H}^{(2)}) = 0.039,$$
$$d_{hnh}((\overline{H}^{(3)})^{(2)}, \overline{H}^{(2)}) = 0.02$$

Now all $d_{hnh}((\overline{H}^{(k)})^{(2)}, \overline{H}^{(2)}) < 0.1$ $(k = 1, 2, 3)$, then go to Step 5.

Step 5. Let $H = \overline{H}^{(2)}$, and employ the AHFA or AHFG operator to fuse all the hesitant preference values $h_{ij}(j = 1, 2, \ldots, n)$ corresponding to the object C_i into the overall hesitant preference value h_i. We hereby use the AHFA operator to fuse the information. By Eq. (6.44), we have

$$h_1 = \{0.315, 0.458, 0.545\}, \quad h_2 = \{0.485, 0.537, 0.694\}$$
$$h_3 = \{0.444, 0.519, 0.633\}, \quad h_4 = \{0.391, 0.503, 0.597\}$$

Step 6. Using Eq. (1.17), we can get $s(h_1) = 0.439$, $s(h_2) = 0.572$, $s(h_3) = 0.532$, and $s(h_4) = 0.495$. As $s(h_2) > s(h_3) > s(h_4) > s(h_1)$, we can draw a conclusion that $C_2 \succ C_3 \succ C_4 \succ C_1$, which denotes that the e-learning material is the most important factor influencing the affectivity of e-learning.

Surely, in this example, we can use the AHFWG and AHFG operators to fuse the HFEs in Steps 2 and 5, and also can use the hesitant normalized Euclidean distance to calculate the deviation between each individual prefect multiplicative consistent HFPR and the collective prefect multiplicative consistent HFPR in Step 3.

In addition, from Step 3, we can see that if we take the consensus level within the interval $0.07 \leq \tau^* \leq 0.128$, the result will keep the same. In other words, if the consensus level is fixed, small error measurements perhaps do not cause a complete different output, which is to say, our procedures are robust. For example, suppose that the third decision maker gives his/her preferences with another HFPR as:

$$\dot{H}_3 = \begin{pmatrix} \{0.5\} & \{0.2, 0.4\} & \{0.4, 0.7\} & \{0.3, 0.6, 0.7\} \\ \{0.6, 0.8\} & \{0.5\} & \{0.6, 0.8\} & \{0.3, 0.6\} \\ \{0.3, 0.6\} & \{0.2, 0.4, \} & \{0.5\} & \{0.4, 0.6\} \\ \{0.3, 0.4, 0.7\} & \{0.4, 0.7\} & \{0.4, 0.6\} & \{0.5\} \end{pmatrix}$$

In the following, we begin to check whether the output will be changed or not.

Step 1. The prefect multiplicative consistent HFPRs of the third decision maker can be calculated easily, which is

$$(\dot{\overline{H}}^{(3)})^{(1)} = \begin{pmatrix} \{0.5\} & \{0.2,0.4\} & \{0.2,0.609\} & \{0.202,0.639\} \\ \{0.6,0.8\} & \{0.5\} & \{0.6,0.8\} & \{0.4,0.778\} \\ \{0.391,0.8\} & \{0.2,0.4\} & \{0.5\} & \{0.4,0.6\} \\ \{0.361,0.798\} & \{0.222,0.6\} & \{0.4,0.6\} & \{0.5\} \end{pmatrix}$$

Step 2. The collective prefect HFPR can be derived by the AHFWA operator as:

$$\dot{\overline{H}}^{(1)} = \begin{pmatrix} \{0.5\} & \{0.242,0.372,0.472\} & \{0.209,0.447,0.579\} & \{0.27,0.506,0.617\} \\ \{0.532,0.633,0.765\} & \{0.5\} & \{0.497,0.710\} & \{0.442,0.771\} \\ \{0.426,0.571,0.792\} & \{0.332,0.543\} & \{0.5\} & \{0.49,0.619\} \\ \{0.394,0.514,0.745\} & \{0.256,0.563\} & \{0.396,0.53\} & \{0.5\} \end{pmatrix}$$

Step 3. Calculate the deviation between each individual prefect multiplicative consistent HFPR and the collective HFPR $\dot{\overline{H}}^{(1)}$:

$$d_{hnh}((\overline{H}^{(1)})^{(1)},\dot{\overline{H}}^{(1)}) = 0.162, \quad d_{hnh}((\overline{H}^{(2)})^{(1)},\dot{\overline{H}}^{(1)}) = 0.130,$$
$$d_{hnh}((\dot{\overline{H}}^{(3)})^{(1)},\dot{\overline{H}}^{(1)}) = 0.093$$

Since the consensus level $\tau^* = 0.1$, then we can see that both $d_{hnh}((\overline{H}^{(1)})^{(1)},\dot{\overline{H}}^{(1)})$ and $d_{hnh}((\overline{H}^{(2)})^{(1)},\dot{\overline{H}}^{(1)})$ are bigger than 0.1. Thus, we need to improve these individual prefect multiplicative consistent HFPRs.

Step 4. Let $\xi = 0.7$, then we can construct respectively the new individual HFPRs $(\dot{\overline{H}}^{(k)})^{(2)}(k = 1,2,3)$ as:

$$(\dot{\overline{H}}^{(1)})^{(2)} = \begin{pmatrix} \{0.5\} & \{0.229,0.35,0.45\} & \{0.206,0.401,0.556\} & \{0.248,0.461,0.583\} \\ \{0.553,0.654,0.78\} & \{0.5\} & \{0.498,0.679\} & \{0.429,0.725\} \\ \{0.445,0.612,0.794\} & \{0.332,0.512\} & \{0.5\} & \{0.463,0.584\} \\ \{0.425,0.552,0.762\} & \{0.296,0.574\} & \{0.427,0.551\} & \{0.5\} \end{pmatrix}$$

$$(\dot{\overline{H}}^{(2)})^{(2)} = \begin{pmatrix} \{0.5\} & \{0.259,0.38,0.48\} & \{0.213,0.463,0.588\} & \{0.296,0.533,0.634\} \\ \{0.522,0.623,0.747\} & \{0.5\} & \{0.468,0.707\} & \{0.459,0.796\} \\ \{0.412,0.55,0.788\} & \{0.303,0.542\} & \{0.5\} & \{0.523,0.644\} \\ \{0.374,0.481,0.715\} & \{0.222,0.544\} & \{0.366,0.491\} & \{0.5\} \end{pmatrix}$$

$$(\dot{\overline{H}}^{(3)})^{(2)} = \begin{pmatrix} \{0.5\} & \{0.229,0.38,0.48\} & \{0.206,0.463,0.588\} & \{0.248,0.504,0.624\} \\ \{0.522,0.623,0.776\} & \{0.5\} & \{0.528,0.739\} & \{0.429,0.773\} \\ \{0.412,0.55,0.794\} & \{0.270,0.482\} & \{0.5\} & \{0.463,0.613\} \\ \{0.384,0.51,0.762\} & \{0.246,0.574\} & \{0.397,0.551\} & \{0.5\} \end{pmatrix}$$

Let $p = 2$, then go back to Step 2. The individual HFPRs $(\dot{\overline{H}}^{(k)})^{(2)}$ $(k = 1,2,3)$ can be fused into a collective HFPR $(\dot{\overline{H}})^{(2)}$ by the AHFWA operator:

$$\dot{\overline{H}}^{(2)} = \begin{pmatrix} \{0.5\} & \{0.241, 0.371, 0.471\} & \{0.209, 0.445, 0.579\} & \{0.268, 0.504, 0.616\} \\ \{0.532, 0.633, 0.766\} & \{0.5\} & \{0.496, 0.709\} & \{0.442, 0.77\} \\ \{0.422, 0.57, 0.792\} & \{0.302, 0.516\} & \{0.5\} & \{0.488, 0.618\} \\ \{0.393, 0.512, 0.744\} & \{0.252, 0.562\} & \{0.394, 0.528\} & \{0.5\} \end{pmatrix}$$

Then, the hesitant normalized Hamming distance between each individual prefect multiplicative consistent HFPR $(\dot{\overline{H}}^{(k)})^{(2)}$ and the collective HFPR $\dot{\overline{H}}^{(2)}$ can be calculated as:

$$d_{hnh}((\dot{\overline{H}}^{(1)})^{(2)}, \dot{\overline{H}}^{(2)}) = 0.056, \quad d_{hnh}((\dot{\overline{H}}^{(2)})^{(2)}, \dot{\overline{H}}^{(2)}) = 0.039,$$
$$d_{hnh}((\dot{\overline{H}}^{(3)})^{(2)}, \dot{\overline{H}}^{(2)}) = 0.023$$

Now all $d_{hnh}((\dot{\overline{H}}^{(k)})^{(2)}, \dot{\overline{H}}^{(2)}) < 0.1$ $(k = 1, 2, 3)$, then go to Step 5.

Step 5. Let $\dot{H} = \dot{\overline{H}}^{(2)}$, and employ the AHFA operator to fuse all the hesitant preference values $h_{ij}(j = 1, 2, \ldots, n)$ corresponding to the object C_i into the overall hesitant preference value h_i:

$$h_1 = \{0.315, 0.458, 0.545\}, \quad h_2 = \{0.494, 0.537, 0.703\}$$
$$h_3 = \{0.433, 0.519, 0.628\}, \quad h_4 = \{0.391, 0.503, 0.597\}$$

Step 6. Using the score function Eq. (1.17), we can get $s(h_1) = 0.439$, $s(h_2) = 0.578$, $s(h_3) = 0.527$, and $s(h_4) = 0.495$. As $s(h_2) > s(h_3) > s(h_4) > s(h_1)$, we can draw a conclusion that $C_2 \succ C_3 \succ C_4 \succ C_1$, which denotes that the e-learning material is also the most important factor influencing the affectivity of e-learning.

From the above example, it can be seen that although the third decision maker slightly changes his/her preference, the output of our procedure also keeps the same. Thus, our algorithm is robust and practicable.

6.3.2 Interactive Consensus Reaching Process

From Example 6.6, we can see Algorithm 6.4 is automatic and easy to implement. However, since it does not interact with the experts when changing the preference values in Step 4, this method sometimes may reach a false or unrealistic consensus degree due to that the experts could not agree with the changes proposed by the system. Indeed, the most directive method for repairing the consensus is returning the preference relations to the experts to reconsider constructing new preference relations according to their new comparison until they have acceptable consensus. Therefore, it is reasonable for us to investigate some interactive consensus reaching processes for HFPRs.

Although Algorithm 6.4 does not focus on feedback mechanism, it provides a good way to reach group consensus. Inspired by this, we can develop an interactive consensus reaching procedure.

Algorithm 6.5

Step 1. See Algorithm 6.4.

Step 2. See Algorithm 6.4.

Step 3. See Algorithm 6.4.

Step 4. Return the inconsistent multiplicative HFPRs $(H^{(k)})^{(p)}$ to the experts to reconsider constructing a new HFPR $(H^{(k)})^{(p+1)}$ according to their new judgments. In this case, they can refer to our new constructed individual HFPRs $(\overline{H}^{(k)})^{(p)} = ((\bar{h}_{ij}^{(k)})_{n \times n})^{(p)}$ $(k = 1, 2, \ldots, m)$. Let $p = p + 1$, then go to Step 2.

Step 5. See Algorithm 6.4.

Step 6. See Algorithm 6.4.

Here we do not want to illustrate this algorithm by numerical examples. The schematic diagram of this algorithm is provided in Fig. 6.1.

Comparing the automatic consensus reaching process (Algorithm 6.4) and interactive consensus reaching process (Algorithm 6.5), we can find both of the two algorithms have advantages and disadvantages:

(1) The automatic Algorithm 6.4 is easy to implement and can save a lot of time. It also can give a quick response to urgent situations. In some settings, if the decision makers do not want to interact with the experts, or if they cannot find the initial experts to re-evaluate and alter their preferences, or if consensus must be urgently obtained, the automatic Algorithm 6.4 is a good choice to derive a consensus solution for group decision making, which involves most initial information. But, sometimes, the results may not reflect the realistic opinions of

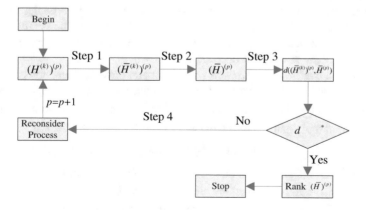

Fig. 6.1 Schematic diagram of interactive procedure

the experts due to that the experts may not agree with the changes proposed by the automatic system.

(2) Interacting with the decision makers frequently during the consensus reaching process is very reliable and accurate. However, the feedback mechanism wastes a lot of time and sometimes this ideal consensus is just a utopian consensus which is difficult to achieve.

6.4   Group Decision Making with Interval-Valued Hesitant Fuzzy Preference Relations

6.4.1   Interval-Valued Hesitant Fuzzy Preference Relation

Xu (2004) introduced the concept of interval-valued fuzzy preference relation to express the uncertainty and vagueness.

Definition 6.8 (*Xu* 2004). Let $A = \{A_1, A_2, \ldots, A_n\}$ be a discrete set of alternatives. An interval-valued fuzzy preference relation (IVFPR) $\tilde{R}$ on the set A is defined as $\tilde{R} = (\tilde{r}_{ij})_{n \times n}$, which satisfies

$$\tilde{r}_{ij} = [\tilde{r}_{ij}^L, \tilde{r}_{ij}^U], \ \tilde{r}_{ij}^U \geq \tilde{r}_{ij}^L \geq 0, \ \tilde{r}_{ij}^L + \tilde{r}_{ji}^U = \tilde{r}_{ij}^U + \tilde{r}_{ji}^L = 1, \ \tilde{r}_{ii}^L = \tilde{r}_{ii}^U = 0.5$$

$$\text{for all } i, j = 1, 2, \ldots, n \tag{6.46}$$

where $\tilde{r}_{ij}$ shows the interval-valued preference degree of the alternative A_i over A_j, and $\tilde{r}_{ij}^L$ and $\tilde{r}_{ij}^U$ are the lower and upper bounds of $\tilde{r}_{ij}$, respectively.

The IVFPRs can be used in group decision making by aggregating the individual IVFPRs into collective one. However, this may lead to the loss of information. As the IVFPRs are unable to directly incorporate the different opinions of different decision makers, it is adequate to introduce the interval-valued HFPR (IVHFPR), whose elements are characterized by several interval values, to depict the group decision making problems under hesitant fuzzy environments. The IVHFPR avoids performing information aggregation and directly reflects the different preference information between different decision makers.

Consider a decision organization with m experts provides some interval-valued fuzzy preference values to describe the degrees that the alternative A_i is superior to A_j, which are denoted as $\tilde{h}_{ij}^1, \tilde{h}_{ij}^2, \tilde{h}_{ij}^3, \ldots, \tilde{h}_{ij}^m$. Then the preference information $\tilde{h}_{ij}$, i.e., A_i is preferred to A_j, can be considered as an IVHFE $\tilde{h}_{ij} = \{\tilde{h}_{ij}^k, k = 1, 2, \ldots, m\}$. All $\tilde{h}_{ij}(i, j = 1, 2, \ldots, n)$ constitute an IVHFPR, which is defined as follows:

Definition 6.9 (*Chen et al.* 2013). Let $A = \{A_1, A_2, \ldots, A_n\}$ be a fixed set. An IVHFPR on A is denoted by a matrix $\tilde{H} = (\tilde{h}_{ij})_{n \times n}$, where $\tilde{h}_{ij} = \{\tilde{h}_{ij}^k, k = 1, 2, \ldots, l_{\tilde{r}_{ij}}\}$ is an IVHFE, indicating all possible degrees to which A_i is preferred to

A_j with $l_{\tilde{r}_{ij}}$ representing the number of intervals in an IVHFE. Moreover, $\tilde{h}_{ij}$ should satisfy

$$\inf \tilde{h}_{ij}^{\sigma(t)} + \sup \tilde{h}_{ji}^{\sigma(l_{\tilde{r}_{ji}} - t + 1)} = \sup \tilde{h}_{ij}^{\sigma(t)} + \inf \tilde{h}_{ji}^{\sigma(l_{\tilde{r}_{ji}} - t + 1)} = 1$$
$$\tilde{h}_{ii} = [0.5, 0.5], \; i, j = 1, 2, \ldots, n$$
(6.47)

where the elements in $\tilde{h}_{ij}$ are arranged in ascending order, and $\tilde{h}_{ij}^{\sigma(t)}$ is the tth smallest value in $\tilde{h}_{ij}$. $\inf \tilde{h}_{ij}^{\sigma(t)}$ and $\sup \tilde{h}_{ij}^{\sigma(t)}$ denote the lower and upper bounds of $\tilde{h}_{ij}^{\sigma(t)}$, respectively.

We use an example originated from a practical decision making problem to illustrate how to construct the IVHFPRs,

Example 6.7 (Chen et al. 2013). Supply Chain Management (SCM) has received considerable attention in industry. Establishing an effective SCM system can reduce supply chain risk, maximize revenue, optimize business processes, etc. Therefore, how to determine suitable suppliers in the supply chain has become a crucial issue. Suppose that a high technique company which manufactures electronic products intends to select the most appropriate supplier of USB connectors. To evaluate three alternative suppliers A_1, A_2 and A_3, a committee composed of three decision makers e_1, e_2 and e_3 is formed. These decision makers provide their evaluation information from three aspects, i.e., finance, performance and technique. The decision makers provide their pairwise preference degrees over the alternatives by interval-valued fuzzy values, which are shown as:

$e_1: \tilde{h}_{11}^1 = \tilde{h}_{22}^1 = \tilde{h}_{33}^1 = [0.5, 0.5], \; \tilde{h}_{12}^1 = [0.4, 0.7], \; \tilde{h}_{21}^1 = [0.3, 0.6], \; \tilde{h}_{13}^1 = [0.5, 0.6],$
$\quad \tilde{h}_{31}^1 = [0.4, 0.5], \; \tilde{h}_{23}^1 = [0.4, 0.5], \; \tilde{h}_{32}^1 = [0.5, 0.6];$
$e_2: \tilde{h}_{11}^2 = \tilde{h}_{22}^2 = \tilde{h}_{33}^2 = [0.5, 0.5], \; \tilde{h}_{12}^2 = [0.2, 0.3], \; \tilde{h}_{21}^2 = [0.7, 0.8], \; \tilde{h}_{13}^2 = [0.3, 0.4],$
$\quad \tilde{h}_{31}^2 = [0.6, 0.7], \; \tilde{h}_{23}^2 = [0.5, 0.8], \; \tilde{h}_{32}^2 = [0.2, 0.5];$
$e_3: \tilde{h}_{11}^3 = \tilde{h}_{22}^3 = \tilde{h}_{33}^3 = [0.5, 0.5], \; \tilde{h}_{12}^3 = [0.4, 0.5], \; \tilde{h}_{21}^3 = [0.5, 0.6], \; \tilde{h}_{13}^3 = [0.6, 0.7],$
$\quad \tilde{h}_{31}^3 = [0.3, 0.4], \; \tilde{h}_{23}^3 = [0.4, 0.6], \; \tilde{h}_{32}^3 = [0.4, 0.6];$

Using the data $\tilde{h}_{ij}^k \left(\tilde{h}_{ij}^k = \left[\left(\tilde{h}_{ij}^k \right)^L, \left(\tilde{h}_{ij}^k \right)^U \right] \right)$, $i, j, k = 1, 2, 3$, the IVFPRs $\tilde{H}^k = (\tilde{h}_{ij}^k)_{3 \times 3}$ are got, which satisfy the complementary properties defined in Eq. (6.46). Moreover, from $\tilde{H}^1$, $\tilde{H}^2$ and $\tilde{H}^3$, we construct the IVHFPR $\tilde{H} = (\tilde{h}_{ij})_{3 \times 3}$:

$$\tilde{H} = \begin{pmatrix} \{[0.5, 0.5]\} & \{[0.2, 0.3], [0.4, 0.5], [0.4, 0.7]\} & \{[0.3, 0.4], [0.5, 0.6], [0.6, 0.7]\} \\ \{[0.3, 0.6], [0.5, 0.6], [0.7, 0.8]\} & \{[0.5, 0.5]\} & \{[0.4, 0.5], [0.4, 0.6], [0.5, 0.8]\} \\ \{[0.3, 0.4], [0.4, 0.5], [0.6, 0.7]\} & \{[0.2, 0.5], [0.4, 0.6], [0.5, 0.6]\} & \{[0.5, 0.5]\} \end{pmatrix}$$

where $\tilde{h}_{ij}$ denotes the group preference degree that the alternative A_i is superior to the alternative A_j. Here we take $\tilde{h}_{12}$ as an example. Since $\tilde{h}_{12}$ represents all possible interval-valued fuzzy preference degrees to which A_1 is preferred to A_2, its values come from $\tilde{h}_{12}^1(= [0.4, 0.7])$, $\tilde{h}_{12}^2(= [0.2, 0.3])$ and $\tilde{h}_{12}^3(= [0.4, 0.5])$ which is provided by the decision makers e_1, e_2 and e_3, respectively. So, we can denote $\tilde{h}_{12}$ by $\{\tilde{h}_{12}^1, \tilde{h}_{12}^2, \tilde{h}_{12}^3\}$ and consider it as an IVHFE. Similarly, we can denote the symmetric element of $\tilde{h}_{12}$, i.e., $\tilde{h}_{21}$, by $\{\tilde{h}_{21}^1, \tilde{h}_{21}^2, \tilde{h}_{21}^3\}$. Other symmetric elements $\tilde{h}_{ij}$ and $\tilde{h}_{ji}$ in $\tilde{H}$ are got in an analogous way and they satisfy the complementary properties defined in Eq. (6.47). Also, when $i = j$, $\tilde{h}_{ii}$ represents the preference degree to which A_i is preferred to itself; that is, it is equally preferred, so $\tilde{h}_{ii} = \{[0.5, 0.5]\}(i = 1, 2, 3)$. Through the above procedure, we can construct the IVHFPR $\tilde{H}$.

It needs to be mentioned that if accounting for the weight vector $\lambda = (\lambda_1, \lambda_2, \lambda_3)^T$ of the decision makers e_1, e_2 and e_3 with the conditions $\lambda_k > 0$ and $\sum_{k=1}^3 \lambda_k = 1$, we can still construct the IVHFPR by replacing the data of $\tilde{h}_{ij}^k(k = 1, 2, 3)$ given in the previous matrix $\tilde{H}$ by $\lambda_k \tilde{h}_{ij}^k \left(\lambda_k \tilde{h}_{ij}^k = \left[\lambda_k \left(\tilde{h}_{ij}^k \right)^L, \lambda_k \left(\tilde{h}_{ij}^k \right)^U \right] \right)$, $k = 1, 2, 3$.

6.4.2 Operators to Aggregate Interval-Valued Hesitant Fuzzy Preference Information

In the following, we introduce a series of specific aggregation operators for IVHFEs.

Definition 6.10 (*Chen et al.* 2013). Let $\tilde{h}_j(j = 1, 2, \ldots, n)$ be a collection of IVHFEs, $\omega = (\omega_1, \omega_2 \cdots \omega_n)^T$ be the weight vector of $\tilde{h}_j(j = 1, 2, \ldots, n)$ with $\omega_j \in [0, 1]$, $\sum_{j=1}^n \omega_j = 1$ and $\lambda > 0$, then

(1) An interval-valued hesitant fuzzy weighted averaging (IVHFWA) operator is a mapping IVHFWA: $\tilde{H}^n \to \tilde{H}$, where

$$\text{IVHFWA}(\tilde{h}_1, \tilde{h}_2, \ldots, \tilde{h}_n) = \bigoplus_{j=1}^n (\omega_j \tilde{h}_j)$$
$$= \left\{ \left[1 - \prod_{j=1}^n (1 - \tilde{\gamma}_j^L)^{\omega_j}, 1 - \prod_{j=1}^n (1 - \tilde{\gamma}_j^U)^{\omega_j} \right] \Big| \tilde{\gamma}_1 \in \tilde{h}_1, \tilde{\gamma}_2 \in \tilde{h}_2, \ldots, \tilde{\gamma}_n \in \tilde{h}_n \right\}$$

$$(6.48)$$

(2) An interval-valued hesitant fuzzy weighted geometric (IVHFWG) operator is a mapping IVHFWG: $\tilde{H}^n \to \tilde{H}$, where

$$\text{IVHFWG}\left(\tilde{h}_1, \tilde{h}_2, \ldots, \tilde{h}_n\right) = \overset{n}{\underset{j=1}{\otimes}}\, \tilde{h}_j^{\omega_j}$$

$$= \left\{ \left[\prod_{j=1}^{n} (\tilde{\gamma}_j^L)^{\omega_j}, \prod_{j=1}^{n} (\tilde{\gamma}_j^U)^{\omega_j} \right] \Big| \tilde{\gamma}_1 \in \tilde{h}_1, \tilde{\gamma}_2 \in \tilde{h}_2, \ldots, \tilde{\gamma}_n \in \tilde{h}_n \right\} \tag{6.49}$$

(3) A generalized interval-valued hesitant fuzzy weighted averaging (GIVHFWA) operator is a mapping GIVHFWA: $\tilde{H}^n \to \tilde{H}$, where

$$\text{GIVHFWA}_\lambda\left(\tilde{h}_1, \tilde{h}_2, \ldots, \tilde{h}_n\right) = \left(\overset{n}{\underset{j=1}{\oplus}} \left(\omega_j \tilde{h}_j^\lambda \right) \right)^{1/\lambda}$$

$$= \left\{ \left[\left(1 - \prod_{j=1}^{n} (1 - (\tilde{\gamma}_j^L)^\lambda)^{\omega_j} \right)^{1/\lambda}, \left(1 - \prod_{j=1}^{n} (1 - (\tilde{\gamma}_j^U)^\lambda)^{\omega_j} \right)^{1/\lambda} \right] \Big| \tilde{\gamma}_1 \in \tilde{h}_1, \tilde{\gamma}_2 \in \tilde{h}_2, \ldots, \tilde{\gamma}_n \in \tilde{h}_n \right\} \tag{6.50}$$

(4) A generalized interval-valued hesitant fuzzy weighted geometric (GIVHFWG) operator is a mapping $\tilde{H}^n \to \tilde{H}$, where

$$\text{GIVHFWG}_\lambda\left(\tilde{h}_1, \tilde{h}_2, \ldots, \tilde{h}_n\right) = \frac{1}{\lambda}\left(\overset{n}{\underset{j=1}{\otimes}} (\lambda \tilde{h}_j)^{\omega_j} \right)$$

$$= \left\{ \left[1 - \left(1 - \prod_{j=1}^{n} \left(1 - (1 - \tilde{\gamma}_j^L)^\lambda \right)^{\omega_j} \right)^{1/\lambda}, 1 - \left(1 - \prod_{j=1}^{n} \left(1 - (1 - \tilde{\gamma}_j^U)^\lambda \right)^{\omega_j} \right)^{1/\lambda} \right] \Big| \tilde{\gamma}_1 \in \tilde{h}_1, \tilde{\gamma}_2 \in \tilde{h}_2, \ldots, \tilde{\gamma}_n \in \tilde{h}_n \right\} \tag{6.51}$$

If $\lambda = 1$, then the GIVHFWA and GIVHFWG operators reduce to the IVHFWA and IVHFWG operators, respectively. In particular, if $\omega = (1/n, 1/n, \cdots, 1/n)^T$, the IVHFWA and IVHFWG operators respectively become the interval-valued hesitant fuzzy averaging (IVHFA) operator:

$$\text{IVHFA}\left(\tilde{h}_1, \tilde{h}_2, \ldots, \tilde{h}_n\right) = \overset{n}{\underset{j=1}{\oplus}} \left(\frac{1}{n} \tilde{h}_j \right)$$

$$= \left\{ \left[1 - \prod_{j=1}^{n} (1 - \tilde{\gamma}_j^L)^{1/n}, 1 - \prod_{j=1}^{n} (1 - \tilde{\gamma}_j^U)^{1/n} \right] \Big| \tilde{\gamma}_1 \in \tilde{h}_1, \tilde{\gamma}_2 \in \tilde{h}_2, \ldots, \tilde{\gamma}_n \in \tilde{h}_n \right\} \tag{6.52}$$

and the interval-valued hesitant fuzzy geometric (IVHFG) operator:

$$\text{IVHFG}\left(\tilde{h}_1, \tilde{h}_2, \ldots, \tilde{h}_n\right) = \overset{n}{\underset{j=1}{\otimes}}\, \tilde{h}_j^{1/n}$$

$$= \left\{ \left[\prod_{j=1}^{n} (\tilde{\gamma}_j^L)^{1/n}, \prod_{j=1}^{n} (\tilde{\gamma}_j^U)^{1/n} \right] \Big| \tilde{\gamma}_1 \in \tilde{h}_1, \tilde{\gamma}_2 \in \tilde{h}_2, \ldots, \tilde{\gamma}_n \in \tilde{h}_n \right\} \tag{6.53}$$

Example 6.8 (Chen et al. 2013). Let $\tilde{h}_1 = \{[0.1, 0.3], [0.4, 0.5]\}$ and $\tilde{h}_2 = \{[0.3, 0.4], [0.4, 0.6], [0.5, 0.7]\}$ be two IVHFEs, and $\omega = (0.2, 0.8)^T$ be the weight vector of them. Using Definition 6.10, we have

$$\text{GIVHFWA}_1\left(\tilde{h}_1, \tilde{h}_2\right) = \text{IVHFWA}\left(\tilde{h}_1, \tilde{h}_2\right)$$

$$= \overset{2}{\underset{j=1}{\oplus}}\left(\omega_j \tilde{h}_j\right) = \left\{\left[1 - \prod_{j=1}^{2}(1 - \tilde{\gamma}_j^L)^{\omega_j}, 1 - \prod_{j=1}^{2}(1 - \tilde{\gamma}_j^U)^{\omega_j}\right] \middle| \tilde{\gamma}_1 \in \tilde{h}_1, \tilde{\gamma}_2 \in \tilde{h}_2\right\}$$

$$= \{[0.2639, 0.3812], [0.3493, 0.5526], [0.4376, 0.6446], [0.3213, 0.4215],$$
$$[0.4, 0.5817], [0.4814, 0.6677]\}$$

$$\text{GIVHFWG}_1\left(\tilde{h}_1, \tilde{h}_2\right) = \text{IVHFWG}\left(\tilde{h}_1, \tilde{h}_2\right)$$

$$= \overset{2}{\underset{j=1}{\otimes}} \tilde{h}_j^{\omega_j} = \left\{\left[\prod_{j=1}^{2}(\tilde{\gamma}_j^L)^{\omega_j}, \prod_{j=1}^{2}(\tilde{\gamma}_j^U)^{\omega_j}\right] \middle| \tilde{\gamma}_1 \in \tilde{h}_1, \tilde{\gamma}_2 \in \tilde{h}_2\right\}$$

$$= \{[0.2408, 0.3776], [0.3031, 0.5223], [0.3624, 0.5909], [0.3178, 0.4183],$$
$$[0.4, 0.5785], [0.4782, 0.6544]\}$$

$$\text{GIVHFWA}_4\left(\tilde{h}_1, \tilde{h}_2\right) = \left(\overset{2}{\underset{j=1}{\oplus}}\left(\omega_j \tilde{h}_j^4\right)\right)^{1/4}$$

$$= \left\{\left[\left(1 - \prod_{j=1}^{2}(1 - (\tilde{\gamma}_j^L)^4)^{\omega_j}\right)^{1/4}, \left(1 - \prod_{j=1}^{2}(1 - (\tilde{\gamma}_j^U)^4)^{\omega_j}\right)^{1/4}\right] \middle| \tilde{\gamma}_1 \in \tilde{h}_1, \tilde{\gamma}_2 \in \tilde{h}_2\right\}$$

$$= \{[0.2840, 0.3857], [0.3786, 0.5713], [0.4737, 0.6675], [0.3284, 0.4265],$$
$$[0.4, 0.5843], [0.4848, 0.6749]\}$$

$$\text{GIVHFWG}_4\left(\tilde{h}_1, \tilde{h}_2\right) = \frac{1}{4}\left(\overset{2}{\underset{j=1}{\otimes}}(4\tilde{h}_j)^{\omega_j}\right)$$

$$= \left\{\left[1 - \left(1 - \prod_{j=1}^{2}\left(1 - (1 - \tilde{\gamma}_j^L)^4\right)^{\omega_j}\right)^{\frac{1}{4}}, 1 - \left(1 - \prod_{j=1}^{2}\left(1 - (1 - \tilde{\gamma}_j^U)^4\right)^{\omega_j}\right)^{\frac{1}{4}}\right] \middle| \tilde{\gamma}_1 \in \tilde{h}_1, \tilde{\gamma}_2 \in \tilde{h}_2\right\}$$

$$= \{[0.23, 0.3747], [0.2744, 0.4804], [0.3053, 0.5060], [0.3158, 0.4157],$$
$$[0.4, 0.5735], [0.4744, 0.6276]\}$$

From Example 6.8, we can find that

$$\text{GIVHFWG}_4\left(\tilde{h}_1, \tilde{h}_2\right) \le \text{IVHFWG}\left(\tilde{h}_1, \tilde{h}_2\right) \le \text{IVHFWA}\left(\tilde{h}_1, \tilde{h}_2\right) \le \text{GIVHFWA}_4\left(\tilde{h}_1, \tilde{h}_2\right)$$

In fact, the above fact can be revealed by Theorem 6.8.

Theorem 6.8 (Chen et al. 2013). *Assume that $\tilde{h}_j(j = 1, 2, \ldots, n)$ are a collection of IVHFEs, $\omega = (\omega_1, \omega_2 \cdots \omega_n)^T$ is the weight vector of them with $\omega_j \in [0, 1]$ and $\sum_{j=1}^{n} \omega_j = 1$, then*

$$\text{IVHFWG}\left(\tilde{h}_1, \tilde{h}_2, \ldots, \tilde{h}_n\right) \le \text{IVHFWA}\left(\tilde{h}_1, \tilde{h}_2, \ldots, \tilde{h}_n\right) \qquad (6.54)$$

Proof For any $\tilde{\gamma}_1 \in \tilde{h}_1$, $\tilde{\gamma}_2 \in \tilde{h}_2, \ldots, \tilde{\gamma}_n \in \tilde{h}_n$, based on Lemma 3.1, we have

$$\prod_{j=1}^{n} (\tilde{\gamma}_j^L)^{\omega_j} \le \sum_{j=1}^{n} \omega_j(\tilde{\gamma}_j^L) = 1 - \sum_{j=1}^{n} \omega_j(1 - \tilde{\gamma}_j^L) \le 1 - \prod_{j=1}^{n} (1 - \tilde{\gamma}_j^L)^{\omega_j}$$

$$(6.55)$$

Similarly, we have

$$\prod_{j=1}^{n} (\tilde{\gamma}_j^U)^{\omega_j} \le \sum_{j=1}^{n} \omega_j(\tilde{\gamma}_j^U) = 1 - \sum_{j=1}^{n} \omega_j(1 - \tilde{\gamma}_j^U) \le 1 - \prod_{j=1}^{n} (1 - \tilde{\gamma}_j^U)^{\omega_j}$$

$$(6.56)$$

Given Eq. (1.23), Eqs. (6.55) and (6.56) imply $\overset{n}{\underset{j=1}{\otimes}} \left(\tilde{h}_j^{\omega_j} \right) \le \overset{n}{\underset{j=1}{\oplus}} (\omega_j \tilde{h}_j)$, which completes the proof of Theorem 6.8.   □

Theorem 6.9 (Chen et al. 2013). *Let $\tilde{h}_j(j = 1, 2, \ldots, n)$ be a collection of IVHFEs, whose weight vector is $\omega = (\omega_1, \omega_2 \cdots \omega_n)^T$ with $\omega_j \in [0, 1]$ and $\sum_{j=1}^{n} \omega_j = 1$, $\lambda > 0$, then*

$$\text{IVHFWG}(\tilde{h}_1, \tilde{h}_2, \ldots, \tilde{h}_n) \le \text{GIVHFWA}_\lambda(\tilde{h}_1, \tilde{h}_2, \ldots, \tilde{h}_n) \quad (6.57)$$

Proof For any $\tilde{\gamma}_1 \in \tilde{h}_1, \tilde{\gamma}_2 \in \tilde{h}_2, \ldots, \tilde{\gamma}_n \in \tilde{h}_n$, by Lemma 3.1, we have

$$\prod_{j=1}^{n} (\tilde{\gamma}_j^L)^{\omega_j} = \left(\prod_{j=1}^{n} ((\tilde{\gamma}_j^L)^\lambda)^{\omega_j} \right)^{1/\lambda} \le \left(\sum_{j=1}^{n} \omega_j(\tilde{\gamma}_j^L)^\lambda \right)^{1/\lambda}$$
$$= \left(1 - \sum_{j=1}^{n} \omega_j(1 - (\tilde{\gamma}_j^L)^\lambda) \right)^{1/\lambda} \le \left(1 - \prod_{j=1}^{n} (1 - (\tilde{\gamma}_j^L)^\lambda)^{\omega_j} \right)^{1/\lambda}$$

$$(6.58)$$

Likewise, we have

$$\prod_{j=1}^{n} (\tilde{\gamma}_j^U)^{\omega_j} = \left(\prod_{j=1}^{n} ((\tilde{\gamma}_j^U)^\lambda)^{\omega_j} \right)^{1/\lambda} \le \left(\sum_{j=1}^{n} \omega_j(\tilde{\gamma}_j^U)^\lambda \right)^{1/\lambda}$$
$$= \left(1 - \sum_{j=1}^{n} \omega_j(1 - (\tilde{\gamma}_j^U)^\lambda) \right)^{1/\lambda} \le \left(1 - \prod_{j=1}^{n} (1 - (\tilde{\gamma}_j^U)^\lambda)^{\omega_j} \right)^{1/\lambda}$$

$$(6.59)$$

Combining Eqs. (6.58) and (6.59), it follows $\overset{n}{\underset{j=1}{\otimes}} \left(\tilde{h}_j^{\omega_j} \right) \le \left(\overset{n}{\underset{i=1}{\oplus}} \left(\omega_j \tilde{h}_j^\lambda \right) \right)^{1/\lambda}$. This completes the proof of Theorem 6.9.   □

Theorem 6.10 (Chen et al. 2013). *For a collection of IVHFEs $\tilde{h}_j(j = 1, 2, \ldots, n)$, $\omega = (\omega_1, \omega_2 \ldots \omega_n)^T$ is the weight vector with $\omega_j \in [0, 1]$ and $\sum_{j=1}^{n} \omega_j = 1$, $\lambda > 0$, then*

$$\text{GIVHFWG}_\lambda \left(\tilde{h}_1, \tilde{h}_2, \ldots, \tilde{h}_n \right) \leq \text{IVHFWA} \left(\tilde{h}_1, \tilde{h}_2, \ldots, \tilde{h}_n \right) \tag{6.60}$$

Proof Let $\tilde{\gamma}_1 \in \tilde{h}_1$, $\tilde{\gamma}_2 \in \tilde{h}_2, \ldots$, $\tilde{\gamma}_n \in \tilde{h}_n$, then based on Lemma 3.1, we obtain

$$1 - \left(1 - \prod_{j=1}^{n} \left(1 - (1 - \tilde{\gamma}_j^L)^\lambda \right)^{\omega_j} \right)^{1/\lambda} \leq 1 - \left(1 - \sum_{j=1}^{n} \omega_j \left(1 - (1 - \tilde{\gamma}_j^L)^\lambda \right) \right)^{1/\lambda}$$
$$= 1 - \left(\sum_{j=1}^{n} \omega_j (1 - \tilde{\gamma}_j^L)^\lambda \right)^{1/\lambda} \leq 1 - \left(\prod_{j=1}^{n} \left((1 - \tilde{\gamma}_j^L)^\lambda \right)^{\omega_j} \right)^{1/\lambda} = 1 - \prod_{j=1}^{n} (1 - \tilde{\gamma}_j^L)^{\omega_j} \tag{6.61}$$

Similarly,

$$1 - \left(1 - \prod_{j=1}^{n} \left(1 - (1 - \tilde{\gamma}_j^U)^\lambda \right)^{\omega_j} \right)^{1/\lambda} \leq 1 - \left(1 - \sum_{j=1}^{n} \omega_j \left(1 - (1 - \tilde{\gamma}_j^U)^\lambda \right) \right)^{1/\lambda}$$
$$= 1 - \left(\sum_{j=1}^{n} \omega_j (1 - \tilde{\gamma}_j^U)^\lambda \right)^{1/\lambda} \leq 1 - \left(\prod_{j=1}^{n} \left((1 - \tilde{\gamma}_j^U)^\lambda \right)^{\omega_j} \right)^{1/\lambda} = 1 - \prod_{j=1}^{n} (1 - \tilde{\gamma}_j^U)^{\omega_j} \tag{6.62}$$

We can see Theorem 6.10 still holds with Eq. (1.23). □

Based on Definition 6.10 and motivated by the idea of the OWA operator (Yager 1988), the OWA operators for IVHFEs can be defined.

Definition 6.11 (*Chen et al.* 2013). Let $\tilde{h}_j (j = 1, 2, \ldots, n)$ be a collection of IVHFEs, $\tilde{h}_{\sigma(j)}$ be the jth largest of them, $\omega = (\omega_1, \omega_2 \cdots \omega_n)^T$ be the associated vector such that $\omega_j \in [0, 1]$, $\sum_{j=1}^{n} \omega_j = 1$ and $\lambda > 0$, then

(1) An interval-valued hesitant fuzzy ordered weighted averaging (IVHFOWA) operator is a mapping IVHFOWA: $\tilde{H}^n \to \tilde{H}$, where

$$\text{IVHFOWA} \left(\tilde{h}_1, \tilde{h}_2, \ldots, \tilde{h}_n \right) = \bigoplus_{j=1}^{n} \left(\omega_j \tilde{h}_{\sigma(j)} \right)$$
$$= \left[1 - \prod_{j=1}^{n} (1 - \tilde{\gamma}_{\sigma(j)}^L)^{\omega_j}, 1 - \prod_{j=1}^{n} (1 - \tilde{\gamma}_{\sigma(j)}^U)^{\omega_j} \middle| \tilde{\gamma}_{\sigma(1)} \in \tilde{h}_{\sigma(1)}, \tilde{\gamma}_{\sigma(2)} \in \tilde{h}_{\sigma(2)}, \ldots, \tilde{\gamma}_{\sigma(n)} \in \tilde{h}_{\sigma(n)} \right] \tag{6.63}$$

(2) An interval-valued hesitant fuzzy ordered weighted geometric (IVHFOWG) operator is a mapping IVHFOWG: $\tilde{H}^n \to \tilde{H}$, where

$$\text{IVHFOWG} \left(\tilde{h}_1, \tilde{h}_2, \ldots, \tilde{h}_n \right) = \bigotimes_{j=1}^{n} \tilde{h}_{\sigma(j)}^{\omega_j}$$
$$= \left\{ \left[\prod_{j=1}^{n} (\tilde{\gamma}_{\sigma(j)}^L)^{\omega_j}, \prod_{j=1}^{n} (\tilde{\gamma}_{\sigma(j)}^U)^{\omega_j} \right] \middle| \tilde{\gamma}_{\sigma(1)} \in \tilde{h}_{\sigma(1)}, \tilde{\gamma}_{\sigma(2)} \in \tilde{h}_{\sigma(2)}, \ldots, \tilde{\gamma}_{\sigma(n)} \in \tilde{h}_{\sigma(n)} \right\} \tag{6.64}$$

(3) A generalized interval-valued hesitant fuzzy ordered weighted averaging (GIVHFOWA) operator is a mapping GIVHFOWA: $\tilde{H}^n \to \tilde{H}$, where

$$\mathrm{GIVHFOWA}_\lambda(\tilde{h}_1, \tilde{h}_2, \ldots, \tilde{h}_n) = \left(\overset{n}{\underset{j=1}{\oplus}} \left(\omega_j \tilde{h}_{\sigma(j)}^\lambda \right) \right)^{1/\lambda}$$
$$= \left\{ \left[\left(1 - \prod_{j=1}^n (1 - (\tilde{\gamma}_{\sigma(j)}^L)^\lambda)^{\omega_j} \right)^{1/\lambda}, \left(1 - \prod_{j=1}^n (1 - (\tilde{\gamma}_{\sigma(j)}^U)^\lambda)^{\omega_j} \right)^{1/\lambda} \right] \Big| \tilde{\gamma}_{\sigma(1)} \in \tilde{h}_{\sigma(1)}, \ldots, \tilde{\gamma}_{\sigma(n)} \in \tilde{h}_{\sigma(n)} \right\}$$

$$(6.65)$$

(4) A generalized interval-valued hesitant fuzzy ordered weighted geometric (GIVHFOWG) operator is a mapping GIVHFOWG: $\tilde{H}^n \to \tilde{H}$, where

$$\mathrm{GIVHFOWG}_\lambda(\tilde{h}_1, \tilde{h}_2, \ldots, \tilde{h}_n) = \frac{1}{\lambda} \left(\overset{n}{\underset{j=1}{\otimes}} (\lambda \tilde{h}_{\sigma(j)})^{\omega_j} \right)$$
$$= \left\{ \left[1 - \left(1 - \prod_{j=1}^n \left(1 - (1 - \tilde{\gamma}_{\sigma(j)}^L)^\lambda \right)^{\omega_j} \right)^{1/\lambda}, 1 - \left(1 - \prod_{j=1}^n \left(1 - (1 - \tilde{\gamma}_{\sigma(j)}^U)^\lambda \right)^{\omega_j} \right)^{1/\lambda} \right] \right.$$
$$\left. \Big| \tilde{\gamma}_{\sigma(1)} \in \tilde{h}_{\sigma(1)}, \ldots, \tilde{\gamma}_{\sigma(n)} \in \tilde{h}_{\sigma(n)} \right\}$$

$$(6.66)$$

In the case where $\omega = (1/n, 1/n, \ldots, 1/n)^T$, the IVHFOWA and IVHFOWG operators reduce to the IVHFA and IVHFG operators, respectively. For the case where $\lambda = 1$, the GIVHFOWA and GIVHFOWG operators become the IVHFOWA and IVHFOWG operators.

Example 6.9 (Chen et al. 2013). Let $\tilde{h}_1 = \{[0.1, 0.3], [0.4, 0.5]\}$ and $\tilde{h}_2 = \{[0.3, 0.4], [0.4, 0.6], [0.5, 0.7]\}$ be two IVHFEs, and $\omega = (0.4, 0.6)^T$ be the weight vector of them.

We firs compute the score values of $\tilde{h}_1$ and $\tilde{h}_2$ using Eq. (1.24) and obtain $s(\tilde{h}_1) = [0.25, 0.4]$, $s(\tilde{h}_2) = [0.4, 0.85]$. Since $s(\tilde{h}_2) > s(\tilde{h}_1)$, then we have $\tilde{h}_{\sigma(1)} = \tilde{h}_2 = \{[0.3, 0.4], [0.4, 0.6], [0.5, 0.7]\}$, and $\tilde{h}_{\sigma(2)} = \tilde{h}_1 = \{[0.1, 0.3], [0.4, 0.5]\}$. According to Definition 6.11, we get

$$\mathrm{GIVHFOWA}_1(\tilde{h}_1, \tilde{h}_2) = \mathrm{IVHFOWA}(\tilde{h}_1, \tilde{h}_2) = \overset{2}{\underset{j=1}{\oplus}} (\omega_j \tilde{h}_{\sigma(j)})$$
$$= \left\{ \left[1 - \prod_{j=1}^2 (1 - \tilde{\gamma}_{\sigma(j)}^L)^{\omega_j}, 1 - \prod_{j=1}^2 (1 - \tilde{\gamma}_{\sigma(j)}^U)^{\omega_j} \right] \Big| \tilde{\gamma}_{\sigma(1)} \in \tilde{h}_{\sigma(1)}, \tilde{\gamma}_{\sigma(2)} \in \tilde{h}_{\sigma(2)} \right\}$$
$$= \{[0.1861, 0.3419], [0.3618, 0.4622], [0.2347, 0.4404], [0.4, 0.5427],$$
$$[0.2886, 0.5012], [0.4422, 0.5924]\}.$$

$$\mathrm{GIVHFOWA}_4(\tilde{h}_1, \tilde{h}_2) = \left(\overset{2}{\underset{j=1}{\oplus}} \left(\omega_j \tilde{h}_{\sigma(j)}^4 \right) \right)^{1/4}$$
$$= \left\{ \left[\left(1 - (1 - (\tilde{\gamma}_2^L)^4)^{0.4} (1 - (\tilde{\gamma}_1^L)^4)^{0.6} \right)^{\frac{1}{4}}, \left(1 - (1 - (\tilde{\gamma}_2^U)^4)^{0.4} (1 - (\tilde{\gamma}_1^U)^4)^{0.6} \right)^{\frac{1}{4}} \right] \Big| \tilde{\gamma}_1 \in \tilde{h}_1, \tilde{\gamma}_2 \in \tilde{h}_2 \right\}$$
$$= \{[0.2398, 0.3508], [0.3695, 0.4679], [0.3192, 0.4920], [0.4, 0.5476],$$
$$[0.3998, 0.5738], [0.4487, 0.6095]\}.$$

$$\text{GIVHFOWG}_1\left(\tilde{h}_1, \tilde{h}_2\right) = \text{IVHFOWG}\left(\tilde{h}_1, \tilde{h}_2\right) = \overset{2}{\underset{j=1}{\otimes}} \tilde{h}_{\sigma(j)}^{\omega_j}$$

$$= \{[0.1552, 0.3366], [0.3565, 0.4573], [0.1741, 0.3959], [0.4, 0.5378],$$
$$[0.1904, 0.4210], [0.4373, 0.5720]\}.$$

$$\text{GIVHFOWG}_4\left(\tilde{h}_1, \tilde{h}_2\right) = \frac{1}{4}\left(\overset{2}{\underset{j=1}{\otimes}}(4\tilde{h}_{\sigma(j)})^{\omega_j}\right)$$

$$= \{[0.1477, 0.3332], [0.3527, 0.4524], [0.1585, 0.3669], [0.4, 0.5321],$$
$$[0.1649, 0.3729], [0.4330, 0.5497]\}.$$

It is observed that the IVHFWA, IVHFWG, GIVHFWA and GIVHFWG operators weight the IVHFEs, and the IVHFOWA, IVHFOWG, GIVHFOWA and GIVHFOWG operators only weight the ordered position of each given IVHFE. But the hybrid aggregation operators developed here for IVHFEs weight both all the given IVHFEs and their ordered positions.

Definition 6.12 (*Chen et al.* 2013). For a collection of IVHFEs $\tilde{h}_j(j = 1, 2, \ldots, n)$, $\lambda = (\lambda_1, \lambda_2 \cdots \lambda_n)^T$ is the weight vector of them with $\lambda_j \in [0, 1]$ and $\sum_{j=1}^n \lambda_j = 1$, n is a balancing factor. Aggregation operators are the mapping $\tilde{H}^n \to \tilde{H}$ with an associated vector $\omega = (\omega_1, \omega_2 \cdots \omega_n)^T$ such that $\omega_j \in [0, 1]$ and $\sum_{j=1}^n \omega_j = 1$:

(1) The interval-valued hesitant fuzzy hybrid averaging (IVHFHA) operator:

$$\text{IVHFHA}\left(\tilde{h}_1, \tilde{h}_2, \ldots, \tilde{h}_n\right) = \overset{n}{\underset{j=1}{\oplus}}\left(\omega_j \dot{\tilde{h}}_{\sigma(j)}\right)$$

$$= \left\{\left[1 - \prod_{j=1}^n (1 - \dot{\tilde{\gamma}}_{\sigma(j)}^L)^{\omega_j}, 1 - \prod_{j=1}^n (1 - \dot{\tilde{\gamma}}_{\sigma(j)}^U)^{\omega_j}\right] \Big| \dot{\tilde{\gamma}}_{\sigma(1)} \in \dot{\tilde{h}}_{\sigma(1)}, \ldots, \dot{\tilde{\gamma}}_{\sigma(n)} \in \dot{\tilde{h}}_{\sigma(n)}\right\}$$

$$(6.67)$$

where $\dot{\tilde{h}}_{\sigma(j)}$ is the jth largest of $\dot{\tilde{h}} = n\lambda_k \tilde{h}_k (k = 1, 2, \ldots, n)$.

(2) The interval-valued hesitant fuzzy hybrid geometric (IVHFHG) operator:

$$\text{IVHFHG}\left(\tilde{h}_1, \tilde{h}_2, \ldots, \tilde{h}_n\right) = \overset{n}{\underset{j=1}{\otimes}} \ddot{\tilde{h}}_{\sigma(j)}^{\omega_j}$$

$$= \left\{\left[\prod_{j=1}^n (\ddot{\tilde{\gamma}}_{\sigma(j)}^L)^{\omega_j}, \prod_{j=1}^n (\ddot{\tilde{\gamma}}_{\sigma(j)}^U)^{\omega_j}\right] \Big| \ddot{\tilde{\gamma}}_{\sigma(1)} \in \ddot{\tilde{h}}_{\sigma(1)}, \ldots, \ddot{\tilde{\gamma}}_{\sigma(n)} \in \ddot{\tilde{h}}_{\sigma(n)}\right\}$$

$$(6.68)$$

where $\ddot{\tilde{h}}_{\sigma(j)}$ is the jth largest of $\ddot{\tilde{h}}_k = \tilde{h}_k^{n\lambda_k}(k = 1, 2, \ldots, n)$.

(3) The generalized interval-valued hesitant fuzzy hybrid averaging (GIVHFHA) operator:

$$\text{GIVHFHA}_\lambda(\tilde{h}_1, \tilde{h}_2, \ldots, \tilde{h}_n) = \left(\overset{n}{\underset{j=1}{\oplus}} \left(\omega_j \dot{\tilde{h}}_{\sigma(j)}^\lambda \right) \right)^{1/\lambda}$$

$$= \left\{ \left[\left(1 - \prod_{j=1}^n (1 - (\dot{\tilde{\gamma}}_{\sigma(j)}^L)^\lambda)^{\omega_j} \right)^{1/\lambda}, \left(1 - \prod_{j=1}^n (1 - (\dot{\tilde{\gamma}}_{\sigma(j)}^U)^\lambda)^{\omega_j} \right)^{1/\lambda} \right] \Big| \dot{\tilde{\gamma}}_{\sigma(1)} \in \dot{\tilde{h}}_{\sigma(1)}, \ldots, \dot{\tilde{\gamma}}_{\sigma(n)} \in \dot{\tilde{h}}_{\sigma(n)} \right\}$$

$$(6.69)$$

where $\lambda > 0$, $\dot{\tilde{h}}_{\sigma(j)}$ is the jth largest of $\dot{\tilde{h}} = n\lambda_k \tilde{h}_k (k = 1, 2, \ldots, n)$.

(4) The generalized interval-valued hesitant fuzzy hybrid geometric (GIVHFHG) operator:

$$\text{GIVHFHG}_\lambda(\tilde{h}_1, \tilde{h}_2, \ldots, \tilde{h}_n) = \frac{1}{\lambda} \left(\overset{n}{\underset{j=1}{\otimes}} \left(\lambda \ddot{\tilde{h}}_{\sigma(j)} \right)^{\omega_j} \right)$$

$$= \left\{ \left[1 - \left(1 - \prod_{j=1}^n \left(1 - (1 - \ddot{\tilde{\gamma}}_{\sigma(j)}^L)^\lambda \right)^{\omega_j} \right)^{1/\lambda}, 1 - \left(1 - \prod_{j=1}^n \left(1 - (1 - \ddot{\tilde{\gamma}}_{\sigma(j)}^U)^\lambda \right)^{\omega_j} \right)^{1/\lambda} \right] \right.$$
$$\left. \Big| \ddot{\tilde{\gamma}}_{\sigma(1)} \in \ddot{\tilde{h}}_{\sigma(1)}, \ldots, \ddot{\tilde{\gamma}}_{\sigma(n)} \in \ddot{\tilde{h}}_{\sigma(n)} \right\}.$$

$$(6.70)$$

where $\lambda > 0$, $\ddot{\tilde{h}}_{\sigma(j)}$ is the jth largest of $\ddot{\tilde{h}}_k = \tilde{h}_k^{n\lambda_k} (k = 1, 2, \ldots, n)$.

Example 6.10 (Chen et al. 2013). Let $\tilde{h}_1 = \{[0.1, 0.3], [0.4, 0.5]\}$ and $\tilde{h}_2 = \{[0.3, 0.4], [0.4, 0.6], [0.5, 0.7]\}$ be two IVHFEs. Suppose that their weight vector and position vector are $\lambda = (0.2, 0.8)^T$ and $\omega = (0.4, 0.6)^T$, respectively. With Definition 6.12, we obtain

$$\dot{\tilde{h}}_1 = \left\{ \left[1 - (1 - 0.1)^{2 \times 0.2}, 1 - (1 - 0.3)^{2 \times 0.2} \right], \left[1 - (1 - 0.4)^{2 \times 0.2}, 1 - (1 - 0.5)^{2 \times 0.2} \right] \right\}$$
$$= \{[0.0413, 0.1330], [0.1848, 0.2421]\}$$

$$\dot{\tilde{h}}_2 = \left\{ \left[1 - (1 - 0.3)^{2 \times 0.8}, 1 - (1 - 0.4)^{2 \times 0.8} \right], \left[1 - (1 - 0.4)^{2 \times 0.8}, 1 - (1 - 0.6)^{2 \times 0.8} \right], \right.$$
$$\left. \left[1 - (1 - 0.5)^{2 \times 0.8}, 1 - (1 - 0.7)^{2 \times 0.8} \right] \right\}$$
$$= \{[0.4349, 0.5584], [0.5584, 0.7692], [0.6701, 0.8543]\}$$

Since $s\left(\dot{\tilde{h}}_1\right) = [0.1131, 0.1876]$, $s\left(\dot{\tilde{h}}_2\right) = [0.5545, 0.7273]$, we have $s\left(\dot{\tilde{h}}_2\right) > s\left(\dot{\tilde{h}}_1\right)$. Thus, $\dot{\tilde{h}}_{\sigma(1)} = \dot{\tilde{h}}_2 = \{[0.4349, 0.5584], [0.5584, 0.7692], [0.6701, 0.8543]\}$, $\dot{\tilde{h}}_{\sigma(2)} = \dot{\tilde{h}}_1 = \{[0.0413, 0.1330], [0.1848, 0.2421]\}$.

Similarly, we have

$$\text{GIVHFHA}_1\left(\tilde{h}_1, \tilde{h}_2\right) = \text{IVHFHA}\left(\tilde{h}_1, \tilde{h}_2\right) = \overset{2}{\underset{j=1}{\oplus}}\left(\omega_j \dot{\tilde{h}}_{\sigma(j)}\right)$$

$$= \left\{\left[1 - (1 - \dot{\tilde{\gamma}}_2^L)^{0.4}(1 - \dot{\tilde{\gamma}}_1^L)^{0.6}, 1 - (1 - \dot{\tilde{\gamma}}_2^U)^{0.4}(1 - \dot{\tilde{\gamma}}_1^U)^{0.6}\right]\Big|\dot{\tilde{\gamma}}_2 \in \dot{\tilde{h}}_2, \dot{\tilde{\gamma}}_1 \in \dot{\tilde{h}}_1\right\}$$

$$= \{[0.2240, 0.3381], [0.2959, 0.3894], [0.2969, 0.4894], [0.3621, 0.5290],$$
$$[0.3743, 0.5752], [0.4323, 0.6081]\}.$$

$$\text{GIVHFHA}_4\left(\tilde{h}_1, \tilde{h}_2\right) = \left(\overset{2}{\underset{j=1}{\oplus}}\left(\omega_j \dot{\tilde{h}}_{\sigma(j)}^4\right)\right)^{1/4}$$

$$= \left\{\left[\left(1 - (1 - (\dot{\tilde{\gamma}}_2^L)^4)^{0.4}(1 - (\dot{\tilde{\gamma}}_1^L)^4)^{0.6}\right)^{1/4}, \left(1 - (1 - (\dot{\tilde{\gamma}}_2^U)^4)^{0.4}(1 - (\dot{\tilde{\gamma}}_1^U)^4)^{0.6}\right)^{1/4}\right]\Big|\dot{\tilde{\gamma}}_2 \in \dot{\tilde{h}}_2, \dot{\tilde{\gamma}}_1 \in \dot{\tilde{h}}_1\right\}$$

$$= \{[0.3468, 0.4480], [0.3509, 0.4529], [0.4475, 0.6309], [0.4493, 0.6325],$$
$$[0.5417, 0.7158], [0.6791, 0.7167]\}.$$

If we make use of the GIVHFHG operators to aggregate the IVHFEs $\tilde{h}_1$, $\tilde{h}_2$, then

$$\ddot{\tilde{h}}_1 = \left\{[0.1^{2\times0.2}, 0.3^{2\times0.2}], [0.4^{2\times0.2}, 0.5^{2\times0.2}]\right\} = \{[0.3981, 0.6178], [0.6931, 0.7579]\}$$

$$\ddot{\tilde{h}}_2 = \left\{[0.3^{2\times0.8}, 0.4^{2\times0.8}], [0.4^{2\times0.8}, 0.6^{2\times0.8}], [0.5^{2\times0.8}, 0.7^{2\times0.8}]\right\}$$
$$= \{[0.1457, 0.2308], [0.2308, 0.4416], [0.3299, 0.5651]\}$$

As $s\left(\ddot{\tilde{h}}_1\right) = [0.5456, 0.6879]$, $s\left(\ddot{\tilde{h}}_2\right) = [0.2355, 0.4125]$, we have $s\left(\ddot{\tilde{h}}_1\right) > s\left(\ddot{\tilde{h}}_2\right)$. Thus,

$$\ddot{\tilde{h}}_{\sigma(1)} = \ddot{\tilde{h}}_1 = \{[0.3981, 0.6178], [0.6931, 0.7579]\}$$
$$\ddot{\tilde{h}}_{\sigma(2)} = \ddot{\tilde{h}}_2 = \{[0.1457, 0.2308], [0.2308, 0.4416], [0.3299, 0.5651]\}$$

Using Definition 6.12, we have

$$\text{GIVHFHG}_1\left(\tilde{h}_1, \tilde{h}_2\right) = \text{IVHFHG}\left(\tilde{h}_1, \tilde{h}_2\right) = \overset{2}{\underset{j=1}{\otimes}}\ddot{\tilde{h}}_{\sigma(j)}^{\omega_j}$$

$$= \left\{\left[(\ddot{\tilde{\gamma}}_1^L)^{0.4}(\ddot{\tilde{\gamma}}_2^L)^{0.6}, (\ddot{\tilde{\gamma}}_1^U)^{0.4}(\ddot{\tilde{\gamma}}_2^U)^{0.6}\right]\Big|\ddot{\tilde{\gamma}}_1 \in \ddot{\tilde{h}}_1, \ddot{\tilde{\gamma}}_2 \in \ddot{\tilde{h}}_2\right\}$$

$$= \{[0.2178, 0.3422], [0.2870, 0.5051], [0.3557, 0.5856], [0.2719, 0.3714],$$
$$[0.3583, 0.5481], [0.4440, 0.6355]\}.$$

$$\text{GIVHFHG}_4\left(\tilde{h}_1, \tilde{h}_2\right) = \frac{1}{4}\left(\overset{2}{\underset{j=1}{\otimes}}\left(4\ddot{\tilde{h}}_{\sigma(j)}\right)^{\omega_j}\right)$$

$$= \{[0.2042, 0.3042], [0.2791, 0.4901], [0.3540, 0.5837], [0.2208, 0.3083],$$
$$[0.3071, 0.5034], [0.4001, 0.6107]\}.$$

6.4.3 An Approach to Group Decision Making with Interval-Valued Hesitant Fuzzy Preference Relations

Suppose that the elements in IVHFE are arranged in ascending order. Let $\tilde{h}_{\sigma(i)}(j = 1, 2, \ldots, n)$ be the ith smallest value in $\tilde{h}$. The distance measures for IVHFEs are given by

$$d_1(\tilde{\alpha}, \tilde{\beta}) = \frac{1}{2l} \sum_{i=1}^{l} \left(\left| \tilde{\alpha}_{\sigma(i)}^L - \tilde{\beta}_{\sigma(i)}^L \right| + \left| \tilde{\alpha}_{\sigma(i)}^U - \tilde{\beta}_{\sigma(i)}^U \right| \right) \tag{6.71}$$

$$d_2(\tilde{\alpha}, \tilde{\beta}) = \sqrt{\frac{1}{2l} \sum_{i=1}^{l} \left(\left| \tilde{\alpha}_{\sigma(i)}^L - \tilde{\beta}_{\sigma(i)}^L \right|^2 + \left| \tilde{\alpha}_{\sigma(i)}^U - \tilde{\beta}_{\sigma(i)}^U \right|^2 \right)} \tag{6.72}$$

Equations (6.71) and (6.72) can be considered as the extension of the well-known Hamming distance and Euclidean distance under the interval-valued hesitant fuzzy environment. They satisfy the following properties:

(1) $0 \leq d(\tilde{\alpha}, \tilde{\beta}) \leq 1$.
(2) $d(\tilde{\alpha}, \tilde{\beta}) = 0$ if and only if $\tilde{\alpha} = \tilde{\beta}$.
(3) $d(\tilde{\alpha}, \tilde{\beta}) = d(\tilde{\beta}, \tilde{\alpha})$.

Based on the above analysis, below we develop a group decision making approach with IVHFPRs:

Algorithm 6.6

Step 1. Let $A = \{A_1, A_2, \ldots, A_n\}$ be a discrete set of alternatives, $O = \{O_1, O_2, \ldots, O_m\}$ be the set of decision organizations composed of several experts, and $\omega = (\omega_1, \omega_2, \ldots, \omega_m)$ be the weight vector of the organizations with $\sum_{k=1}^{m} \omega_k = 1$ and $\omega_k \geq 0$, $k = 1, 2, \ldots, m$. Each expert in an organization provides interval preference for each pair of alternatives, so for each organization its IVHFPR is constructed with $\tilde{H}^{(k)} = (\tilde{h}_{ij}^{(k)})_{n \times n}$. We utilize the GIVHFA (or GIVHFG) operator to aggregate all $\tilde{h}_{ij}^{(k)}(j = 1, 2, \ldots, n)$ that correspond to the alternative A_i. We get the IVHFE $\tilde{h}_i^{(k)}$ of the alternative A_i over all the other alternatives for the organization O_k.

Step 2. To make our approach have large feasibility and wide practicability, the weights of decision organizations are also properly incorporated into the group decision making problem. If the weight vector $\omega = (\omega_1, \omega_2, \ldots, \omega_m)^T$ of all organizations is known, then go to Step 3; Otherwise, the weight of each organization needs to be accounted for. The smaller the difference between the preference information offered by one organization with that offered by the rest organizations, the more precise the preference information. Consequently, a larger weight is assigned to the organization. Calculate the weights of decision organizations proceeds as follows. We first compute the differences between any two organizations O_l and O_k using distance measure given by Eq. (6.71) or Eq. (6.72), and obtain

$$D_{lk} = \left(d_{ij}^{(lk)} \right)_{n \times n} = \left(d\left(\tilde{h}_{ij}^{(l)}, \tilde{h}_{ij}^{(k)} \right) \right)_{n \times n}, \; l, k = 1, 2, \ldots, m \qquad (6.73)$$

where (1) $d_{ij}^{(lk)} \geq 0$, especially, if $l = k$, then $d_{ij}^{(lk)} = 0, i, j = 1, 2, \ldots, n$; (2) $d_{ij}^{(lk)} = 0$, $i = j$; (3) $d_{ij}^{(lk)} = d_{ji}^{(lk)}$. D_{lk} is a symmetric matrix and the values of diagonal elements are zero. Then, we compute the average value of the matrix D_{lk} by

$$\bar{d}_{lk} = \frac{1}{n^2} \sum_{i=1}^{n} \sum_{j=1}^{n} d_{ij}^{(lk)} \qquad (6.74)$$

Afterwards, let $\bar{d}_l = \sum\limits_{k=1, k \neq l}^{m} \bar{d}_{lk} = \frac{1}{n^2} \sum\limits_{k=1, k \neq l}^{m} \sum\limits_{i=1}^{n} \sum\limits_{j=1}^{n} d_{ij}^{(lk)}$, which denotes the deviation of the organization O_l from the rest organizations. The smaller the $\bar{d}_l$, the closer the preference information given by O_l and the rest organizations, and hence, the more valuable the preference information coming from the organization O_l. It means that O_l should be given a large weight w_l, which is written as

$$w_l = \frac{(\bar{d}_l)^{-1}}{\sum\limits_{l=1}^{m} (\bar{d}_l)^{-1}}, \quad l = 1, 2, \ldots, m \qquad (6.75)$$

Step 3. Utilize the GIVHFWA (or the GIVHFWG) operator to aggregate all $\tilde{h}_i^{(k)} (k = 1, 2, \ldots, m)$ into a collective IVHFE $\tilde{h}_i$ for the alternative A_i.

Step 4. Compute the score functions of $\tilde{h}_i (i = 1, 2, \ldots, n)$ by Eq. (1.24), and rank all the alternatives $A_i (i = 1, 2, \ldots, n)$ according to $s(A_i) (i = 1, 2, \ldots, n)$.

In the following, a large project of Jiudianxia reservoir operation is employed to demonstrate the validity of our approach:

Example 6.11 (Chen et al. 2013). The reservoir is designed for many purposes, such as power generation, irrigation, total water supply for industry, agriculture, residents and environment, etc. Because of different requirements for the partition of the amount of water, four reservoir operation schemes A_1, A_2, A_3 and A_4 are suggested.

A_1: maximum plant output, enough supply of water used in the Tao River basin, higher and lower supply for society and economy.

A_2: maximum plant output, enough supply of water used in the Tao River basin, higher and lower supply for society and economy, lower supply for ecosystem.

A_3: maximum plant output, enough supply of water used in the Tao River basin, higher and lower supply for society and economy, total supply for ecosystem and environment, whose 90 % is used for flushing sands at low water period.

A_4: maximum plant output, enough supply of water used in the Tao River basin, higher and lower supply for society and economy, total supply for ecosystem and environment, whose 50 % is used for flushing sands at low water period.

To select the best scheme, the government assigns three decision organizations $O_k(k = 1, 2, 3)$ to evaluate the four competing schemes. Due to uncertainties, the decision makers in each organization give their preference information over alternatives in the form of interval values. Taking O_2 as an example, the decision makers evaluate the degrees to which A_1 is preferred to A_2. Some give $[0.2, 0.3]$ and the others give $[0.5, 0.6]$. Consider that these decision makers in the organization O_2 cannot be persuaded each other, the preference information that A_1 is preferred to A_2 provided by the decision organization O_2 can be considered as an IVHFE, i.e., $\{[0.2, 0.3], [0.5, 0.6]\}$. The preference information of these three organizations is listed in Tables 6.4, 6.5 and 6.6, respectively.

To get the optimal alternative, the following steps are adopted:

Step 1. Compute the averaged IVHFE $\tilde{h}_i^{(k)}$ of the alternative A_i over all the other alternatives for the organization $O_k(k = 1, 2, 3)$ by the GIVHFA (let $\lambda = 1$) operator. All the aggregation results are listed in Table 6.7.

Table 6.4 The preference relation of the decision organization O_1

	A_1	A_2	A_3	A_4
A_1	{[0.5,0.5]}	{[0.4,0.5], [0.7,0.9]}	{[0.5,0.6], [0.8,0.9]}	{[0.3,0.5]}
A_2	{[0.1,0.3], [0.5,0.6]}	{[0.5,0.5]}	{[0.4,0.5]}	{[0.6,0.8]}
A_3	{[0.1,0.2], [0.4,0.5]}	{[0.5,0.6]}	{[0.5,0.5]}	{[0.3,0.4], [0.5,0.6]}
A_4	{[0.5,0.7]}	{[0.2,0.4]}	{[0.4,0.5], [0.6,0.7]}	{[0.5,0.5]}

Table 6.5 The preference relation of the decision organization O_2

	A_1	A_2	A_3	A_4
A_1	{[0.5,0.5]}	{[0.2,0.3], [0.5,0.6]}	{[0.5,0.6], [0.7,0.9]}	{[0.2,0.4]}
A_2	{[0.4,0.5], [0.7,0.8]}	{[0.5,0.5]}	{[0.5,0.8]}	{[0.3,0.5], [0.6,0.7], [0.8,0.9]}
A_3	{[0.1,0.3], [0.4,0.5]}	{[0.2,0.5]}	{[0.5,0.5]}	{[0.4,0.5], [0.7,0.8]}
A_4	{[0.6,0.8]}	{[0.1,0.2], [0.3,0.4], [0.5,0.7]}	{[0.2,0.3], [0.5,0.6]}	{[0.5,0.5]}

Table 6.6 The preference relation of the decision organization O_3

	A_1	A_2	A_3	A_4
A_1	{[0.5,0.5]}	{[0.4,0.5], [0.7,0.8]}	{[0.6,0.7]}	{[0.3,0.5], [0.6,0.7]}
A_2	{[0.2,0.3], [0.5,0.6]}	{[0.5,0.5]}	{[0.4,0.6]}	{[0.7,0.8]}
A_3	{[0.3,0.4]}	{[0.4,0.6]}	{[0.5,0.5]}	{[0.3,0.4], [0.5,0.7], [0.8,0.9]}
A_4	{[0.3,0.4], [0.5,0.7]}	{[0.2,0.3]}	{[0.1,0.2], [0.3,0.5], [0.6,0.7]}	{[0.5,0.5]}

Step 2. Derive the weights of the decision organizations. We first use Eq. (6.71) to compute $d\left(\tilde{h}_{ij}^{(l)}, \tilde{h}_{ij}^{(k)}\right)$, $i,j = 1, 2, \ldots, 4$; $l, k = 1, 2, 3$. The difference matrix $D_{lk} = \left(d_{ij}^{(lk)}\right)_{n \times n}$ can thus be obtained:

$$D_{12} = D_{21} = \begin{pmatrix} 0 & 0.225 & 0.025 & 0.1 \\ 0.225 & 0 & 0.2 & 0.1667 \\ 0.025 & 0.2 & 0 & 0.15 \\ 0.1 & 0.1667 & 0.15 & 0 \end{pmatrix},$$

$$D_{13} = D_{31} = \begin{pmatrix} 0 & 0.025 & 0.15 & 0.125 \\ 0.025 & 0 & 0.05 & 0.05 \\ 0.15 & 0.05 & 0 & 0.1167 \\ 0.125 & 0.05 & 0.1167 & 0 \end{pmatrix}$$

$$D_{23} = D_{32} = \begin{pmatrix} 0 & 0.2 & 0.125 & 0.225 \\ 0.2 & 0 & 0.15 & 0.1833 \\ 0.125 & 0.15 & 0 & 0.1167 \\ 0.225 & 0.1833 & 0.1167 & 0 \end{pmatrix}, \quad D_{11} = D_{22} = D_{33} = \begin{pmatrix} 0 & 0 & 0 & 0 \\ 0 & 0 & 0 & 0 \\ 0 & 0 & 0 & 0 \\ 0 & 0 & 0 & 0 \end{pmatrix}$$

Compute the average values of the difference matrix by Eq. (6.74):

$$\bar{d}_{12} = \bar{d}_{21} = 0.1083, \ \bar{d}_{13} = \bar{d}_{31} = 0.0646, \ \bar{d}_{23} = \bar{d}_{32} = 0.125$$

Using Eq. (6.75), we obtain $\omega_1 = 0.38$, $\omega_2 = 0.28$, $\omega_3 = 0.34$.

Step 3. Compute a collective IVHFE $\tilde{h}_i(i = 1, 2, 3)$ of the alternative A_i over all the other alternatives by using the GIVHFWA (let $\lambda = 1$) operator $\tilde{h}_i = \text{IVHFWA}(\tilde{h}_i^{(1)}, \tilde{h}_i^{(2)}, \tilde{h}_i^{(3)})$.

Step 4. Compute the score functions of $\tilde{h}_i(i = 1, 2, \ldots, n)$, and rank all the alternatives $A_i(i = 1, 2, \ldots, n)$ according to the values of $s(\tilde{h}_i)(i = 1, 2, \ldots, n)$:

$$s(\tilde{h}_1) = [0.5105, 0.6324], \ s(\tilde{h}_2) = [0.5063, 0.6359]$$
$$s(\tilde{h}_3) = [0.4304, 0.5368], \ s(\tilde{h}_4) = [0.4247, 0.5462]$$

Table 6.7 The aggregation results of the decision organization $O_k(k = 1, 2, 3)$

		The aggregation results of the decision organization
O_1	$\tilde{h}_1^{(1)}$	{[0.4308,0.5271], [0.5473,0.6656], [0.5213,0.6838], [0.6193,0.7764]}
	$\tilde{h}_2^{(1)}$	{[0.4267,0.5675], [0.5051,0.6239]}
	$\tilde{h}_3^{(1)}$	{[0.3700,0.4434], [0.4209,0.4970], [0.4308,0.5051], [0.4767,0.5528]}
	$\tilde{h}_4^{(1)}$	{[0.4114,0.5394], [0.4682,0.5946]}
O_2	$\tilde{h}_1^{(2)}$	{[0.3675,0.4616], [0.4434,0.6193], [0.4377,0.5319], [0.5051,0.6690]}
	$\tilde{h}_2^{(2)}$	{[0.4308,0.6024], [0.5051,0.6500], [0.5838,0.7341], [0.5213,0.6838], [0.5838,0.7217], [0.6500,0.7885]}
	$\tilde{h}_3^{(2)}$	{[0.3183,0.4561], [0.4267,0.5675], [0.3840,0.5000}, [0.4820,0,6024]}
	$\tilde{h}_4^{(2)}$	{[0.3840,0.5135], [0.4523,0.5771], [0.4215,0.5473], [0.4856,0.6064], [0.4682,0.6193], [0.5271,0.6690]}
O_3	$\tilde{h}_1^{(3)}$	{[0.4616,0.5599], [0.5319,0.6127], [0.5473,0.6500], [0.6064,0.6920]}
	$\tilde{h}_2^{(3)}$	{[0.4820,0.5909], [0.5394,0.6443]}
	$\tilde{h}_3^{(3)}$	{[0.3808,0.4820], [0.4308,0.5644], [0.5473,0.6690]}
	$\tilde{h}_4^{(3)}$	{[0.2915,0.3598], [0.3346,0.4308], [0.4215,0.4990], [0.3486,0.4616], [0.3883,0.5213], [0.4682,0.5787]}

Using Eq. (1.23), we get $s(\tilde{h}_1) > s(\tilde{h}_2) > s(\tilde{h}_4) > s(\tilde{h}_3)$, then $A_1 \succ A_2 \succ A_4 \succ A_3$, which shows the scheme A_1 is the best among these four schemes.

From Example 6.11, we can see that the IVHFPRs are useful in tackling large group decision making problems, because they intuitively express the uncertain preference information provided by the decision makers in each decision organization.

In the approach to group decision making with IVFPRs, the pairwise comparison values on alternatives are first aggregated and correspondingly only the average interval-valued preference information is obtained. However, with IVHFPRs, there is no need to perform such an aggregation process. Thus, we can provide more comprehensive description on the opinions of these decision makers. In order to compare the results obtained by IVHFPRs and those by IVFPRs, we present the detailed calculation process of solving the problem in Example 6.11 with IVFPR.

Step 1. Compute interval-valued fuzzy preferences by averaging the individual decision makers' preference opinions in each decision organization. The results obtained on the basis of the data in Tables 6.4, 6.5 and 6.6 are summarized in Tables 6.8, 6.9 and 6.10, respectively. For example, for O_1,
$$\tilde{h}_{12}^{(1)} = \tfrac{1}{2}([0.4, 0.5] + [0.7, 0.9]) = [0.55, 0.7].$$

Step 2. Compute the averaged $\tilde{h}_i^{(k)}$ of the alternative A_i over all the other alternatives corresponding to the organization $O_k(k = 1, 2, 3)$ by the arithmetic average (AA) operator:

Table 6.8 The interval-valued fuzzy preference relation of the decision organization O_1

	A_1	A_2	A_3	A_4
A_1	[0.5,0.5]	[0.55,0.7]	[0.65,0.75]	[0.3,0.5]
A_2	[0.3,0.45]	[0.5,0.5]	[0.4,0.5]	[0.6,0.8]
A_3	[0.25,0.35]	[0.5,0.6]	[0.5,0.5]	[0.4,0.5]
A_4	[0.5,0.7]	[0.2,0.4]	[0.5,0.6]	[0.5,0.5]

Table 6.9 The interval-valued fuzzy preference relation of the decision organization O_2

	A_1	A_2	A_3	A_4
A_1	[0.5,0.5]	[0.35,0.45]	[0.6,0.75]	[0.2,0.4]
A_2	[0.55,0.65]	[0.5,0.5]	[0.5,0.8]	[0.5667,0.7]
A_3	[0.25,0.4]	[0.2,0.5]	[0.5,0.5]	[0.55,0.65]
A_4	[0.6,0.8]	[0.3,0.4333]	[0.35,0.45]	[0.5,0.5]

Table 6.10 The interval-valued fuzzy preference relation of the decision organization O_3

	A_1	A_2	A_3	A_4
A_1	[0.5,0.5]	[0.55,0.65]	[0.6,0.7]	[0.45,0.6]
A_2	[0.35,0.45]	[0.5,0.5]	[0.4,0.6]	[0.7,0.8]
A_3	[0.3,0.4]	[0.4,0.6]	[0.5,0.5]	[0.5333,0.6667]
A_4	[0.4,0.55]	[0.2,0.3]	[0.3333,0.4667]	[0.5,0.5]

$$\tilde{h}_i^{(k)} = \mathrm{AA}(\tilde{h}_{i1}^{(k)}, \tilde{h}_{i2}^{(k)}, \tilde{h}_{i3}^{(k)}, \tilde{h}_{i4}^{(k)}) = \frac{1}{n}\sum_{j=1}^{n} \tilde{h}_{ij}^{(k)} \tag{6.76}$$

The aggregation results of the decision organization $O_k(k = 1, 2, 3)$ are:

O_1: $\tilde{h}_1^{(1)} = [0.5, 0.6125]$, $\tilde{h}_2^{(1)} = [0.45, 0.5625]$, $\tilde{h}_3^{(1)} = [0.4125, 0.4875]$, $\tilde{h}_4^{(1)} = [0.425, 0.55]$;

O_2: $\tilde{h}_1^{(2)} = [0.4125, 0.525]$, $\tilde{h}_2^{(2)} = [0.5292, 0.6625]$, $\tilde{h}_3^{(2)} = [0.375, 0.5125]$, $\tilde{h}_4^{(2)} = [0.4375, 0.5458]$;

O_3: $\tilde{h}_1^{(3)} = [0.525, 0.6125]$, $\tilde{h}_2^{(3)} = [0.4875, 0.5875]$, $\tilde{h}_3^{(3)} = [0.4333, 0.5417]$, $\tilde{h}_4^{(3)} = [0.3583, 0.4542]$.

Step 3. Utilize the weighted average (WA) operator to aggregate all $\tilde{h}_i^{(k)}(k = 1, 2, 3)$ into a collective $\tilde{h}_i$ of the alternative A_i over all the other alternatives:

$$\tilde{h}_i = \text{WA}(\tilde{h}_i^{(1)}, \tilde{h}_i^{(2)}, \tilde{h}_i^{(3)}) = \sum_{k=1}^{3} \omega_i \tilde{h}_i^{(k)} \qquad (6.76)$$

In order to be consistent with Example 6.11, the same weights for the decision organizations are assigned as $\omega_1 = 0.38$, $\omega_2 = 0.28$, $\omega_3 = 0.34$. Then, the calculated results are:

$$\tilde{h}_1 = [0.484, 0.588], \ \tilde{h}_2 = [0.4849, 0.599],$$
$$\tilde{h}_3 = [0.4091, 0.5129], \ \tilde{h}_4 = [0.4058, 0.5163]$$

Step 4. Rank these interval numbers by Eq. (1.23), and get $A_2 \succ A_1 \succ A_4 \succ A_3$. From the results of calculations, one can find that the ranking results derived by these two approaches are different. The reason is that in the later approach, the group members' preference values are aggregated and such an aggregation process actually amounts to perform a transformation of IVHFE into interval-valued fuzzy numbers. As a result, it leads to the loss of information, which affects the final ranking results. In other words, the comparison shows the benefits of the proposed group decision making approaches based on IVHFPRs.

References

Alonso S, Herrera-Viedma E, Chiclana F, Herrera F (2009) Individual and social strategies to deal with ignorance situations in multi-person decision making. Int J Inf Technol Decis Making 8:313–333

Cabrerizo FJ, Alonso S, Herrera-Viedma E (2009) A consensus model for group decision making problems with unbalanced fuzzy linguistic information. Int J Inf Technol Decis Making 8:109–131

Chen N, Xu ZS, Xia MM (2013) Interval-valued hesitant preference relations and their applications to group decision making. Knowl-Based Syst 37:528–540

Chiclana F, Herrera-Viedma E, Alonso S (2009a) A note on two methods for estimating missing pairwise preference values. IEEE Trans Syst Man Cybern Part B 39:1628–1633

Chiclana F, Herrera-Viedma E, Alonso S, Herrera F (2009b) Cardinal consistency of reciprocal preference relation: a characterization of multiplicative transitivity. IEEE Trans Fuzzy Syst 17:14–23

Fan ZP, Ma J, Jiang YP, Sun YH, Ma L (2006) A goal programming approach to group decision making based on multiplicative preference relations and fuzzy preference relations. Eur J Oper Res 174:311–321

Gong ZW, Li LS, Cao J, Zhou FX (2010) On additive consistent properties of the intuitionistic fuzzy preference relation. Int J Inf Technol Decis Making 9:1009–1025

Herrera F, Herrera-Viedma E (2000) Choice functions and mechanisms for linguistic preference relations. Eur J Oper Res 120:144–161

Herrera-Viedma E, Herrera F, Chiclana F, Luque M (2004) Some issues on consistency of fuzzy preference relations. Eur J Oper Res 154:98–109

Herrera-Viedma E, Chiclana F, Herrera F, Alonso S (2007) Group decision-making model with incomplete fuzzy preference relations based on additive consistency. IEEE Trans Syst Man Cybern Part B 37:176–189

Liao HC, Xu ZS (2014a) Some algorithms for group decision making with intuitionistic fuzzy preference information. Int J Uncertainty Fuzziness Knowl-Based Syst 22(4):505–529

Liao HC, Xu ZS (2014b) Intuitionistic fuzzy hybrid weighted aggregation operators. Int J Intell Syst 29(11):971–993

Liao HC, Xu ZS (2015) Consistency of the fused intuitionistic fuzzy preference relation in group intuitionistic fuzzy analytic hierarchy process. Appl Soft Comput 35:812–826

Liao HC, Xu ZS, Xia MM (2014a) Multiplicative consistency of hesitant fuzzy preference relation and its application in group decision making. Int J Inf Technol Decis Making 13(1):47–76

Liao HC, Xu ZS, Xia MM (2014b) Multiplicative consistency of interval-valued intuitionistic fuzzy preference relations. J Intell Fuzzy Syst 27(6):2969–2985

Liao HC, Xu ZS, Zeng XJ, Merigó JM (2015) Framework of group decision making with intuitionistic fuzzy preference information. IEEE Trans Fuzzy Syst 23(4):1211–1227

Liu HF, Xu ZS, Liao HC (2016) The multiplicative consistency index of hesitant fuzzy preference relation. IEEE Trans Fuzzy Syst 24(1):82–93

Ma J, Fan ZP, Jiang YP, Mao JY, Ma L (2006) A method for repairing the inconsistency of fuzzy preference relations. Fuzzy Sets Syst 157:20–33

Mata F, Martínez L, Herrera-Viedma E (2009) An adaptive consensus support model for group decision making problems in a multi-granular fuzzy linguistic context. IEEE Trans Fuzzy Syst 17:279–290

Ngwenyama O, Bryson N (1999) Eliciting and mapping qualitative preferences to numeric rankings in group decision-making. Eur J Oper Res 116:487–497

Saaty TL (1980) The analytic hierarchy process. McGraw-Hill, New York

Tanino T (1984) Fuzzy preference orderings in group decision making. Fuzzy Sets Syst 12:117–131

Tapia García JM, del Moral MJ, Martinez MA, Herrera-Viedma E (2012) A consensus model for group decision making problems with interval fuzzy preference relations. Int J Inf Technol Decis Making 11(4):709–725

Tzeng GH, Chiang CH, Li CW (2007) Evaluating intertwined effects in e-learning program: A novel hybrid MCDM model based on factor analysis and DEMATEL. Expert Syst Appl 32:1028–1044

Xia MM, Xu ZS (2011a) Methods for fuzzy complementary preference relations based on multiplicative consistency. Comput Ind Eng 61(4):930–935

Xia MM, Xu ZS (2011b) Some issues on multiplicative consistency of interval reciprocal relations. Int J Inf Technol Decis Making 10:1043–1065

Xia MM, Xu ZS, Liao HC (2013) Preference relations based on intuitionistic multiplicative information. IEEE Trans Fuzzy Syst 21:113–133

Xu ZS (2004) On compatibility of interval fuzzy preference matrices. Fuzzy Optim Decis Making 3:217–225

Xu ZS (2005) An approach to group decision making based on incomplete linguistic preference relations. Int J Inf Technol Decis Making 4:153–160

Xu ZS (2006) Incomplete linguistic preference relations and their fusion. Inf Fusion 7:331–337

Xu ZS (2007) Intuitionistic preference relations and their application in group decision making. Inf Sci 177:2363–2379

Xu ZS, Cai XQ (2011) Group consensus algorithms based on preference relations. Inf Sci 181:150–162

Xu ZS, Liao HC (2014) Intuitionistic fuzzy analytic hierarchy process. IEEE Trans Fuzzy Syst 22(4):749–761

Xu ZS, Liao HC (2015) A survey of approaches to decision making with intuitionistic fuzzy preference relations. Knowl-Based Syst 80:131–142

Xu ZS, Yager RR (2009) Intuitionistic and interval-valued intuitionistic fuzzy preference relations and their measures of similarity for the evaluation of agreement within a group. Fuzzy Optim Decis Making 8:123–139

Yager RR (1988) On ordered weighted averaging aggregation operators in multi-criteria decision making. IEEE Trans Syst Man Cybern 18(1):183–190

Printed in the United States
By Bookmasters